Communications in Computer and Information Science　　1661

More information about this series at https://link.springer.com/bookseries/7899

Yury Kochetov · Anton Eremeev ·
Oleg Khamisov · Anna Rettieva (Eds.)

Mathematical Optimization Theory and Operations Research

Recent Trends

21st International Conference, MOTOR 2022
Petrozavodsk, Russia, July 2–6, 2022
Revised Selected Papers

 Springer

Editors
Yury Kochetov 🆔
Sobolev Institute of Mathematics
Novosibirsk, Russia

Anton Eremeev 🆔
Sobolev Institute of Mathematics
Novosibirsk, Russia

Oleg Khamisov 🆔
Melentiev Energy Systems Institute
Irkutsk, Russia

Anna Rettieva 🆔
Institute of Applied Mathematical Research
Petrozavodsk, Russia

ISSN 1865-0929   ISSN 1865-0937 (electronic)
Communications in Computer and Information Science
ISBN 978-3-031-16223-7   ISBN 978-3-031-16224-4 (eBook)
https://doi.org/10.1007/978-3-031-16224-4

This Springer imprint is published by the registered company Springer Nature Switzerland AG
The registered company address is: Gewerbestrasse 11, 6330 Cham, Switzerland

Preface

This volume contains refereed selected papers presented at the 21st international conference on Mathematical Optimization Theory and Operations Research (MOTOR 2022), which was held July 2–6, 2022, in the Karelia region, Russia. MOTOR is the annual scientific conference of the Operational Research Society of Russia. MOTOR 2022 was the fourth joint scientific event (http://motor2022.krc.karelia.ru/en/section/1) unifying four well-known conferences that had been held in Ural, Siberia, and the Russian Far East for a long time. The next conference, MOTOR 2023, will be organized in Ekaterinburg.

As per tradition, the main conference scope included, but was not limited to, mathematical programming, bi-level and global optimization, integer programming and combinatorial optimization, approximation algorithms with theoretical guarantees and approximation schemes, heuristics and meta-heuristics, game theory, optimal control, optimization in machine learning and data analysis, and their valuable applications in operations research and economics. To accelerate the cooperation of academia and industry, we organized round table sessions and invited Russian and international companies to discuss business applications of optimization methods and operations research.

In response to the call for papers, MOTOR 2022 received 161 submissions. Out of 88 full papers considered for review (73 abstracts and short communications were excluded on formal reasons) only 21 papers were selected by the Program Committee (PC) for publication in the first volume of the proceedings, published in Springer LNCS, vol. 13367. Out of the remaining papers, the PC selected 24 revised contributions, including three short communications and two invited reviews by plenary speakers Rentsen Enkhbat and Sergey Sevastyanov.

Each submission was reviewed by at least three PC members or invited reviewers, experts in their fields, in order to supply detailed and helpful comments. In addition, the PC recommended publishing 13 papers in the Journal of Applied and Industrial Mathematics and the journal Siberian Electronic Mathematical Reports after their revision with respect to the reviewers' comments.

We thank the authors for their submissions, the members of the Program Committee, and all the external reviewers for their efforts in providing exhaustive reviews. We thank our sponsors and partners: the Institute of Applied Mathematical Research (IAMR) of the Karelian Research Centre of RAS, the Sobolev Institute of Mathematics of SB RAS, Novosibirsk State University, the International Mathematical Center in Akademgorodok, the Krasovsky Institute of Mathematics and Mechanics of UB RAS, the Ural Mathematical Center, the Center for Research and Education in Mathematics, the Higher

School of Economics (Nizhny Novgorod), and the Matrosov Institute for System Dynamics and Control Theory of SB RAS. We are grateful to the colleagues from the Springer LNCS and CCIS editorial boards for their kind and helpful support.

July 2022

Yury Kochetov
Anton Eremeev
Oleg Khamisov
Anna Rettieva

School of Economics (Nizhny Novgorod), and the Matrosov Institute for System Dynamics and Control Theory of SB RAS. We are grateful to the colleagues from the Springer LNCS and CCIS editorial boards for their kind and helpful support.

July 2022

Yury Kochetov
Anton Eremeev
Oleg Khamisov
Anna Rettieva

Preface

This volume contains refereed selected papers presented at the 21st international conference on Mathematical Optimization Theory and Operations Research (MOTOR 2022), which was held July 2–6, 2022, in the Karelia region, Russia. MOTOR is the annual scientific conference of the Operational Research Society of Russia. MOTOR 2022 was the fourth joint scientific event (http://motor2022.krc.karelia.ru/en/section/1) unifying four well-known conferences that had been held in Ural, Siberia, and the Russian Far East for a long time. The next conference, MOTOR 2023, will be organized in Ekaterinburg.

As per tradition, the main conference scope included, but was not limited to, mathematical programming, bi-level and global optimization, integer programming and combinatorial optimization, approximation algorithms with theoretical guarantees and approximation schemes, heuristics and meta-heuristics, game theory, optimal control, optimization in machine learning and data analysis, and their valuable applications in operations research and economics. To accelerate the cooperation of academia and industry, we organized round table sessions and invited Russian and international companies to discuss business applications of optimization methods and operations research.

In response to the call for papers, MOTOR 2022 received 161 submissions. Out of 88 full papers considered for review (73 abstracts and short communications were excluded on formal reasons) only 21 papers were selected by the Program Committee (PC) for publication in the first volume of the proceedings, published in Springer LNCS, vol. 13367. Out of the remaining papers, the PC selected 24 revised contributions, including three short communications and two invited reviews by plenary speakers Rentsen Enkhbat and Sergey Sevastyanov.

Each submission was reviewed by at least three PC members or invited reviewers, experts in their fields, in order to supply detailed and helpful comments. In addition, the PC recommended publishing 13 papers in the Journal of Applied and Industrial Mathematics and the journal Siberian Electronic Mathematical Reports after their revision with respect to the reviewers' comments.

We thank the authors for their submissions, the members of the Program Committee, and all the external reviewers for their efforts in providing exhaustive reviews. We thank our sponsors and partners: the Institute of Applied Mathematical Research (IAMR) of the Karelian Research Centre of RAS, the Sobolev Institute of Mathematics of SB RAS, Novosibirsk State University, the International Mathematical Center in Akademgorodok, the Krasovsky Institute of Mathematics and Mechanics of UB RAS, the Ural Mathematical Center, the Center for Research and Education in Mathematics, the Higher

Organization

General Chair

Vladimir Mazalov — Institute of Applied Mathematical Research, Russia

Honorary Chair

Panos Pardalos — University of Florida, USA

Program Committee Chairs

Michael Khachay — Krasovsky Institute of Mathematics and Mechanics, Russia

Oleg Khamisov — Melentiev Institute of Energy Systems, Russia

Yury Kochetov — Sobolev Institute of Mathematics, Russia

Anton Eremeev — Omsk Branch of Sobolev Institute of Mathematics, Russia

Program Committee

Anatoly Antipin — Dorodnicyn Computing Centre, FRC, CSC, RAS, Russia

René Van Bevern — Huawei Cloud Technologies Co., Ltd., Russia

Maxim Buzdalov — ITMO University

Igor Bykadorov — Sobolev Institute of Mathematics SB RAS, Russia

Tatjana Davidović — Mathematical Institute SANU, Serbia

Stephan Dempe — TU Bergakademie Freiberg, Germany

Adil Erzin — Sobolev Institute of Mathematics, Russia

Stefka Fidanova — Institute of Information and Communication Technologies, Bulgaria

Fedor Fomin — University of Bergen, Norway

Eduard Gimadi — Sobolev Institute of Mathematics, Russia

Evgeny Gurevsky — Université de Nantes, France

Feng-Jang Hwang — University of Technology Sydney, Australia

Sergey Ivanov — Moscow Aviation Institute, Russia

Milojica Jaćimović — Montenegrin Academy of Sciences and Arts, Montenegro

Valeriy Kalyagin — Higher School of Economics, Russia

Vadim Kartak	Ufa State Aviation Technical University, Russia
Alexander Kazakov	Matrosov Institute for System Dynamics and Control Theory SB RAS, Russia
Lev Kazakovtsev	Siberian State Aerospace University, Russia
Igor Konnov	Kazan University, Russia
Alexander Kononov	Sobolev Institute of Mathematics, Russia
Dmitri Kvasov	DIMES, University of Calabria, Italy
Bertrand M. T. Lin	National Yang Ming Chiao Tung University, Taiwan
Vittorio Maniezzo	University of Bologna, Italy
Nenad Mladenović	Khalifa University, United Arab Emirates
Yury Nikulin	University of Turku, Finland
Evgeni Nurminski	Far Eastern Federal University, Russia
Nicholas Olenev	Doronicyn Computing Centre FRC CSC RAS, Russia
Leon Petrosyan	Saint Petersburg State University, Russia
Alex Petunin	Ural Federal University, Russia
Leonid Popov	Krasovsky Institute of Mathematics and Mechanics, Russia
Mikhail Posypkin	Dorodnicyn Computing Centre, FRC CSC RAS, Russia
Artem Pyatkin	Sobolev Institute of Mathematics, Russia
Soumyendu Raha	Indian Institute of Science, India
Yaroslav Sergeyev	University of Calabria, Italy
Sergey Sevastyanov	Sobolev Institute of Mathematics, Russia
Natalia Shakhlevich	University of Leeds, UK
Aleksandr Shananin	Moscow Institute Physics and Technology, Russia
Angelo Sifaleras	University of Macedonia, Greece
Vladimir Skarin	Krasovsky Institute of Mathematics and Mechanics, Russia
Alexander Strekalovskiy	Matrosov Institute for System Dynamics and Control Theory SB RAS, Russia
Tatiana Tchemisova	University of Aveiro, Portugal
Raca Todosijevic	Université Polytechnique Hauts-de-France, France
Alexey Tret'yakov	Dorodnicyn Computing Centre, FRC CSC RAS, Russia

Additional Reviewers

Artemyeva, Liudmila
Baklanov, Artem
Buzdalov, Maxim
Chernykh, Ilya

Chirkova, Julia
Dang, Duc-Cuong
Gasnikov, Alexander
Gluschenko, Konstantin

Gribanov, Dmitriy
Gusev, Mikhail
Il'ev, Victor
Ivashko, Anna
Jelisavcic, Vladisav
Khachay, Daniel
Khlopin, Dmitry
Khutoretskii, Alexander
Kononova, Polina
Konovalchikova, Elena
Kulachenko, Igor
Lebedev, Pavel
Lempert, Anna
Levanova, Tatyana
Lushchakova, Irina
Melnikov, Andrey
Nikitina, Natalia
Ogorodnikov, Yuri
Pankratova, Yaroslavna
Parilina, Elena
Plotnikov, Roman

Plyasunov, Alexander
Potapov, Mikhail
Pyatkin, Artem
Rettieva, Anna
Rybalov, Alexander
Sedakov, Artem
Shkaberina, Guzel
Simanchev, Ruslan
Sopov, Evgenii
Stanimirovic, Zorica
Stanovov, Vladimir
Takhonov, Ivan
Tarasyev, Alexander
Tovbis, Elena
Tur, Anna
Ushakov, Anton
Vasilyev, Igor
Vasin, Alexandr
Yanovskaya, Elena
Zakharova, Yulia
Zubov, Vladimir

Industry Session Chair

Vasilyev Igor — Matrosov Institute for System Dynamics and Control Theory, Russia

Organizing Committee

Chair

Vladimir Mazalov — Institute of Applied Mathematical Research, Russia

Deputy Chair

Anna Rettieva — Institute of Applied Mathematical Research, Russia

Scientific Secretary

Yulia Chirkova — Institute of Applied Mathematical Research, Russia

Anna Ivashko — Institute of Applied Mathematical Research, Russia

Elena Parilina	Saint-Petersburg State University, Russia
Polina Kononova	Sobolev Institute of Mathematics, Russia
Timur Medvedev	Higher School of Economics, Russia
Yuri Ogorodnikov	Krasovsky Institute of Mathematics and Mechanics, Russia

Organizers

Institute of Applied Mathematical Research, Russia
Sobolev Institute of Mathematics, Russia
Mathematical Center in Akademgorodok, Russia
Krasovsky Institute of Mathematics and Mechanics, Russia
Higher School of Economics (Campus Nizhny Novgorod), Russia

Sponsors

Higher School of Economics (Campus Nizhny Novgorod), Russia
Mathematical Center in Akademgorodok, Novosibirsk, Russia
Ural Mathematical Center, Russia

Contents

Operational Research Applications

Invited Talks

Three Efficient Methods of Approximate Solution of NP-Hard Discrete Optimization Problems. (Illustrated by Applications to Scheduling Problems)

Sergey Sevastyanov$^{(\boxtimes)}$ (iD)

Sobolev Institute of Mathematics, Koptyug Avenue 4, Novosibirsk 630090, Russia
seva@math.nsc.ru
http://math.nsc.ru

Abstract. Three fairly universal and efficient methods of finding near-optimal solutions for NP-hard problems will be discussed and illustrated by examples of their application to scheduling problems (although, they are surely applicable to a much wider area of discrete optimization problems): (1) Methods of *compact vector summation* in a ball of minimum radius (of any norm), and of non-strict summation of vectors in a given area of d-dimensional space. These methods are applicable to problems of finding uniform distributions of multi-dimensional objects and to multi-stage scheduling problems. (2) Searching for the maximal flow most uniform in sinks, as a method of finding uniform distributions of one-component items under distribution restrictions. (3) The method of finding balanced restrictions on the set of feasible solutions.

Keywords: Efficient methods · Approximation algorithms · Discrete optimization · Scheduling · Uniform distributions · Compact vector summation

1 Introduction

More than half a century has passed since the day I proved my first theorem. Over such a long period of my scientific activity, I came to the conclusion that the most valuable "products" in mathematics are not individual results (no matter how bright they sound and no matter how long standing open problems they close), but *Ideas and Methods* to solve such problems. Indeed, as a rule, the solution of long standing open problems is impossible without the development of some **new methods**. On the other hand, some **old methods** (perhaps undeservedly forgotten by now) may turn out to be quite efficient for solving new problems

The research was supported by the program of fundamental scientific researches of the SB RAS No I.5.1., project No 0314-2019-0014, and by the Russian Foundation for Basic Research, projects 20-07-00458.

(arising in some recently emerging fields of science or in some new research directions). As a result of such oblivion, "bicycles" are invented and results are published that are much weaker than the ones that could be obtained by means of "old" (but forgotten) methods.

Two examples of such an "oblivion" I encountered quite recently, this year, in two papers published in reputable international journals. One of these papers was written by Biró and McDermid, and published in 2014 in *Discrete Applied Mathematics* [8]. The second paper [9] by four authors was published in March of this year in the *Journal of Scheduling*.

The authors of the first paper considered the problem $\langle P \mid \mathcal{M}_k \mid C_{\max} \rangle$, in which jobs from a given set $\{J_1, \ldots, J_n\}$ with given durations $\{p_k\}$ must be distributed among m identical machines $\mathcal{M} = \{M_1, \ldots, M_m\}$ subject to the restrictions given by sets $\mathcal{M}_k \subseteq \mathcal{M}$ of *admissible executors* of jobs J_k, and in such a way that the whole set of jobs is performed within the time interval of minimum length. It is assumed that each machine M_i executes the jobs from $\mathcal{J}_i$ (allocated to M_i) sequentially, no matter in which order, which requires time $\sum_{J_k \in \mathcal{J}_i} p_k$.

For a special case of this problem, when all job processing times $\{p_k\}$ are **integer powers of two**, the authors presented a pseudo-polynomial ρ-approximation algorithm with $\rho = 2 - 1/p_{\max}$. At the same time, if the authors had used the method described in 1986 in the "Cybernetics" journal [22] (or in its English translation published in the same year), they would be able to get a much prettier result, namely: for a problem with **arbitrary integer** job processing times, a **completely polynomial** ρ-approximation algorithm with the same value of ρ could be constructed. And moreover, after citing [22], they would only need to add a few lines of additional explanation on how to obtain such a result.

The authors of the second paper considered the problems $\langle P \mid \mathcal{M}_k(nested), p\text{-}batch(B_i) \mid C_{\max} \rangle$ and $\langle Q \mid \mathcal{M}_k(tree), p\text{-}batch(B_i) \mid C_{\max} \rangle$, for which they constructed polynomial-time 2-approximation algorithms. I should explain now the settings of both problems. Letters P and Q denote *identical* and *uniform* machines, *i.e.*, machines with identical speeds and different speeds, respectively. At that, the machines are not "ordinary" ones (those processing the jobs "one by one"), but are *p-batch machines*, able to process the jobs "by batches" (several jobs in parallel), which is denoted in the problem notation (given above) as "*p-batch* (B_i)". Here B_i specifies the maximum possible number of jobs that could be included in each batch on machine M_i. The process of performing the jobs in a batch is represented as if all these jobs are started and completed simultaneously (as if they all represent a single "batch-job"), and the processing time of that "batch-job" is equal to that of the longest job in the batch.

And again, if the authors of the paper had acknowledged the paper from "Cybernetics" (mentioned above), they, probably, would manage to obtain a stronger result, namely, a polynomial-time 2-approximation algorithm for a much more general problem $\langle Q \mid \mathcal{M}_k, p\text{-}batch(B_i) \mid C_{\max} \rangle$ with **arbitrary processing set restrictions**.

Thus, it turned out that the method developed nearly 40 years ago (see [21,22]) for solving the problem of "even distribution of jobs among machines subject to distribution restrictions" remains relevant to these days. As a result, I came to the conclusion that it is necessary to "refresh" the memory on this method in the minds of discrete optimizers (or to make a "remake" of the good old movie called "Uniform (over all sinks) maximal flows in networks and their application to scheduling problems"). I admit that many of the younger scientists (discrete optimizers) may hear about this method for the first time.

Other "uniformization" methods, which I would also like to recall today and which are also far from having exhausted their potential for application to discrete optimization problems, are the methods of *compact vector summation* in a finite-dimensional space within a ball (of some norm) of minimum radius, as well as the methods of *nonstrict vector summation* in some fixed (or optimizable) domains of d-dimensional space. These methods were developed quite actively by me and a few my colleagues from Ukraine in the 70s–90s of the last century, and certainly require further development in order to apply them to other problems.

Finally, the third method, which I invented quite recently, has also got, in my opinion, a great potential for application to discrete optimization problems. And perhaps, already while reading this paper, some of the readers will come up with the idea of using a similar method for solving problems relevant to them. The name of this method has not been settled yet. For now, as a "working choice", I will call it a *method of finding balanced constraints on the set of feasible solutions*.

To describe the methods listed above, first of all, it is required to determine the areas of their application, *i.e.*, to formulate problems that can be successfully solved using these methods. So, we will pass on to formulation of the problems.

2 Uniformization Problems

To emphasize that this kind of problems has been relevant since immemorial time, I will tell you the following "real-life" story.

There lived an old miller. His wife died a long time ago from overwork and endless childbirth. Children also tended to die at an early age. Only three survived to adulthood: two sons and a daughter. And he had some property acquired by a hard work: a small mill on the river, a haymaker, a cow, two horses, two dozens of sheep, old-fashioned watch (made of silver), Aivazovsky's painting "The Ninth Wave" (a copy, of course), and further down the list. And once the old miller felt that his strength has been run out (he was already in his fifties). So, he decided to prepare for the inevitable. He thought about writing a will, dividing his entire property among his three children "most fairly". The last circumstance (the requirement for a "fair" partition) caused him the greatest difficulty (as is often the case in our modern world, as well). First of all, it was clear that not every item of his household is suitable for everyone. (For example, he understood that it would be a bad idea to give his daughter a mill—she would overstrain. The idea of giving an Aivazovsky painting to his eldest son is just as bad.) After sitting all the night long at this problem, he came to the following mathematical setting.

The Problem of a Fair Sharing of an Inheritance. There is an *inheritance* $\{J_1, \ldots, J_n\}$ consisting of n indivisible items, and there is a set $\mathcal{M} = \{M_1, \ldots, M_m\}$ of *equal-right heirs*. The "equality of rights" of heirs is expressed in terms that the heirs claim for total shares **as equal as possible**. However, when moving on to the "details" of the sharing, significant differences between the contenders for the inheritance become clear. For each inheritance item J_k, a price p_k and a set $\mathcal{M}_k \subseteq \mathcal{M}$ of heirs claiming it are given. It is required to distribute the inheritance in the "most fair" way, namely: so that the largest total "jackpot" that goes to one of the heirs is minimal.

Perhaps, for some readers, such a mathematical setting of the problem may seem not quite adequately reflecting the real life situation, and the whole "prehistory" of the appearance of this problem may seem doubtful either. (Where have you ever seen a miller solving a mathematical problem all the night long?) Someone else might think that this optimality criterion does not fit quite well the notion of a "fair partition". Both those and others will be absolutely right, since real everyday problems, as a rule, do not lend themselves well to mathematical formalization. But we will not discuss this and dwell on the postulate that mathematical formulations in general are very rarely adequate to reality. Next, we will focus on the mathematical problem itself.

As you can see, it can be easily reformulated in other terms so that it becomes quite a "recognizable" problem of finding the optimal schedule, denoted in standard notation as $\langle P \,|\, \mathcal{M}_k \,|\, C_{\max}\rangle$. (We already mentioned this problem above.) Indeed, once we rename "inherited items" to "*jobs*", "heirs" to "*machines*", and the "cost" of an inherited item to the "*processing time of a job*" (or the *duration* of the job), then the discrete optimization problem formulated above immediately transforms into a typical scheduling problem of performing n jobs on m equal-speed machines, with the *minimum makespan* criterion. The difference between the resulting scheduling problem and the (even more classical) problem $\langle P \,\|\, C_{\max}\rangle$ on identical parallel machines (also known to the reader under the "folklore name" as a "problem on stones" that should be evenly distributed among m "heaps") is the presence of processing set restrictions specified by sets $\mathcal{M}_k$.

Another generalization of the above problem is the assumption that machines have *individual speeds* $\{s_i \,|\, i = 1, \ldots, m\}$. (Such machines are called *uniform* ones.) In this case, the processing time of job J_k, when performed by machine M_i, is determined by the formula $p_{i,k} = p_k/s_i$, where p_k is the *amount of work* of job J_k specified at the input. It is clear that in order to find an optimal solution to such a problem, it is no longer necessary to distribute the total amount of work "uniformly" among the machines, but a distribution with a "given unevenness", proportionate to speeds of machines is needed. (According to the principle: "Those whom more is given to, more will be asked".) If we return to the terminology of the problem of "fair sharing of inheritance", then "fair sharing" will no longer be an absolute, but a relative concept, and will be governed by the initially given "merit coefficients" of each heir.

We just encountered above a situation where essentially different physical quantities, such as (a) the *cost* of the kth item of the inheritance and (b) the *duration* of the *k*th job, were implemented in the mathematical model in exactly the same way (using the same variable p_k). This fact can be "reversed" by saying that the same variable p_k in the mathematical model can be interpreted in two ways: both as "duration" and as "cost". But what if in the problem of uniform distribution of some objects, each object has both characteristics: both duration and cost, and besides, many other ones, **independent of each other**, and it is required in the problem that the uniformity should be achieved **for each of those characteristics**[1]?!

Indeed, in real-life problems of finding "uniform distributions", distributed objects usually have several independent characteristics, for each of which it is desirable to achieve some kind of "uniformity". I encountered one of such real problems in 70s, when our laboratory was making a report for the *Iskitim cement plant*. In that time, the *Planned Economy* operated in USSR, and each enterprise had an *annual plan*.

The annual plan of the *Iskitim cement plant* consisted of a list of products planned for releasing over the coming year. Each product was characterized by many parameters, of which a dozen of the main ones was chosen. What are those parameters? First, the parameters that characterize the consumption of the most important resources used to manufacture the product. Here are the main consumables, the time of occupying the machines of various types (lathes, milling machines, etc.), the "labor force" (measured in man-hours). Another important parameter was the cost of the end-product. And this annual plan, represented by a list of products with given characteristics, was to be divided into 4 quarters (and ultimately, into 12 months) as evenly as possible for each parameter.

The solution of this problem (called in the literature a *Volumetric-Calendar Planning problem*, or *VCP*-problem) was intended to solve a long overdue problem of the enterprise associated with the **irregularity of production**. For example, an excessive consumption of some consumables in certain periods led to the need to allocate additional space for the storage of those resources; excessive consumption of machines of a certain type (again, in certain periods) created bottlenecks in production, which led to delays and disruption of the planned production schedule; the uneven loading of the "labor force" led to the need in using the (more expensive) overtime. All these led to disruptions in the implementation of the annual plan. Besides, the enterprise had to incur additional, unplanned expenses: the pay for the creation and the service of an excess space for excessive consumption of resources, the pay for the overtime work, the pay for extra work (that could be dispensed with in a well-established systematic production).

The methods used to solve the uniformization problems mentioned above will be discussed in the following sections.

---

[1] I should note right away that we are going to talk neither about *multi-criteria optimization*, nor about finding complete sets of representatives of Pareto-optimal solutions.

3 Two Algorithms of Finding Uniform Distributions of One-Component Items Under Distribution Restrictions

In this section, I will tell about two algorithms for solving such DO-problems (of finding uniform distributions of one-component items) which can be modeled by scheduling problems $\langle P \mid \mathcal{M}_k \mid C_{\max} \rangle$ and $\langle Q \mid \mathcal{M}_k \mid C_{\max} \rangle$. In the latter, each machine M_i has an individual *speed* s_i of processing jobs. The "speed" of machine M_i means that it takes time $t = p/s_i$ to complete the amount of work p. Thus, to minimize the objective function of this problem ($C_{\max}$), instead of achieving the "maximum homogeneity" of the distribution of jobs among machines, one should achieve the "maximum proportionality" of the loads of the machines to their speeds, which will ensure the arrival of all machines to the "finish line" in the minimum time (measured by the last participant).

For solving the latter problem approximately, I developed two versions of the algorithm: a "short" and a "long" one. The result of the "short algorithm" is a schedule S with a guaranteed upper bound on its **absolute error**:

$$C_{\max}(S) \leq LB + \left(1 - \frac{1}{m}\right) \hat{p}_{\max}, \tag{1}$$

where LB is a lower bound on the optimum, and $\hat{p}_{\max} = \max_{(v_k, w_i) \in E_0} p_{i,k}$ is the maximum *job processing time* (the maximum is taken over all arcs of the bipartite graph $G_0 = (V_0, W_0; E_0)$, where the vertices from V_0 correspond to jobs, vertices from W_0 to machines, and $(v_k, w_i) \in E_0 \Leftrightarrow M_i \in \mathcal{M}_k$). The result of the "long algorithm" is a schedule S with an upper bound on the **worst-case performance ratio** of the algorithm:

$$\frac{C_{\max}(S)}{OPT} \leq 2 - \frac{1}{m}. \tag{2}$$

As one can see, for problem $\langle P \mid \mathcal{M}_k \mid C_{\max} \rangle$ (in which the speeds of all machines are identical and are taken for 1), bound (2) follows from (1), since $\hat{p}_{\max} = p_{\max} \leq OPT$. Thus, the "short algorithm" is sufficient to achieve both bounds in the case of problem $\langle P \mid \mathcal{M}_k \mid C_{\max} \rangle$.

This is not the case for problem $\langle Q \mid \mathcal{M}_k \mid C_{\max} \rangle$, since the value of $\hat{p}_{\max}$ can be much higher than the optimum. To solve this problem with bound (2), we will need to use the "long algorithm", implementing the method of "finding balanced restrictions on the set of feasible solutions" (method no. 3, mentioned in the Abstract) used in conjunction with the "short algorithm" procedures.

3.1 The "short" Approximation Algorithm for Solving the Problems $\langle P \mid \mathcal{M}_k \mid C_{\max} \rangle$ and $\langle Q \mid \mathcal{M}_k \mid C_{\max} \rangle$

The algorithm consists of three stages.

At **Stage 1**, we solve a relaxed problem $Z(G_0)$: with the abolition of the requirement to select a unique executor for each job. In this problem, we allow each job to be divided into parts to be processed on different machines. (And we even allow parts of one job to be processed simultaneously.) Thus, the set of feasible solutions of the original problem is significantly expanded, which makes it possible to reduce the value of the optimum. As a result, the exact solution of the relaxed problem gives us a lower bound (LB) on the optimum of the original problem.

The next two stages are aimed at getting rid of the fragmentation of jobs. For this end, we leave in graph G_0 only the arcs containing **fractional parts** of jobs. The resulting sub-graph will be denoted by G_{fr}.

At **Stage 2**, we transform the distribution of jobs on the arcs of graph G_{fr} so that the **total load** of each machine **remains unchanged**, and after the deletion from G_{fr} of all arcs with "integral" shares of jobs (equal to zero or to the whole amount of work of the corresponding job), the graph becomes **acyclic**.

At **Stage 3**, the fractional distribution of jobs on the acyclic graph G_{fr} is **rounded off** to a fully integral one. At that, the load of some machines will increase, but not more than by the amount $\left(1 - \frac{1}{m}\right) \hat{p}_{\max}$ (defined above).

Next, we will describe the methods for solving the problems posed for each stage.

A Solution Method for the Relaxed Problem (Stage 1 of the "short algorithm").

Evidently, the relaxed problem can be modeled by a problem of finding an optimal flow in the directed bipartite graph $G_0 = (V_0, W_0; E_0)$ (defined above), where the "liquid substance" flowing through the arcs of the graph is the *amount of work* that is to be performed on the given set of machines. In any *feasible flow* Π, the total amount of flow issuing from each vertex $v_k \in V_0$ must be equal to p_k (which corresponds in the original scheduling problem to the requirement of the complete performing of job J_k). Denoting by $L_i(\Pi) \doteq \sum_{\{(v_k, w_i) \in E_0\}} \Pi(v_k, w_i)$ the *load* of sink $w_i \in W_0$ (equal to the total amount of flow Π entering vertex w_i) and by $\tilde{L}_i(\Pi) \doteq L_i(\Pi)/s_i$ the *relative load* of sink w_i, we define the objective function of this problem as the **maximum relative load** over sinks ($\tilde{L}_{\max}(\Pi) \doteq \max_{w_i \in W_0} \tilde{L}_i(\Pi)$), that should be minimized over the set $\mathcal{P}$ of feasible flows. It can be easily seen that this objective function coincides with the objective function $C_{\max}(S)$ of the relaxed scheduling problem. Thus, these two problems are equivalent, which gives us the right to apply the same name ("the relaxed problem") to both problems.

Let us next define a functional $\mu : \mathcal{P} \mapsto \mathbb{R}^m$ which maps any flow $\Pi \in \mathcal{P}$ to the vector $\mu(\Pi) = (\mu_1(\Pi), \ldots, \mu_m(\Pi))$ whose components are elements of the family $\{\tilde{L}_i(\Pi) \mid i = 1, \ldots, m\}$ sequenced in the non-increasing order: $\mu_1(\Pi) \geq \cdots \geq \mu_m(\Pi)$. Then for any given flow $\Pi \in \mathcal{P}$, we have: $\mu_1(\Pi) = \tilde{L}_{\max}(\Pi)$.

Now, instead of solving the relaxed problem $Z(G_0)$ (of minimizing the function $\mu_1(\Pi)$ over the set of feasible flows $\Pi \in \mathcal{P}$) we will solve a stronger problem (denoted as $Z^*(G_0)$) of **lexicographical minimization** of the vector-function $(\mu_1(\Pi), \ldots, \mu_m(\Pi))$. The reader may decide that we have overcomplicated our solution method for the relaxed problem. But in fact, it is the other way around! By imposing in this problem higher demands on the optimal solution, we provide its additional remarkable *flow segregation* property, *i.e.*, that of splitting the optimal flow into independent parallel streams flowing in sub-graphs of strictly smaller sizes, which enables us to apply for its solution a recursive routine. As a result, the optimal solution of problem $Z^*(G_0)$ (and with it, the solution of the relaxed problem $Z(G_0)$) can be found in time not greater than $O(T \min\{m, n\})$, where T is the running time of any algorithm for finding the maximal flow/minimal cut in a "nearly bipartite" graph with $O(n + m)$ vertices. Let us show how we do that.

To begin with, let us think over the question: what the optimal flow might look like? Clearly, an "ideal lexicographical minimum" of the objective vector-function $\mu(\Pi)$ could be the vector $\bar{B} = (B, \ldots, B) \in \mathbb{R}^m$, where $B = P/\tilde{S}$, $P \doteq \sum_{v_k \in V_0} p_k$ is the total amount of work over all jobs, and $\tilde{S} \doteq \sum_{w_i \in W_0} s_i$ is the total speed of all machines. Meanwhile, such an "ideal minimum" is attainable only when there is a feasible flow Π in graph G_0 entering each sink $w_i \in W_0$ at the amount of $L_i(\Pi) = Bs_i$. This provides the idea for solving problem $Z^*(G_0)$.

First, we build on graph G_0 by adding two vertices: a common source $\hat{S}$ and a common sink $\hat{F}$. We connect the common source with all vertices $\{v_k\}$ by arcs $(\hat{S}, v_k)$ with capacities $\gamma(\hat{S}, v_k) = p_k$, while the common sink is connected with all vertices $\{w_i\}$ by arcs $(w_i, \hat{F})$ with capacities $\gamma(w_i, \hat{F}) = Bs_i$. The capacities of the remaining arcs $u \in E_0$ will be assumed to be unlimited (set to infinity). This extended graph (in fact, a network) is denoted as $\widehat{G}_0$.

Now, to solve problem $Z^*(G_0)$, we find the **maximal flow** Π^* in network $\widehat{G}_0$ and compute its power $|\Pi^*|$ (the total amount of flow issuing from the source $\hat{S}$, or entering the sink $\hat{F}$). If $|\Pi^*| = P$, this firstly means that the flow **is feasible** for problem $Z^*(G_0)$, and secondly, that all arcs $\{(w_i, \hat{F}) \,|\, w_i \in W\}$ are saturated with this flow, and so, the "ideal minimum" of the objective vector-function is attained, and flow Π^* is the **optimal solution** of problem $Z^*(G_0)$.

In the opposite case (when $|\Pi^*| < P$) we start with finding the *minimal cut* in network $\widehat{G}_0$ (*i.e.*, a partition of the whole set of vertices into two subsets $\langle X, Y \rangle$ such that $\hat{S} \in X$, $\hat{F} \in Y$, and the total capacity of the arcs going from X to Y is minimum). By the Ford & Fulkerson theorem [11], the capacity of this cut coincides with the power $|\Pi^*|$ of the maximal flow. Let $A \doteq V_0 \cap X$, $B \doteq V_0 \cap Y$, $C \doteq W_0 \cap X$, $D \doteq W_0 \cap Y$ (see Fig. 1). It follows from the Ford & Fulkerson theorem that all "forward arcs" of the cut (*i.e.*, those following from X to Y) must be saturated with the maximal flow, while the flow on all "backward arcs" (from Y to X) must be zero. Thus, the following properties of flow Π^* hold:

(a) all arcs from $C \times \{\hat{F}\}$ are saturated;
(b) all arcs from $\{\hat{S}\} \times B$ are saturated;
(c) the flow on backward arcs of the cut (which can only be the arcs from $B \times C$) is equal to zero.

Furthermore, we have the property of network $\widehat{G}_0$:
(d) the set of arcs $A \times D$ is empty. (All such arcs have to be saturated with flow Π^*. Yet once their capacities are infinite, the existence of at least one such arc would mean that flow Π^* has an infinite power, which cannot be true, since $|\Pi^*| < P < \infty$.)

All together, these properties mean that the maximal flow Π^* in network $\widehat{G}_0$ splits into two independent streams: the *left* one ($\hat{S} \to A \to C \to \hat{F}$) and the *right* one ($\hat{S} \to B \to D \to \hat{F}$; see Fig. 1).

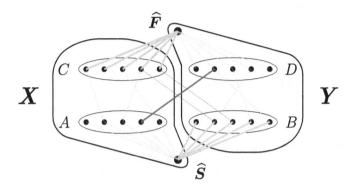

Fig. 1. Splitting of the maximal flow Π^* in network $\widehat{G}_0$ into two streams.

And next we show that the **optimal flow** Π_{opt} in problem $Z^*(G_0)$ has got **exactly the same property**, *i.e.*, it splits into two independent streams: the *left* one (Π'_{opt}), flowing in the sub-graph G_1 defined on the set of vertices $A \cup C$, and the *right* one (Π''_{opt}), flowing in the sub-graph G_2 defined on the set of vertices $B \cup D$. On both sub-graphs G_1, G_2 we can pose similar problems $Z^*(G_1)$ and $Z^*(G_2)$ of finding lexicographic minima of vector-functions $\mu'(\Pi')$ and $\mu''(\Pi'')$ of smaller dimensions equal to $|C|$ and $|D|$, respectively. Clearly, the union of the optimal solutions of these two sub-problems (*i.e.*, the union of the streams Π'_{opt} and Π''_{opt} found in those sub-problems) will give an optimal solution to problem $Z^*(G_0)$, which enables us to speak on a **recursive procedure** for solving problem $Z^*(G_0)$.

Now, concerning the running time of this recursive algorithm. Since, each splitting of the optimal flow is accompanied by dividing the bipartite graph into two sub-graphs defined on nonempty subsets of vertices (A, C) and (B, D), it becomes clear that there may be no more than $(\alpha - 1)$ such splittings, where $\alpha \doteq \min\{n, m\}$. Each of those splittings is accompanied by solving the problems of finding max flow/min cut on sub-graphs of graph G_0. Furthermore, for each of the final (non-divisible) sub-graphs $G_1, \ldots, G_{\alpha'}$ (where $\alpha' \leq \alpha$), into which graph G_0 will be divided as an ultimate result of splitting the optimal flow, we will need to solve the max flow problem. In total, no more than 2α such problems should be solved, which yields the (mentioned above) bound $O(T \min\{m, n\})$ on the running time of the algorithm for solving the relaxed problem at Stage 1.

A Solution Method for the Problem at Stage 2 of the "short algorithm". I remind the reader that the goal of Stages 2 and 3 is to get rid of the fractionality of the flow Π_{opt} obtained by solving the relaxed problem at

Stage 1. First, in graph G_0 we get rid of arcs that already contain "integral" flow values (equal to 0 or p_k). When an arc with flow $\Pi_{opt}(v_k, w_i) = p_k$ is found, we add job J_k to the list $\mathcal{J}_i$ (of jobs assigned to run on machine M_i), and then we remove the arc from graph G_0. Arcs with $\Pi_{opt}(v_k, w_i) = 0$ are just deleted. The bipartite graph with the remaining arcs containing "fractional" flow values only is denoted by G_{fr}; the fractional flow itself is denoted by Π_{fr}. We count the number (K) of arcs of graph G_{fr}.

At Stage 2, while treating the arcs as undirected edges, we aim to **get rid of the cyclability** of graph G_{fr}. To that end, we iteratively repeat procedure $CycleSearch(G_{fr})$ until graph G_{fr} becomes empty:

$$\textbf{repeat } CycleSearch(G_{fr}) \textbf{ until } K = 0.$$

You may ask me: What will we work with at Stage 3, if graph G_{fr} becomes empty already at Stage 2? Actually, it is not all that tragic, because we act as follows.

The search for a cycle is performed in the $CycleSearch$ procedure by constructing a simple path $\mathcal{P}$ in graph G_{fr} until it intersects itself. We start the path from an arbitrary vertex $w_i \in W_0$, and take the first edge in the list $E(w_i)$ of edges incident to it. Having come to some vertex x along an edge e and having made sure that up to this moment vertex x did not appear in the path $\mathcal{P}$ (for otherwise, the desired cycle is found), we simply take the next edge (after e) in list $E(x)$. Such a "blind" way of searching for a cycle can sometimes lead us to a "dead end", to a pendant vertex x (of degree 1). In this case, we remove from G_{fr} the terminal edge incident to x. Yet we remove it not "permanently", but rather transfer it to a "temporary" graph G'_{fr}, initially empty. At the end of the repeat...until cycle, when graph G_{fr} becomes empty, we return all those edges by assigning $G_{fr} \leftarrow G'_{fr}$. (As can be shown, the resulting graph must be acyclic.)

After each such "failed" attempt to build a cycle (ending in a dead end), we start building it again, from the very beginning,—but already in a graph with one less edge. Assume now that the attempt was "successful": the path $\mathcal{P}$ has self-intersected, which enables us to easily detect a cycle $\mathcal{C}$ in it, and thereby complete **Step 1** of procedure $CycleSearch$. Further work of the procedure consists of the following three steps.

Step 2. Clearly, every cycle in graph G_{fr} contains an even number of edges (2β), non-greater than 2α. Starting from an arbitrary vertex $w_i \in \mathcal{C}$, we go around the cycle, numbering its edges in the order of their running: $e_1, e_2, \ldots, e_{2\beta}$. Let E_{odd} and E_{even} denote the sets of edges $\{e_k\}$ of cycle $\mathcal{C}$ with odd and even indices $\{k\}$, respectively. Compute $\delta \doteq \min_{e_k \in E_{odd}} \Pi_{fr}(e_k)$ (where under the "amount of flow on an edge" we assume that amount on the corresponding arc).

Step 3. Transform flow Π_{fr} on the edges of cycle $\mathcal{C}$ by decreasing the flow on all "odd edges" by δ, and increasing it on all edges from E_{even} by the same δ. Flow Π_{fr} obtained after such transformation meets the properties:

(a) The balance of the flow at vertices $v_k \in V_0$ has not changed: the amount of flow outgoing from v_k remains equal to p_k. Thus, the flow **remains feasible**.

(b) The balance of the flow at vertices $w_i \in W_0$ has not changed, as well. Hence, the total load of each sink w_i, as well as its relative load remain "optimal" (like in the optimal flow Π_{opt}).

(c) On at least one edge $e_j \in C$ we have $\Pi_{fr}(e_j) = 0$, and thus, we can do

Step 4. Delete the edge $e_j = (v_k, w_i) \in C$ with flow $\Pi_{fr}(e_j) = 0$ from graph G_{fr}. If the degree of vertex v_k in graph G_{fr} becomes equal to 1 (which means, vertex v_k becomes pendant), delete the last edge incident to v_k from G_{fr}. (Every time when we delete an edge, we decrease K by 1.)

This completes the description of procedure *CycleSearch* and of the algorithm at Stage 2.

A Solution Method for the Problem at Stage 3 of the "short algorithm". The goal of the third (the last) stage of the "short algorithm" is to find such a "rounding" of the fractional flow Π_{fr} (specified on the arcs of the acyclic graph G_{fr}) that the load of any sink w_i increases by at most $\left(1 - \frac{1}{m}\right)\hat{p}_{max}$.

It is easy to propose a simple rounding algorithm with a slightly coarser accuracy bound ($\hat{p}_{max}$). Here it is.

A simple rounding algorithm

Iteratively repeat the following procedure until the number of edges in graph G_{fr} becomes zero.

Procedure *Delete a Terminal Edge*

We look through the vertices $w_i \in W_0$ and find the vertex w_{i^*} of degree 1 and the terminal edge (v_{k^*}, w_{i^*}) incident to it. (If graph G_{fr} still contains $K > 0$ edges, such a vertex and an edge exist for sure.) Add job J_{k^*} to set $\mathcal{J}_{i^*}$ and remove from G_{fr} all edges of list $E(v_{j^*})$. ∎

Justification of the accuracy bound of such an algorithm, as well as that of the polynomial bound on its running time is not difficult. Yet it is of more interest to build an algorithm with the accuracy bound $\left(1 - \frac{1}{m}\right)\hat{p}_{max}$ guaranteed. The first rounding algorithm with such bound was described in [26] by Shchepin and Vakhania in 2005.

I note right away that the accuracy bound is unimprovable. It is achieved in the instance with one job equally distributed in a "fractional distribution" among m machines. Next, I will describe the idea of one of possible algorithms for obtaining such a rounding.

The **key idea** of such an algorithm is to represent the fractional flow not in absolute terms (expressed in fractional parts of either the amounts of work, p_k, or the values of job durations, $p_{i,k}$), but in relative values $\delta(v_k, w_i)$ showing the *share* of job J_k being performed on machine M_i. Thus, for all vertices $v_k \in V_0$, the fractional flow Π_{fr} will meat the equalities

$$\sum_{w_i \in \mathcal{M}_k} \Pi_{fr}(v_k, w_i) = 1, \quad v_k \in V_0 \qquad (3)$$

(which means just the requirement to complete each job J_k). Let us show that any fractional "share flow" Π_{fr} defined on the arcs of the acyclic graph G_{fr} and satisfying (3) can be transformed into an integral flow Π (with shares $\delta(v_k, w_i) \in \{0,1\}$) so as at each vertex $w_i \in W_0$ the sum (a_i) of "additional" shares of jobs executed on machine M_i (that is, the sum of shares received from other machines) will not exceed $\left(1 - \frac{1}{m}\right)$.

Let us refine the definition of quantities a_i. On each arc $u \in E$ we define a non-negative function: $\Delta^+(u) \doteq (\Pi(u) - \Pi_{\mathrm{fr}}(u))^+$ (where Π_{fr} is the original fractional flow, and Π is the integral flow). Then $a_i = \sum_{u \in E(w_i)} \Delta^+(u)$.

It is easy to guess that the fulfillment of the inequalities $a_i \leq 1 - \frac{1}{m}$ over all $w_i \in W_0$ provides the required absolute accuracy bound (1) of the "short algorithm".

The new algorithm (let us call it a "Smart Algorithm") for transforming a given "share flow" Π_{fr} into an integral flow Π also consists of iterations of repeating the procedure "Iteration", until $E_{\mathrm{fr}} = \varnothing$. Before starting the cycle on iterations, we compute the degrees $d(x)$ of all vertices of graph G_{fr} and put $a_i \equiv 0$ for all $w_i \in W_0$.

Iteration

Looking through first all vertices $x \in W_0$, and then $x \in V_0$, we find vertices of degree $d(x) = 1$. (These can always be found as long as there are edges in the graph G_{fr}.) For each vertex x found, we find the (unique) edge $(x, y) \in E_{\mathrm{fr}}$.

In case $x = v_k \in V_0$, $y = w_i \in W_0$ (when a "job vertex" v_k, incident to the only "machine vertex" w_i turned out to be pendant), we **do**:
$\{AddJob(J_k, M_i); DelEdge(v_k, w_i)\}$,
where the procedure $AddJob(J_k, M_i)$ performs all actions associated with adding job J_k to list $\mathcal{J}_i$ (in particular, it defines the integral flow on the arc (v_k, w_i): $\Pi(v_k, w_i) \leftarrow 1$, and adjusts the value of a_i: $a_i \leftarrow a_i + 1 - \Pi_{\mathrm{fr}}(v_k, w_i)$).

The $DelEdge(x, y)$ procedure performs actions related to removing an edge (x, y) from G_{fr} (while not forgetting to decrease the degrees of its end vertices).

In case $x = w_i \in W_0, y = v_k \in V_0$ do **begin**
if $a_i + 1 - \Pi_{\mathrm{fr}}(v_k, w_i) \leq 1 - \frac{1}{m}$ **then** $AddJob(J_k, M_i)$ **else** $\Pi(v_k, w_i) \leftarrow 0$;
$DelEdge(x, y)$
end

The running time of the algorithm is $O(mn)$. The proof of the accuracy bound is not that trivial, yet still possible.

3.2   The Method of Finding Balanced Restrictions on the Set of Feasible Solutions. ("Long Approximation" for $\langle Q \,|\, \mathcal{M}_k \,|\, C_{\max}\rangle$.)

Problem $\langle Q \,|\, \mathcal{M}_k \,|\, C_{\max}\rangle$ being discussed in this section (with restrictions to the job distribution specified by an arbitrary bipartite graph $G = (V, W; E)$) will also be called "Problem $X(G)$", for brevity.

In the previous section we found out that applying the three-stage "short algorithm" to problem $X(G)$ with an input graph G enables us to construct a schedule S with bound

$$C_{\max}(S) \leq \tilde{L}_{\max}(G) + \left(1 - \frac{1}{m}\right) \hat{p}_{\max}(G), \tag{4}$$

where $\tilde{L}_{\max}(G) \doteq \tilde{L}_{\max}(\Pi_{opt}(G))$ is the optimum of the relaxed problem and $\hat{p}_{\max}(G) = \max p_{i,k}$ is the maximum job length reached on the arcs $(v_k, w_i) \in E$ of graph G. As was noted above, for graph G_0 specifying the original distribution restrictions in problem $X(G_0)$, the first summand in bound (4) provides a lower bound on the optimum: $L_{\max}(G_0) \leq OPT$ (which is good for us). At the same time, the second summand may attain quite large values on some "long" arcs of graph G_0, and those values may even exceed the optimum. But we can simply not use such arcs in our solution (since we know that no arc with length $p_{i,k} > OPT$ is used in the optimal schedule)! Let us also not use such arcs in our approximate solution (built by an efficient algorithm), just by removing them from graph G_0. The remaining graph will be denoted as G.

As you might guess, the optimum $(\tilde{L}_{\max}(G))$ of the relaxed problem on the resulting (reduced) graph G may increase slightly (compared to $\tilde{L}_{\max}(G_0)$), but not critical—it will be still not greater than OPT. (Indeed, since the optimal schedule S_{opt} of the original problem $X(G_0)$ does not use any of the removed arcs, it defines an **admissible flow** Π' in network G, which gives an approximate solution for the relaxed problem $Z(G)$: $\tilde{L}_{\max}(G) \leq \tilde{L}_{\max}(\Pi')$, which, due to relations $\tilde{L}_{\max}(\Pi') = C_{\max}(S_{opt}) = OPT$, gives the required inequality: $\tilde{L}_{\max}(G) \leq OPT$.) At that, we will have $\hat{p}_{\max}(G) \leq OPT$, which, due to (4), yields the desired bound (2): $C_{\max}(S) \leq \left(2 - \frac{1}{m}\right) OPT$.

It is O'k, but the question arises: shouldn't we be able to "guess" the value of the optimum (OPT), in order to get the desired bound? No, we will do it another way. We will set various upper bounds $\hat{p}$ on the allowed job lengths $p_{i,k}$, and will remove from graph G_0 all arcs (v_k, w_i) that do not meat the inequality $p_{i,k} \leq \hat{p}$. The remaining graph and the set of its edges will be denoted by $G(\hat{p})$ and $E(\hat{p})$, respectively. Then, denoting $\tilde{L}_{\max}(G(\hat{p})) \doteq \tilde{L}_{\max}(\hat{p})$, we get bound (4) in the form:

$$C_{\max}(S(\hat{p})) \leq \tilde{L}_{\max}(\hat{p}) + \left(1 - \frac{1}{m}\right) \hat{p}. \tag{5}$$

By varying (but within reasonable limits) the values of parameter $\hat{p}$, we will follow what is happening in the right side of bound (5). "Reasonable limits" of parameter $\hat{p}$ are specified by an interval $[\hat{p}_1, \hat{p}_2]$, which initially is set by $\hat{p}_1 \leftarrow \hat{p}_{\min} \doteq \max_{v_k \in V_0} \min_{(v_k, w_i) \in E_0} p_{i,k}$ and $\hat{p}_2 \leftarrow \hat{p}_{\max} \doteq \max_{(v_k, w_i) \in E_0} p_{i,k}$. Indeed, setting too small values of parameter $\hat{p}$ may lead to the removal of too many edges, resulting in the appearance of isolated vertices in graph G. If it appeared to be a "machine vertex" w_i, this makes no problem. (The remaining machines can handle any load, as long as it is "legitimate".) But if suddenly a node v_k turned out to be isolated, then this means that there are no allowed machines to perform job J_k, and therefore, the problem instance with such restrictions has no solution. These conditions determine the lower bound on the range of admissible values of parameter $\hat{p}$ established above. The upper bound (set in $\hat{p}_{\max}$) is explained by the fact that larger values of parameter $\hat{p}$ restrict yet nothing, and the value of $\tilde{L}_{\max}(\hat{p})$, having reached its minimum at $\hat{p} = \hat{p}_2$ (or even earlier), does not change further, while the second term in bound (5) only increases, worsening the total bound.

Let us try to analyze how bound (5) changes when parameter $\hat{p}$ varies. Note that this bound consists of two terms (summands). The first term (function $\tilde{L}_{\max}(\hat{p})$) has got the following properties.

(a) It is a monotonically non-increasing function of $\hat{p}$. It is undefined for the values $\hat{p} < \hat{p}_{\min} = \hat{p}_1$ (due to the absence of feasible solutions to problem $Z(\hat{p})$); however, for the sake of completeness, we will define it for $\hat{p} \in [0, \hat{p}_1)$ by the infinite value: $\tilde{L}_{\max}(\hat{p}) = \infty$.

(b) Function $\tilde{L}_{\max}(\hat{p})$ is piecewise constant and has a finite number of discontinuities. The breaks occur at points $\hat{p} = p_{i,k}$ and are caused by the removal of arcs from graph G_0. (However, not every removal of an arc causes a jump in the value of $\tilde{L}_{\max}(\hat{p})$.) At that, the function is continuous from the right.

(c) In the interval $\hat{p} \in [\hat{p}_2, \infty)$, it is constant and takes its minimum value not exceeding OPT.

The second term in bound (5) is a monotonically increasing function of $\hat{p}$, taking values from 0 to ∞. At the same time, we can say nothing interesting about the total function of $\hat{p}$ in the right part of (5).

Yet let us draw the graphs of functions $\tilde{L}_{\max}(\hat{p})$ and $f(\hat{p}) = \hat{p}$ (see Fig. 2). Since one of these functions is monotonically non-increasing, and the other one is, vice versa, monotonically increasing from 0 to ∞, they have to inevitably intersect at some point $\hat{p}_0$.[2] At that, for all $\hat{p} \in [0, \hat{p}_0) \doteq \mathcal{P}_N$, the inequalities $\tilde{L}_{\max}(\hat{p}) > \hat{p}$ hold, while for all points $\hat{p} \in [\hat{p}_0, \infty) \doteq \mathcal{P}_Y$ (which will be called "Yes-points") the opposite relations hold:

$$\tilde{L}_{\max}(\hat{p}) \leq \hat{p}. \tag{6}$$

We note that inequality (6) is valid at point $\hat{p} = OPT$, which implies that $OPT \in \mathcal{P}_Y$ and $OPT \geq \hat{p}_0$. It follows that any point $\hat{p}$ from the interval $[\hat{p}_0, OPT]$ is also a "Yes-point", and so, it also satisfies the desired bound (2). **Finding such a point is the ultimate goal of our algorithm.**

Note, however, that a more general sufficient condition will also do for us:

$$(\hat{p} \leq OPT) \ \& \ (\tilde{L}_{\max}(\hat{p}) \leq OPT),$$

under which bound (2) also holds (although $\hat{p}$ may not belong to $\mathcal{P}_Y$). Thus, we will be satisfied with any point of the graph of function $\tilde{L}_{\max}(\hat{p})$ lying inside the square $\mathcal{K}^* \doteq [0, OPT] \times [0, OPT]$ (see Fig. 2).

---

[2] However, we should precise what we call "an intersection point of two functions" in case, when one of the two functions is discontinuous. We say that functions $f'(x)$ and $f''(x)$ intersect at point x_0, if either $f'(x_0) = f''(x_0)$ or, for any $\varepsilon > 0$ there are points $x_1, x_2 \in (x_0 - \varepsilon, x_0 + \varepsilon)$ such that $f'(x_1) < f''(x_1)$ and $f'(x_2) > f''(x_2)$.

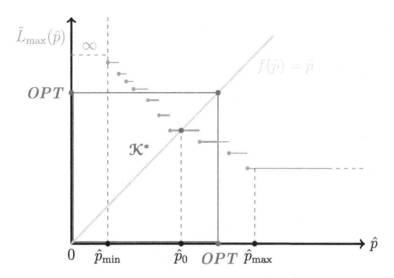

Fig. 2. Graphs of functions $\tilde{L}_{\max}(\hat{p})$ and $f(\hat{p}) = \hat{p}$.

So, let us first make two rough attempts to hit the square $\mathcal{K}^*$ at once. As the first "trial point", we choose the value $\hat{p}_1 = \hat{p}_{\min}$ (for which we know that $\hat{p}_1 \leq OPT$). Solving the problem $Z(\hat{p}_1)$, we find $\tilde{L}_{\max}(\hat{p}_1)$. If it appears that $\tilde{L}_{\max}(\hat{p}_1) \leq \hat{p}_1$ (*i.e.*, $\hat{p}_1$ is a "Yes-point"), than this will mean that $\tilde{L}_{\max}(\hat{p}_1) \leq OPT$, and so, $(\hat{p}_1, \tilde{L}_{\max}(\hat{p}_1)) \in \mathcal{K}^*$, and we are done. Thus, we can further consider the "unfortunate" case, that $\hat{p}_1 \in \mathcal{P}_N$.

As the second "trial point", we take $\hat{p}_2 = \hat{p}_{\max}$, for which we know that $\tilde{L}_{\max}(\hat{p}_2) \leq OPT$. Let us find this value (by solving the $Z(\hat{p}_2)$ problem). If it appears that $\tilde{L}_{\max}(\hat{p}_2) > \hat{p}_2$, then $\hat{p}_2 < OPT$, and we again have a point in $\mathcal{K}^*$ (and we are done). Thus we may further consider only the opposite case: $\hat{p}_2 \in \mathcal{P}_Y$.

So, in the "unfortunate case", when both our attempts to hit square $\mathcal{K}^*$ were unsuccessful, we get a pair of points $\hat{p}_1 \leq \hat{p}_2$ such that $\hat{p}_1 \in \mathcal{P}_N$ and $\hat{p}_2 \in \mathcal{P}_Y$.

And now let us return to our *ultimate goal* declared above: to find (any) point $p^* \in [\hat{p}_0, OPT]$ (*i.e.*, a "Yes-point" lying to the left of OPT). This goal will be achieved efficiently by means of a two-stage algorithm.

The target of **Stage 1** is to find a pair of points $\{\hat{p}_1, \hat{p}_2\}$ such that the following properties hold: (a) $\hat{p}_1 \in \mathcal{P}_N$; (b) $\hat{p}_2 \in \mathcal{P}_Y$; (c) $\hat{p}_2 - \hat{p}_1 \leq \frac{1}{s_{\max}}$. We already have a pair of points $\{\hat{p}_1, \hat{p}_2\}$ satisfying properties (a) and (b). To get a pair that satisfies all three required properties, we apply a *Dichotomy*: we will repeat the following

Iteration

For $p^{\#} \doteq (\hat{p}_1 + \hat{p}_2)/2$, we solve problem $Z(p^{\#})$. If it appears that $\tilde{L}_{\max}(p^{\#}) \leq p^{\#}$ (*i.e.*, $p^{\#}$ is a "Yes-point"), then $\hat{p}_2 \leftarrow p^{\#}$. Otherwise, $\hat{p}_1 \leftarrow p^{\#}$. ∎

After each such iteration, the length of the interval $[\hat{p}_1, \hat{p}_2]$ is halved. Since the length of the original interval $([\hat{p}_1, \hat{p}_2] = [\hat{p}_{\min}, \hat{p}_{\max}])$ is less than $\hat{p}_{\max} \leq p_{\max}/s_{\min}$, in order to obtain an interval of length $\leq \frac{1}{s_{\max}}$, we need to execute no more than $O(\log_2 p_{\max} + \log_2 \frac{s_{\max}}{s_{\min}}) \leq O(|I|)$ iterations, where $|I|$ is the input length of a given instance I of the original problem X.

Note that at each iteration of this dichotomy, point $\hat{p}_0$ (*i.e.*, the left-most point of the "Yes"-interval) belongs to the interval $[\hat{p}_1, \hat{p}_2]$. And since the length of the latter rapidly decreases, point $\hat{p}_2$ is getting closer and closer to point $\hat{p}_0$, and at some moment (as we hope) it will appear **to the left of point** OPT (which stands to the right of $\hat{p}_0$), and this would be exactly what we need. . .

Tr-r-r!!! Not that fast. In fact, Stage 1 may end without happening this happy event. In this case, OPT must be somewhere in the open interval $(\hat{p}_1, \hat{p}_2)$. Note that OPT coincides with $C_{\max}(S_{opt})$, which, in turn, coincides with the completion time of some (critical) machine M_{i^*}, and this time is equal to $\sum_{J_k \in \mathcal{J}_{i^*}} p_k/s_{i^*}$, *i.e.*, is equal to the total processing time of jobs running on machine M_{i^*}. Taking into account that all $\{p_k\}$ are integers, we conclude that OPT is a multiple of $1/s_{i^*}$ for some index $i^* \in \{1, \dots, m\}$. Since we don't know what i^* is equal to (because we don't know which machine will become critical in the optimal schedule S_{opt}), we have to check all possible contenders for the OPT value. As such, we consider **all values** that are multiples of some of m numbers $\{1/s_1, \dots, 1/s_m\}$ and fall into the interval $(\hat{p}_1, \hat{p}_2)$. (Such points will be referred to as "trial points".) Fortunately, there cannot be too many such points: for each $i \in \{1, \dots, m\}$, **no more than one** (since $\hat{p}_2 - \hat{p}_1 \leq 1/s_{\max} \leq 1/s_i$).

We determine all such trial points, and among them, we find the smallest Yes-value $\hat{p}^*$ (if any, of course; it may happen that among trial points there are no Yes-points; it even may be that there are no trial points at all). We note that finding such a value $\hat{p}^*$ requires solving no more than m additional problems $Z(\hat{p})$.

If the value $\hat{p}^*$ appeared to be defined, then we can state with a full confidence that OPT cannot lie strictly to the left of this value, which implies that the Yes-point $\hat{p}^*$ lies to the left of OPT, as required. We take $\hat{p}^*$ for the desired value p^*, and we are done.

In the opposite case, when no Yes-point has appeared among trial points (or there are no trial points at all), we can assert with no less certainty that OPT cannot lie strictly to the left of point $\hat{p}_2$, which implies that the Yes-point $\hat{p}_2$ lies to the left of OPT, and we are done again.

4   Problems of Multidimensional Uniformization and the Method of Compact Vector Summation

Let us present the formal setting of the *VCP-problem* (that was introduced in Sect. 2) in more abstract terms.

The Problem on Uniform Distribution of Items

There are n things (items) $\{J_k \mid k = 1, \dots, n\}$ and m parameters for their characterization; the kth item is characterized by an m-dimensional vector

$p_k = (p_{1,k}, \ldots, p_{m,k})$. It is required to partition the set of items into r subsets "uniformly" in all parameters, so as the total m-dimensional characteristic vectors of all parts were "nearly equal" as much as possible.

For the first time, I got acquainted with this problem in the spring of 1973, when my supervisor Vitaly Afanasyevich Perepelytsa formulated it for me as a "summer vacation task". For this problem at that time neither efficient exact solution algorithms were known (as they are not known to this day), nor any efficient approximation algorithms with accuracy bounds guaranteed. The only exception was a special case of this problem, in which a set of "two-dimensional stones" had to be split into two "heaps". Such a problem can be easily solved by a simple algorithm (in linear time) with an absolute error not greater than $p_{\max} \doteq \max p_{i,k}$.[3] At the same time, for "3D stones" no efficient approximation algorithms were known. This is what I was suggested to do: to come up with an efficient algorithm for approximate solution of the three-dimensional VCP-problem. Thus, "an interesting summer" was guaranteed to me.

Since I was not constrained by the obligation to solve the problem exactly, I could reason quite freely, allowing the solution of the problem to deviate from the "mainstream" optimization in any direction (but within reasonable limits). I reasoned as follows.

The annual plan is represented by the total vector $P = \sum p_k$, which means that the total characteristic vector of the first quarter should be $P/4$ (or so). Assume that we have some "near optimal" solution. Then, summing the vectors of objects that fell into the first quarter, we get the total vector "near the point" $P/4$. If we add the vectors of the second quarter, then the total vector should be near the point $2P/4$, and so on. Moreover, if we try to evenize the solution over months, then the *trajectory of vector summation* (*i.e.*, the sequence of intermediate sums of these vectors) should also pass near the points $P/12, 2P/12, \ldots$. But why, one wonders, should we achieve the uniformity only by the end of each reporting period? After all, the production rhythm is required throughout the planning period, isn't it?! So, let us demand that the summation trajectory of vectors p_k deviates not too far from the straight line $\hat{P} = \{tP \,|\, t \in \mathbb{R}\}$ (determined by vector P) **all the way along**. Then, sooner or later, it will "pass by" point $P/12$, point $2P/12$, *etc.*,$\ldots$, and will pass by all the points we need, and it will only remain to "cut" this sequence at right places into pieces of "about the same length" (see Fig. 3).

Yet how to ensure that the entire trajectory is close to the straight line $\hat{P}$? It can be observed that if we project all vectors $\{p_k\}$ onto the hyperplane H orthogonal to line $\hat{P}$, then the sum of the projections will be equal to zero, and the requirement to the summation trajectory "not to deviate too much" from line $\hat{P}$ transforms into the requirement to find such a summation order of vectors-projections that all their intermediate sums (or their entire "summation trajectory") fit in an $(m-1)$-dimensional ball (in hyperplane H) of the smallest possible radius. Indeed, if we derive a simple formula for an upper bound on inaccuracy of our approximate solution to the VCP-problem, then its dependence

---

[3] I recommend it to readers as a simple exercise.

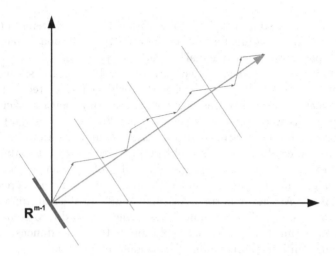

Fig. 3. A geometrical method for solving the VCP-problem.

on the radius of projections summation will be easily seen. From this formula, one will be able to see that the smaller is the radius of projections summation, the smaller is the bound on the inaccuracy of the resulting solution, estimated by the norm of the vector-difference between the total vector of each quarter and vector $P/4$ corresponding to the quarterly load in the "ideal" (absolutely uniform) quarterly partition of the annual plan.

And then I suggest the reader to go beyond the framework of this certain discrete optimization problem (so as not to bind ourselves with the need to elaborate efficient algorithms) and to formulate a more **generally valid mathematical problem**, trying to answer the question: by which amount can be estimated (in the worst case) the summation radius of a given family of vectors with zero sum? Obviously, **by nothing** except the infinity, if we allow the family contain vectors of an arbitrarily large norm. On the other hand, it is clear that any given instance of the problem **can be scaled** (by dividing every vector by the amount equal to the maximum norm among the vectors of the family), which nohow affects the essence of the problem under solution. As a result, we obtain a finite family of vectors with two properties (let us, so far, call such families as "proper" ones): (a) all these vectors are from the **unit ball** of the norm defined in space $\mathbb{R}^d$; (b) their sum is equal to zero. It is required to estimate the minimum radius within which the vectors of the "worst" such family (if any) can be summed. First of all, it is interesting to find out if there is some finite value of this "minimum radius" or, alternatively, it can take arbitrarily large values (infinitely growing, for example, along with the number of vectors, despite that we have limited their norms by 1)? In case that the radius is limited, what does it depend on? (Does it depend, say, on the dimension of space or on the norm specified in the space?)

Thus, such a seemingly simple problem unexpectedly gave rise to a lot of interesting and rather difficult questions that I had to resolve somehow. One of the questions was solved after I managed to prove that on the Euclidean plane any such family of vectors can be summed up within the ball of radius $\sqrt{3}$. (Some time later, this bound was improved down to $\sqrt{2}$.) Thus, in the case of the plane, it was proved that the summation radius is bounded by a constant. This immediately implied the existence of such a solution in the three-dimensional VCP-problem in which the norm of the vector-difference between the total vector of each month/quarter and the "ideal vector" is not greater than $Cp_{\max}$, where C is a constant. As an "additional bonus", it was shown that the sequence of vectors that provided their summation in a ball of radius $\sqrt{3}$ can be found **in polynomial time**, which gave an **efficient approximation algorithm** for the 3D VCP-problem with an absolute performance guarantee. (So, as mathematicians say: "*quod erat demonstrandum!*")

In fact, Perepelytsa formulated to me (as a "task for the summer") **two problems**. One of them is already known to the reader—this is the VCP-problem. The second one is the even more famous *Johnson problem for three machines*. The polynomial algorithm for exact solution of the Johnson problem for two machines (belonging to Salmer Martin Johnson) was well known in scheduling theory[4]. At the same time, numerous attempts to construct a similar algorithm for the three-machine problem, undertaken over the next 20 years, had not brought any tangible results. No efficient approximation algorithms for its solution with non-trivial accuracy bounds guaranteed were known for it, either. So, Vitaly Afanasyevich also advised me to think on this problem and try to solve it, if not exactly, then at least somehow approximately. (It should be said that approximation algorithms were not respectable enough at that time, and were treated as results of a "second class".)

In total, two completely different (at first glance) problems were presented to me "for my choice", which needed to be solved "somehow approximately". With one of them, the VCP-problem, I successfully coped during my summer vacation, which I spent in my "small Motherland", Petropavlovsk (a city in Northern Kazakhstan). Novosibirsk had to be reached by railway. And when I was already sitting at the station waiting for the train, an unexpected thought came to my mind: "Is it possible to "apply" (such a nice) problem that I recently invented (on vector summation within a ball of minimum radius; hereinafter, it will be called a *Compact Vector Summation* problem, or *CVS*-problem) to the Johnson problem, so as to solve it approximately?" It seemed to have taken me no more than 10 min to conclude: "It works, indeed!"

Indeed, in the Johnson problem, all machines line up in a chain $M_1, \ldots, M_m$ along one conveyor line (see Fig. 4), and each job J_k consists in manufacturing a *product* k on those machines. The whole job J_k consists of m successive operations $O_{1,k}, \ldots, O_{m,k}$ to be performed on machines $M_1, \ldots, M_m$ in this order. Operation durations are given by a matrix $(p_{i,k})_{i=1,\ldots,m}^{k=1,\ldots,n}$.

---

[4] In fact, from this result, published on May 5, 1953 [15], the science called today *Scheduling Theory* takes its origin.

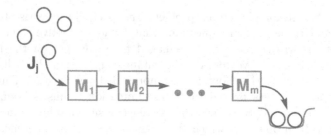

Fig. 4. The Johnson problem.

In addition, the problem has a limitation that is standard for all classical multi-stage scheduling problems: no more than one job can be executed simultaneously on each machine. Thus, all parts have to go through each machine in some order $\pi = (\pi_1, \pi_2, \ldots, \pi_n)$ (that can be chosen). If the job orders on all machines are identical (and can be specified by a single permutation π), then such a schedule S_π is called a *permutation schedule*. Any given permutation of jobs (π) specifies precedence constraints on the set of operations, visually depicted as a directed network whose vertices are operations (see Fig. 5).

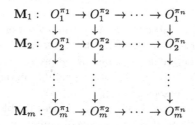

Fig. 5. Precedence constraints on the set of operations in schedule S_π.

In fact, the orders of processing the jobs **may be different** for different machines! Of course, the reader may ask me a question: how to get different job orders on two neighboring machines under pipeline conditions? Indeed, to get such different orders, at least one part has to overtake on the conveyor some other part that goes ahead of it. The answer to this question is that the master running on machine M_i has a "waiting shelf". After processing a part on his machine, he may not immediately send it to the conveyor, but put it on the shelf. Then he processes several parts following the postponed part, and only after that he puts the delayed part on the conveyor and lets it go further. It may seem that such artificial delays are not able to help in shortening a schedule. And this is really true, if the number of machines on the conveyor is not greater than three! But already for four or more machines, problem instances are known when such a delay enables one to build a shorter final schedule. (In other words, for such instances, none of the $n!$ permutation schedules is optimal.)

However, despite of this fact, we will look for an approximate schedule precisely in the class of permutation schedules. As will be seen from the bound on the absolute error of such schedules, the rejection of the wide class of non-permutation schedules (including, maybe, the optimal one) worsens the final result of the problem solution not too much. At that, to find a suitable permutation of jobs π that defines a "nearly optimal" permutation schedule S_π, the solution of the CVS-problem can be used. Let us show how to do this.

Next, let us consider network G_π determining precedence constraints on the set of operations in permutation schedule S_π specified by a job permutation π (see Fig. 5). With respect to this value, we will estimate the length of any schedule S. At that, the difference $C_{\max}(S) - P_{\max}$ will be estimated in units equal to the maximum operation length $(p_{\max} \doteq \max_{i,k} p_{i,k})$. To get an upper bound on the "worst-case" absolute error of the schedule under construction (i.e., a bound that would be valid for the "worst" problem instance), let us "worsen" the original instance as follows: we equalize the loads of all machines up to $P_{\max}$ by increasing (if necessary) the durations of some operations on underloaded machines, but without incrementing the value of $p_{\max}$. (It can always be done, since the number of operations performed on each machine is the same, n.) Thus, we can further assume that

$$P_i = P_{\max}, \quad i = 1, \ldots, m. \tag{7}$$

And now let us take a closer look at network G_π determining precedence constraints on the operations in a permutation schedule S_π, specified by a permutation of jobs π. The nodes of this network (representing operations $O_{i,k}$) have weights $p_{i,k}$. It is well known in scheduling theory that the length of an active schedule specified by such a network is equal to the length of the longest path in that network. Thus, we have: $C_{\max}(S_\pi) = \max_{\mathcal{P} \in \mathcal{P}(G_\pi)} P(\mathcal{P})$, where the "length" of a path $\mathcal{P}$ is defined as the total weight of its nodes: $P(\mathcal{P}) = \sum_{O_{i,k} \in \mathcal{P}} p_{i,k}$.

While looking at network G_π (Fig. 5), we can observe that each longest path in this network goes from operation O_{1,π_1} to operation O_{m,π_n} (top left corner to bottom right corner of the network) and looks like a "staircase" descending to the right, with the ith stair rung representing the operations of some subset of jobs processed by machine M_i (see Fig. 6).

The length of such path $\mathcal{P}$ can be represented in the form:

$$P(\mathcal{P}) = \sum_{k=1}^{k_1} p_{1,\pi_k} + \sum_{k=k_1}^{k_2} p_{2,\pi_k} + \cdots + \sum_{k=k_{m-1}}^{n} p_{m,\pi_k}$$

$$= \sum_{i=1}^{m-1} p_{i,\pi_{k_i}} + \left(\sum_{k=1}^{k_1-1} p_{1,\pi_k} - \sum_{k=1}^{k_1-1} p_{2,\pi_k} \right) + \left(\sum_{k=1}^{k_2-1} p_{2,\pi_k} - \sum_{k=1}^{k_2-1} p_{3,\pi_k} \right)$$

$$+ \cdots + \left(\sum_{k=1}^{k_{m-1}-1} p_{m-1,\pi_k} - \sum_{k=1}^{k_{m-1}-1} p_{m,\pi_k} \right) + P_{\max}.$$

In this expression, we can observe the amount $P_{\max}$ representing a lower bound on the optimum. Beyond this, the total length of the last operations of the

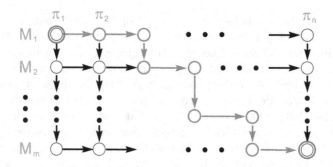

Fig. 6. A critical path in network G_π.

stair rungs from the first to $(m-1)$th (gathered in the first sum) can be simply estimated by $(m-1)p_{\max}$. But what can be said about the expressions in parenthesis?

To estimate these expressions, let us define for each $k = 1,\ldots,n$ vector $p_k = (p_{1,k},\ldots,p_{m,k})$ compound of the lengths of all operations of job J_k. Due to (7), we can say that vector $P = (P_1,\ldots,P_m) \doteq \sum_{k=1}^{n} p_k$ (representing the vector of machine loads) has equal components. This means that if we find an order π of summing the vectors $\{p_k\}$ such that the summation trajectory in its every point is close to the straight line $\hat{P} = \{tP\,|t \in \mathbb{R}\}$, then any node of this trajectory (representing the sum of the first j vectors p_k of the trajectory for some j) will have "approximately equal" components. The difference between any two of them can be estimated in terms of the distance from the node to the straight line $\hat{P}$. From the previous part of this paper we already know that to obtain the summation trajectory of vectors $\{p_k\}$ close (in every its point) to the straight line $\hat{P}$, it is sufficient to solve the CVS-problem for projection vectors in the space of dimension $(m-1)$. Thus, we can derive an upper bound on the length of schedule S_π obtained by means of solving the CVS-problem:

$$C_{\max}(S_\pi) \le P_{\max} + \psi_2(r(m-1), m-1)p_{\max}, \tag{8}$$

where $r(m)$ is the radius of vector summation that we can guarantee for any "proper" family of vectors in $\mathbb{R}^m$, and $\psi_2(r,m)$ is a polynomial function on r and m. And once I already had a polynomial-time algorithm for approximate solution of the two-dimensional CVS-problem (providing the radius $\sqrt{3}$ of vector summation for any "proper" family of vectors in euclidean $\mathbb{R}^2$), this immediately gave a polynomial-time approximation algorithm for the three-machine Johnson problem with a bound on the absolute error: $C_{\max}(S) - OPT \le C_{\max}(S) - P_{\max} \le Cp_{\max}$, where C is a constant. Thus, with one shot, I "killed two birds": the VCP-problem (so far, three-dimensional) and the Johnson problem for three machines.

But the appetite, as we know, comes with eating! I really wanted to build similar approximation algorithms for these two problems in the general case, for arbitrary values of m. And this desire bumped into the lack of an efficient algorithm for solving the VCP-problem in a space of arbitrary (finite) dimension. Of course, first of all, I wanted to get an answer to the fundamental question: is it possible in space $\mathbb{R}^d$ to limit the summation radius of "proper" vectors to some constant, not growing with the number of vectors?

I received the answer in winter of 1973/74, and luckily, it turned out to be **positive**. Although, the bound on the radius obtained ($r(m) \leq 2^m \cdot (1 + \varepsilon)^{m-1}$, attainable for any fixed value of $\varepsilon \in (0, 1)$) was exponential in space dimension m, but a positive bonus was that the bound was achieved constructively, in polynomial time $T \leq O(\varepsilon^{-1} n \log n)$ (for any chosen value of $\varepsilon \in (0, 1)$).

While being quite efficient, that algorithm was rather sophisticated. So, I invented a more straightforward algorithm that guaranteed even better bound:

$$r(m) \leq \sqrt{4^m - 1}/\sqrt{3} < 2^m/\sqrt{3}. \tag{9}$$

Yet, it was also exponential in m (which was the fault of the method of deriving the bound by induction on space dimension). Besides, the running time of the straightforward algorithm was much worse.

Thus, my efficient algorithm of vector summation provided not less efficient algorithms (of running time $O(n \log n)$) for approximate solution of the two problems being of interest to me (the VCP- and the Johnson problem) **in the general case** (for arbitrary values of m), with bounds on their absolute errors not greater than $\varphi_1(m) p_{max}$ and $\varphi_2(m) p_{max}$ for some functions $\varphi_i(m)$ ($i = 1, 2$). Well, those functions were exponential in m so far, yet for any fixed value of m and an increasing n, those bounds guaranteed an **asymptotic optimality** of the solutions obtained, provided that a rather uncomplicated condition is met: $p_{max}/P_{max} \to 0$. (Indeed, the condition is quite natural, since P_{max} is the total length of n operations, while p_{max} is the length of one operation.)

These results were briefly announced [17] in the abstracts of the 3rd All-Union Conference on Theoretical Cybernetics, held in June 1974 in Novosibirsk. (The complete version of these results was submitted to transactions of the Sobolev Institute of mathematics, where they were published in 1975 [18]) And a few months later my supervisor Perepelytsa received a letter from our Ukrainian colleagues, Belov and Stolin (from Kharkov) containing a strange and intriguing question: "What a proof of the *Steinitz Lemma* did your student use?"...

Thus, I suddenly found out that I was far from being the first researcher of this interesting problem (on vector summation within a ball of minimum radius) and not the first one who obtained the result about the possibility of summing the "short" vectors within a ball of limited radius. In whole, this story turned out to be quite rich in interesting results, unexpected applications, and the names of famous mathematicians involved in this fascinating process. A brief background of the events described was as follows.

Apparently, the first guy who investigated this issue was the German mathematician Ernst Steinitz. In 1913, he published the paper [27], in which Riemann's classical theorem on the set of sums of a conditionally convergent number series was generalized to the case of m-dimensional vector series. (Steinitz proved that the set of sums of any m-dimensional series is a k-dimensional linear manifold for some $k \leq m$.) The proof of this fact was essentially based on the lemma (proved in the same paper)[5] which stated that any finite "proper" family of vectors $\{x_1, \ldots, x_n\}$ in m-dimensional Euclidean space (*i.e.*, a family that meets the properties:

$$\|x_i\| \leq 1 \ (i = 1, \ldots, n), \quad \sum x_i = 0)$$

can be "summed within the ball" (with the center in the origin) of a limited radius R_{St}, independent of the number of vectors of the family. The latter means that all partial sums of the first t vectors (for any $t = 1, \ldots, n$) belong to that ball.

Steinitz had no need to minimize the radius R_{St} to obtain his main result (the theorem on the range of sums of a series). He was satisfied by the fact that R_{St} was bounded by a constant (for a given space $\mathbb{R}^m$ with fixed dimension m). However, other researchers were interested in the very question of the minimum value of the radius. (This minimum value of the radius will be further called a *Steinitz radius* or a *Steinitz constant*, and denoted by $R_{St}(s, m)$, where m is the space dimension and s is the norm given in $\mathbb{R}^m$.) As a result, the papers by (Gross, 1917 [13]) and (Bergström, 1931 [7]) appeared in which the following upper bounds on the Steinitz constant in the Euclidean plane were derived:

$$R_{St}(\ell_2, 2) \leq \sqrt{2}; \tag{10}$$

$$R'_{St}(\ell_2, 2) \leq \sqrt{5/4}. \tag{11}$$

True, the value $R'_{St}(\ell_2, 2)$ present in bound (11) does not quite match what we are interested in. Yes, Bergström can fit the vector summation trajectory into a ball of radius $\sqrt{5/4}$, but the center of the ball may not coincide with the origin. (This means that we cannot guarantee bound (11) for all partial sums of the first t vectors.)

In his other paper [6], published in the same issue of the journal in 1931, Bergström proved exactly bound (9).[6]

Apparently, the papers of Bergström were similarly missed by the young Ukrainian mathematician M.I. Kadeč, who also reproved Bergström's bound (9) in 1953 (see [16]). Of course, the "classical theorist" Kadeč in 1953 was not interested in issues of "running time" of algorithms. (This concept acquired such an great significance some time later, along with inventing the computers.) However, in 1974, his (former) students, I.S. Belov and Ya.N. Stolin derived a bound [5] on the complexity of Kadeč's constructive proof. That was $O(n^m)$, which was, clearly, inferior to the $O(n \log n)$ bound of my efficient algorithm. It was probably this difference that led them to the idea that I used "a different proof of the Steinitz Lemma".

As I said above, unlike "pure theorists", I was concerned with applications of the CVS problem to "more real-life" problems. The quality of such an application can be evaluated by two characteristics: the accuracy of the resulting solution and the running time of the algorithm. And while everything was OK with the

---

[5] now referred to in the literature as the *Steinitz Lemma*.

[6] Thus, my second bound turned out to be a repetition of a known result...

second indicator, the first one (exponentially depending on m) still could not satisfy me, and I continued to think over how to solve this problem. As a result, in the autumn of 1977 I made a crucial improvement of the bound on the radius of vector summation:

$$r(s, m) \leq m. \tag{12}$$

This bound was valid for any symmetric norm s in $\mathbb{R}^m$, and the desired order of vector summation could be found in $O(m2^m \cdot n^2)$ time. Thus, it was not only the first bound different from the exponent[7], but also the first bound obtained for norms different from the Euclidean norm.

This bound had interested Professor Kadeč, and in spring of 1978 I was invited to have a talk at his seminar in the Kharkov University. Yet the day before that talk, a detailed analysis of my result took place at the seminar by I.S. Belov and V.S. Greenberg (at the same university). And it should be noted that despite the rather high complexity of that proof (and the complexity of the algorithm for finding the order of vector summation), the seminar coped with this task "for an excellent mark" ! Practically, not a single nuance of that complex result remained incomprehensible to the audience (which is not that common).

Along with the long proof, I had a short proof of the rougher bound $r(s, m) \leq 2m$ (and the corresponding "short" vector summation algorithm). And since Kadeč was primarily interested in the very fact of the existence of a linear bound on the radius, I was recommended to present at Kadeč's seminar only my short proof, what I've done. However, shortly after my returning to Novosibirsk, I received a letter from Greenberg, where he wrote that my long proof was not needed, since the same bound (12) holds for the "short" algorithm. In his letter, he gave a very short proof. This is how our joint paper (of one and a half pages) appeared in the journal "Functional analysis and its applications" [12], containing a complete (although, not quite constructive) proof of bound (12). It was also shown there that the bound is valid for any norm s (including **non-symmetric norms**), and moreover, in the class of non-symmetric norms it was **unimprovable**. We could not prove this for symmetric norms, but an instance was constructed that gave a lower bound on the (minimal) Steinitz radius for the ℓ_1 norm:

$$R_{St}(\ell_1, m) \geq (m + 1)/2. \tag{13}$$

That instance showed that in the class of symmetric norms, bound (12) cannot be improved by the order of magnitude.

In the same 1980, paper [20] was issued, where I designed a fully polynomial algorithm for finding the order of summation of any proper family of vectors with bound (12) on the radius guaranteed. Namely, instead of bound $O(m2^m \cdot n^2)$ on the running time (exponential in m) obtained earlier in [19], the new algorithm had running time $O(m^2 n^2)$.

---

[7] As strange it may seem, by that time it was not known that, in fact, Steinitz's proof of his Lemma yielded a **certain value** of the summation radius, and this value was $2m$ (!) Unfortunately, neither I, nor Kadeč, nor even Bergström had a correct idea of that result.

In addition, the scheme of reducing the Flow Shop problem to the CVS problem was significantly improved, which was facilitated by the use of a special s_0 norm invented by me. The norm is defined by its unit ball $B_{s_0} = \text{conv}\{C^+, C^-\}$, which is the convex hull of two convex bodies: of the "positive unit cube" $C^+ \doteq \{x = (x_1, \ldots, x_m) \in \mathbb{R}^m \mid x_i \in [0,1] \ (i = 1, \ldots, m)\}$ and of the "negative unit cube" $C^- \doteq -C^+ \{x = (x_1, \ldots, x_m) \in \mathbb{R}^m \mid x_i \ in[-1,0] \ (i = 1, \ldots, m)\}$. As a result, a formula was obtained for the dependence of the absolute error on the bound on the summation radius for vectors from $B_{s_0}(m-1)$:

$$C_{\max}(S_\pi) - P_{\max} \leq (m-1)(r(s_0, m-1) + 1)p_{\max},$$

providing (together with bound (12)) an improved bound on the absolute error of the algorithm:

$$C_{\max}(S_\pi) - P_{\max} \leq m(m-1)p_{\max}.$$

Its running time was the running time of the CVS algorithm, namely, $O(m^2 n^2)$.

Although, as mentioned above, it was not possible to improve the order of magnitude of bound (12) in the class of all symmetric norms, one could try to do this with respect to some particular norm. For example, the lower bound on the Steinitz radius

$$R_{St}(\ell_2, m) \geq \sqrt{m+3}/2$$

given in [12] (and still unimproved) provides a ground to assume that the exact value of $R_{St}(\ell_2, m)$ is most likely of the order of $O(\sqrt{m})$. However, this assumption still remains open: it neither has been proven, nor disproved. The exact value of the Steinitz radius in Euclidean space is known so far for only the two-dimensional space $\mathbb{R}^2$, due to the paper of the talented Polish mathematician Wojciech Banaszczyk [3]. He showed that $R_{St}(\ell_2, 2) = \sqrt{5}/2$, thus heaving removed just a "prime" from Bergström's bound (11); but what an important "prime"!

In the same paper, Banaszczyk proved a bound valid for any norm in $\mathbb{R}^2$:

$$R_{St}(s, m) \leq 3/2.$$

The tightness of this bound follows from (13). Thus, we got the second exact value of the Steinitz radius:

$$R_{St}(\ell_1, 2) = 3/2.$$

There is an assumption (also open at the moment) that the value of the Steinitz radius for any symmetric norm in space $\mathbb{R}^2$ is in the interval between these two known values:

$$\sqrt{5}/2 = R_{St}(\ell_2, 2) \leq R_{St}(s, 2) \leq R_{St}(\ell_1, 2) = 3/2.$$

Finally, in [3] Banaszczyk made a seemingly minor but, in fact, quite important advance in the bound of the Steinitz radius for a space of arbitrary dimension m, by announcing the bound (for any symmetric norm s):

$$R_{St}(s, m) \leq m - 1 + 1/m \tag{14}$$

strictly improving bound (12). It wasn't for me a great job to design a polynomial-time algorithm of vector summation within the ball of this radius [23]. It's running time was the same as for bound (12), $O(m^2n^2)$. Besides, I showed that the algorithm also works (with the same bound (14)) with any non-symmetric norm whose unit ball is symmetric with respect to some inner point (not necessarily coinciding with the origin).

Among other interesting results obtained on the CVS problem, we could mention the lower bounds on the Steinitz radius in space $\mathbb{R}^m$ with the norm ℓ_∞, obtained by means of the Hadamard matrices H_n (the existence which for any n multiple of 4 also remains an intriguing open problem since the end of the 19th century).

Yet, it is time to revisit the application problems, for the sake of which the CVS problem was "reinvented". In fact, in parallel with the work on improving the bounds on the Steinitz radius for various norms (and the work on improving vector summation algorithms), an intensive work was also carried out to find new applications of this problem to other scheduling problems. The initiators of this research direction were my Ukrainian colleagues: I.S. Belov, Ya.N. Stolin, B.I. Dushin et al. In particular, they showed that in addition to the "single-route" Flow Shop problem, the CVS problem is also applicable to the (two-route) counter-routes problem [2], to the three-machine Akers-Friedman problem (a special case of the Job Shop problem, in which for each job J_k its route through machines is given and is a permutation of machines) [1], to the Akers-Fridman p-routes problem [10]. At the same time, Bárány and Fiala [4] in 1982 implemented the CVS problem to the Open Shop problem (the scheduling problem with non-fixed routes of jobs through machines), which allowed finding (in polynomial time) optimal schedules for a wide class of instances of the (NP-hard, by the way) Open Shop problem. While keeping in mind that for the CVS problem, bound (12) on the radius of vector summation was already known, this class was defined by the relation:

$$P_{\max} \geq (m^2 + 2m - 1)p_{\max},$$

sufficient to guarantee the equality $C_{\max}(S_{opt}) = P_{\max}$. And moreover, the efficient vector summation algorithm with bound (12) makes it possible to find such an optimal schedule in polynomial time.

However, I have mentioned here only a small part of the results related to this topic. The readers who want to get acquainted in more detail with all the vicissitudes of that exciting race for records in accuracy and running time (the race in which I also happened to take an active part), are referred to my survey paper [24] of 1994. And yet one more certain result should be mentioned apart.

We note that all scheduling problems with fixed job routes through machines mentioned above are special cases of a much more complex problem, known in the literature as the *Job Shop problem*. In this problem, each job J_k represents a chain of successive operations $(O_{1,k}, O_{2,k}, \ldots, O_{\mu_k,k})$, each of which $(O_{q,k})$ is characterized by its duration $p_{q,k}$ and a machine-executor $M_{q,k}$. As a result, each job has an individual (and fixed) route through machines in which the same machine can be repeated any number of times (even at a stretch; see Fig. 7). Thus, the *route length* (μ_k) of each job J_k is a parameter independent of the parameters m (number of machines) and n (number of jobs).

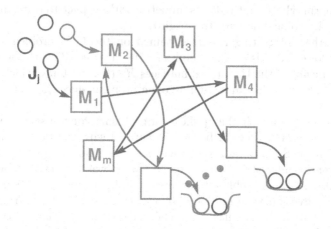

Fig. 7. Job Shop problem.

As you remember, for the Flow Shop problem, we built a *permutation schedule*, in which on each machine the jobs were performed in the same order, namely, in the order that specifies a "uniform summation" of the vectors of operation durations of each job. (We used the CVS problem to find such a "uniform summation".) But for most instances of the Job Shop problem, a construction of such a schedule is impossible in principle. And at first glance, it is absolutely unclear, where (in that confusing "web" of different routes of jobs through machines) a single permutation of vectors may be hidden? (And of which vectors, by the way?). In other words, at first glance, the CVS problem seems to be inapplicable to such a complex problem as Job Shop (in its general form). However, even here, the **uniformization of the loads of machines in time** (achieved by solving the CVS problem) has ultimately played a decisive role.

The application of the CVS problem to the Job Shop problem became possible after implementing several ideas.

Idea N 1 consists in **route unification** of all jobs, by means of which we kind of make a multi-route problem a "single-route" one. To that end, we do **two transformations** of the route of each job J_k. **Firstly,** we lengthen the job route to the maximum length $\mu \doteq \max_{k=1,\ldots,n} \mu_k$. To do this, we just add the missing number ($\mu - \mu_k$) of zero-length dummy operations (with arbitrarily assigned executing machines). **With the second transformation,** we lengthen the route further by a factor of m, assuming that the q-th operation of job ($q = 1, \ldots, \mu$) is just a *representative* of the whole family $\{O_{q,k}^1, \ldots, O_{q,k}^m\}$ of m "equal-rights" operations of the *q-th level* of job J_k. (At that, operation $O_{q,k}^i$ ($i = 1, \ldots, m$) is executed on machine M_i.) It is clear that apart from the real operation $O_{q,k}$, the other operations of the qth level are dummy ones, and initially have zero length. Thus, each job consists of $m\mu$ operations. At that, a "subordination" of operations must hold: each operation of the $(q + 1)$th level of job J_k must be performed after each of its qth level operations, while all operations of the same level of job J_k can be executed independently of each other (in parallel). For each job J_k, we define an $(m\mu)$-dimensional vector x_k of the durations of its operations.

Idea N 2 consists in balancing the loads on all machines by increasing the durations of some operations (both real and dummy) without changing the values of $P_{\max}$ and $p_{\max}$. (We can do this because all machines now have the same number of operations, $n\mu$.)

Idea N 3. We define a vector x'_k for each job J_k, so that, all together, they constitute a proper family of vectors in space $\mathbb{R}^{m\mu}$ with a specific non-symmetric norm $\tilde{s}$, whose unit ball is centrally-symmetric. We apply the compact summation algorithm to this family of vectors, which finds a "suitable" permutation π^* of job indices. This permutation is taken as the basis for determining the order of executing the operations of each machine.

Idea N 4. To adjust the defined above order of processing the operations on each machine (in order to respect the "subordination of operations"), a certain number of dummy operations of each level $q = 2, \ldots, \mu$ are added to each machine. (The larger the value of q, the more such operations, but their total number is limited by a function of m and μ.)

As a result, by means of the compact vector summation algorithm and with the help of the ideas mentioned above, it was possible to construct a polynomial-time approximation algorithm for the Job Shop problem which guaranteed constructing a schedule S' such that

$$C_{\max}(S') - P_{\max} \leq \psi(m, \mu)p_{\max},$$

where $\psi(m, \mu)$ is a polynomial of order 4. The first versions of such algorithm were published in [21, 22]. The algorithm also formed the basis for constructing the first polynomial-time scheme (PTAS) for an approximate solution of the Job Shop problem, presented by Jansen, Solis-Oba and Sviridenko [14] at APPROX99.

Interesting possibilities for constructing near-optimal schedules are also contained in *non-strict vector summation* methods related to compact vector summation. The difference between these methods and the CVS algorithms is that they consider the possibility of summation not only in a ball of some norm, but also in arbitrary (convex) domains of m-dimensional space (including unbounded ones). In this case, the summation trajectory is allowed to go out of the designated area from time to time, but under the condition that already at the next step it will return to the specified area. (The number of such exits of the trajectory beyond the region is not limited.) Such a problem setting allowed us to improve the performance guarantees of polynomial-time algorithms for constructing approximate schedules, based on the vector summation method. Unfortunately, the limited size of this paper gave me small possibility to describe the methods in more detail. Instead, I refer the interested readers to my survey paper [25].

5   Conclusion

This paper presented three methods of approximate solution of the problems in which "uniformization" of one or more indicators of the desired solution was either the direct target of the problem under solution, or helped to find solutions close to the optimum. Although the paper contains a fairly large number of examples of a successful application of these methods, however, I am deeply convinced (and not only convinced, but I know this for sure) that their potential in the field of solving discrete optimization problems is far from being exhausted.

The paper also mentioned a large number of open theoretical questions and problems (which indicates the presence of a great potential for the development of this area of research). The author hopes that these open problems will find their "deliverers" among the readers of this paper. In fact, the most interesting work is just beginning! Have a nice trip!

References

1. Babushkin, A.I., Bashta, A.L., Belov, I.S.: Scheduling a three machine job shop. Avtomatika i Telemekhanika **7**, 154–158 (1976). (in Russian)
2. Babushkin, A.I., Bashta, A.L., Belov, I.S.: Construction of a schedule for the counter-routes problem. Kibernetika **4**, 130–135 (1977). (in Russian)
3. Banaszczyk, W.: The Steinitz constant of the plane. J. Reine und Angew. Math. **373**, 218–220 (1987)
4. Bárány, I., Fiala, T.: Nearly optimum solution of multi-machine scheduling problems. Szigma **15**(3), 177–191 (1982). (in Hungarian)
5. Belov, I.S., Stolin, Ja.N.: An algorithm for the single-route scheduling problem. In: Mathematical Economics and Functional Analysis. "Nauka", Moscou, pp. 248–257 (1974). (in Russian)
6. Bergström, V.: Ein neuer Beweis eines Satzes von E.Steinitz. Abhendlungen aus dem Mathematischen Seminar der Hamburgischen Universitat **8**, 148–152 (1931)
7. Bergström, V.: Zwei Satze über ebene Vektorpolygone. Abhendlungen aus dem Mathematischen Seminar der Hamburgischen Universitat **8**, 206–214 (1931)
8. Biró, P., McDermid, E.: Matching with sizes (or scheduling with processing set restrictions). Discret. Appl. Math. **164**(1), 61–67 (2014)
9. Chai, X., et al.: Approximation algorithms for batch scheduling with processing set restrictions. J. Sched. (2022). https://doi.org/10.1007/s10951-022-00720-2
10. Dushin, B.I.: Algorithm for the p-route Johnson problem. Kibernetika (Kiev) **2**, 119–122 (1989). (in Russian)

11. Ford Jr., L.R., Fulkerson, D.R.: Maximal flow through a network. Can. J. Math. **8**, 399–404 (1956)

12. Grinberg, V.S., Sevast'yanov, S.V.: Value of the Steinitz constant. Functional Anal. Appl. **14**(2), 125–126 (1980). https://doi.org/10.1007/BF01086559

13. Gross, W.: Bedingt Konvergente Reihen. Monatsh. Math. und Physik. **28**, 221–237 (1917)

14. Jansen, K., Solis-Oba, R., Sviridenko, M.: A linear time approximation scheme for the job shop scheduling problem. In: Hochbaum, D.S., Jansen, K., Rolim, J.D.P., Sinclair, A. (eds.) APPROX/RANDOM -1999. LNCS, vol. 1671, pp. 177–188. Springer, Heidelberg (1999). https://doi.org/10.1007/978-3-540-48413-4_19

15. Johnson, S.M.: Optimal two- and three-stage production schedules with setup times included. P-402, 5 May 1953, the RAND Corporation, Santa Monica, California, p. 10 (1953). https://www.rand.org/pubs/papers/P402.html

16. Kadeč, M.I.: On a property of polygonal paths in n-dimensional space. Uspekhi Mat. Nauk **8**(1), 139–143 (1953). (in Russian)

17. Sevastyanov, S.V.: Asymptotic approach to some scheduling problems. In: Third All-Union Conference on Problems of Theoretical Cybernetics, Thes. Dokl., pp. 67–69. Sobolev Institute of mathematics, Novosibirsk, June 1974. (in Russian)

18. Sevastyanov, S.V.: Asymptotic approach to some scheduling problems. Upravlyaemye Sistemy **14**, 40–51 (1975). (in Russian)

19. Sevastyanov, S.V.: On the approximate solution of some problems of scheduling theory. Metody Diskret. Analiz. **32**, 66–75 (1978). (in Russian)

20. Sevastyanov, S.V.: Approximation algorithms for Johnson's and vector summation problems. Upravlyaemye Sistemy **20**, 64–73 (1980). (in Russian)

21. Sevastyanov, S.V.: Efficient construction of near-optimal schedules for the cases of arbitrary and alternative routes of parts. Doklady Akademii Nauk SSSR **276**(1), 46–48 (1984), in Russian. (English translation. In: Sevast'yanov, S.V.: Efficient construction of schedules close to optimal for the cases of arbitrary and alternative routes of parts. Soviet Math. Dokl. **29**, 447–450 (1984))

22. Sevast'yanov, S.V.: Bounding algorithm for the routing problem with arbitrary paths and alternative servers. Cybern. Syst. Anal. **22**, 773–781 (1986). https://doi.org/10.1007/BF01068694. (Translated from "Kibernetika", November–December 1986, vol. 6, pp. 74–79. Original Article Submitted 12 December 1983)

23. Sevastyanov, S.V.: On a compact vector summation. Diskret. Mat. **3**(3), 66–72 (1991). (in Russian)

24. Sevast'janov, S.V.: On some geometric methods in scheduling theory: a survey. Discrete Appl. Math. **55**(1), 59–82 (1994). https://doi.org/10.1016/0166-218X(94)90036-1

25. Sevastianov, S.: Nonstrict vector summation in multi-operation scheduling. Ann. Oper. Res. **83**, 179–211 (1998). https://doi.org/10.1023/A:1018908013582

26. Shchepin, E.V., Vakhania, N.: An optimal rounding gives a better approximation for scheduling unrelated machines. Oper. Res. Lett. **33**(2), 127–133 (2005). https://doi.org/10.1016/j.orl.2004.05.004

27. Steinitz, E.: Bedingt Konvergente Reihen und Convexe Systeme. J. Reine und Angew. Math. **143**, 128–175 (1913)

Recent Advances in Sphere Packing Problems

Enkhbat Rentsen$^{(\boxtimes)}$ (iD)

Institute of Mathematical and Digital Technology, Mongolian Academy of Sciences,
Ulaanbaatar 13330, Mongolia
renkhbat46@yahoo.com

Abstract. We consider a general (constrained) sphere packing problem which is to pack nonoverlapping spheres (balls) with the maximum
volume into a convex set. This problem has important applications in
science, technology and economics. First, we survey existing methods
algorithms for solving the circle packing problems as well as their industrial applications. Second, we formulate a new optimal control problem
based on sphere packing problem. We also illustrate applications of circle
packing problems in mining industry and business.

Keywords: Sphere packing problem · Convex maximization ·
Optimality conditions · Malfatti's problem · Optimal control · Nash
equilibrium · Multiobjective optimization · Game theory

1 Introduction

The sphere packing problem is one of the most applicable areas in mathematics
which finds numerous applications in science and technology. Sphere packing
problem with one sphere is called the design centering problem. A practical
application of the design centering problem in the diamond industry has been
discussed in [21]. The industry needs to cut the largest diamond of a prescribed
form inside a rough stone. This form can often be described by a ball. The
quality control problem which arise in a fabrication process where the quality of
manufactured item is measured can be reduced to the design centering problem
[26]. Two dimensional sphere packing problem is circle packing problem. Circle
packing problems finds applications in circular cutting problems, communication
networks, facility location and dashboard layout. The circular cutting problem
is to cut out from a rectangular plate as many circular pieces as possible of N
different radii. In [16] it has been shown that the circular cutting problem is
NP-hard and the authors propose heuristic methods and algorithms for solving
it. In [15] Fraser and George consider a container loading problem for pulp
industries which reduces to the circle packing problem. For solving this problem,
the authors implemented a heuristic approach.

In [2] Dowsland considers the problem of packing cylindrical units into a rectangular container. Under some assumptions, the author shows that the problem

Y. Kochetov et al. (Eds.): MOTOR 2022, CCIS 1661, pp. 34–52, 2022.
https://doi.org/10.1007/978-3-031-16224-4_2

is equivalent to the circle packing problem and proposes a simulated annealing method to solve it.

In [14] Erkut introduced the p-dispersion problem which consider the optimized location of a set of points that represent facilities to be found. In [3] it is shown that the continuous p-dispersion problem and the circle packing problems are equivalent. A local search methods were used to solve this problem.

Martin in [19] showed that the robot communication problem is equivalent to the circle packing problem. Drezner in [3] considers the facility layout problem reduced to a circle packing problem where facilities are modeled as circles. Different optimization methods and algorithms [4] are used for solving the problem. In general, the circle packing problems are reduced to difficult nonconvex optimization problems which cannot be handled effectively by analytical approaches. Moreover, the complexity of the circle packing problems increases rapidly as a number of circles increases. Thus, only heuristic type methods are available for high dimensional cases ($N > 50$).

The circle packing problem has also important applications in automated radio-surgical treatment planning [27,28].

Gamma-rays are focused on a common center creating high radiation dose spherical volume. The key problem in the gamma knife treatment is how to place spheres in the tumor of arbitrary shape.

The unconstrained sphere packing problem asks for the densest packing of $\mathbb{R}^n$ with congruent balls. What's the largest fraction of $\mathbb{R}^n$ that can be covered by congruent balls with disjoint interiors?

A sphere packing P is a nonempty subset of $\mathbb{R}^n$ consisting of congruent balls with disjoint interiors.

The upper density of P is

$$\limsup_{r \to \infty} \frac{vol(B_r^n(0) \cap P)}{vol(B_r^n(0))},$$

$B_r^n(x)$ denotes the closed ball of radius r about x, the sphere packing density $\Delta_{\mathbb{R}}$ in $\mathbb{R}^n$ is the supremum of all the upper densities of sphere packings. The exist a sphere packing (Groewer theorem) for which

$$\lim_{r \to \infty} \frac{vol(B_r^n(x) \cap P)}{vol(B_r^n(x))} = \tilde{\Delta}$$

uniformly for all $x \in \mathbb{R}^n$

$$\Delta_{R^2} = \frac{\pi}{\sqrt{12}} = 0.906..., \quad \Delta_{\mathbb{R}^3} = \frac{\pi}{\sqrt{18}} = 0.7404...,$$

the best packing densities currently up to 36 dimensions.

The following statement plays an important role in the unconstrained sphere packing problem.

Theorem 1 (Maryna Viazovska-2016). *The E_8 lattice packing in $\mathbb{R}^8$ has density*

$$\Delta_{\mathbb{R}^8} = \frac{\pi^4}{384} = 0.2536....$$

The sphere packing density of the Leech lattice is

$$\frac{vol(B_1^{24})}{vol(\mathbb{R}^{24}/\Lambda_{24})} = 0.0011929...$$

In [8] it has been shown that the constrained sphere packing problem can be treated as the convex maximization problem which belongs to a class of global optimization. Also, necessary and sufficient conditions for inscribing a finite number of balls into a convex compact set have been derived in [8]. In [5], for the first time, two hundred years old Malfatti's problem [18] has been shown as a particular case of the circle packing problem. The hight dimensional sphere packing problem was considered in [7]. Multiobjective optimization approach to sphere packing problem has been proposed in [9]. In [11], the sphere packing problem was treated from a view point of game theory. The sphere packing problem in a Hilbert space has been examined in [12]. Also, DC programming approach has been proposed and tested for the constrained sphere packing problem in [13]. Applications of sphere packing theory in business and mining industry have been given in [10, 25].

2   Formulation of Sphere Packing Problem

Let $B(x^0, r)$ be a ball with a center $x^0 \in \mathbb{R}^n$ and radius $r \in \mathbb{R}$.

$$B(x^0, r) = \{x \in \mathbb{R}^n | \|x - x^0\| \le r\}. \tag{1}$$

here $\langle, \rangle$ denotes the scalar product of two vectors in $\mathbb{R}^n$, and $\|\cdot\|$ is Euclidean norm.

The n-dimensional volume of the Euclidean ball $B(x^0, r)$ is [22, 29]:

$$V(B) = \frac{\pi^{\frac{n}{2}}}{\Gamma(\frac{n}{2} + 1)} r^n. \tag{2}$$

where Γ is Leonhard Euler's gamma function.

Define the set D as follows:

$$D = \{x \in \mathbb{R}^n | g_i(x) \le 0, i = \overline{1, m}\}. \tag{3}$$

where $g_i : \mathbb{R}^n \to \mathbb{R}, i = \overline{1, m}$, are convex functions. Assume that D is a compact set which is not congruent to a sphere and $int D \ne \emptyset$. Clearly, D is a convex set in $\mathbb{R}^n$.

Let us introduce the functions φ_i for $x, h \in \mathbb{R}^n$ and r.

$$\varphi_i(x, r) = \max_{\|h\| \le 1} g_i(x + rh), r > 0, i = \overline{1, m}. \tag{4}$$

Lemma 1. $B(x^0, r) \subset D$ *if and only if*

$$\max_{1 \le i \le m} \varphi_i(x^0, r) \le 0. \tag{5}$$

Proof. Necessity. Let $y \in B(x^0, r)$ and $y \in D$. The point $y \in B(x^0, r)$ can be presented as $y = x^0 + rh, h \in \mathbb{R}^n, \|h\| \leq 1$. Then condition $y \in D$ implies that

$$g_i(x^0 + rh) \leq 0, \forall h \in \mathbb{R}^n : \|h\| \leq 1, i = \overline{1, m}.$$

Consequently, $max_{1 \leq i \leq m} \varphi_i(x^0, r) \leq 0$.

Sufficiency. Let condition (4) be satisfied, and on the contrary, assume that there exists $\widetilde{y} \in B(x^0, r)$ such that $\widetilde{y} \notin D$. Then we have $\widetilde{h} \in \mathbb{R}^n$ so that $\widetilde{y} = x^0 + r\widetilde{h}, \|\widetilde{h}\| \leq 1$. Since $\widetilde{y} \notin D$, there exists $j \in \{1, 2, ..., m\}$ such that $g_j(x^0 + r\widetilde{h}) > 0$ which contradicts (5).

Denote by $u^1, u^2, \ldots u^k$ centers of the spheres inscribed in D defined by (3). Let $r_1, r_2, ..., r_k$ be their corresponding radii. Now we consider a problem of maximizing a total volume of k non-overlapping spheres (balls) inscribed in $D \subset \mathbb{R}^n$. This problem in the literature [15] is often called sphere packing problem for different spheres. Then the sphere packing problem is:

$$max_{(u,r)} \ V = \frac{\pi^{\frac{n}{2}}}{\Gamma(\frac{n}{2}+1)} \sum_{i=1}^{k} r_i^n. \tag{6}$$

subject to:

$$\varphi_i(u^j, r_j) \leq 0, i = \overline{1, m}, j = \overline{1, k}. \tag{7}$$

$$\|u^i - u^j\|^2 \geq (r_i + r_j)^2, i, j = \overline{1, k}, \ i, j \in \binom{k}{2}, \ i \neq j \tag{8}$$

$$r_1 \geq 0, r_2 \geq 0, ..., r_k \geq 0. \tag{9}$$

The function V, which is convex on the positive or than t of $\mathbb{R}^n$, denotes the total volume of all spheres inscribed in D. Conditions (7) describe that all spheres are in D while conditions (8) secure non-overlapping case of balls. Denote by $S \subset R^{(n+1)k}$ the feasible set of problem (6)–(9). $(u, r) \in S \subset R^{(n+1)k}$, where $(u, r) = (u^1, u^2, ..., u^k, r_1, r_2, ..., r_k)$, $u^i = (u_1^i, u_2^i, ..., u_n^i), i = \overline{1, k}$.
Denote problem (6)–(9) by $SP(n, m, k)$, where n, m and k are its parameters.
    Then problem $SP(n, m, k)$ can be written as:

$$max_{(u,r) \in S} V. \tag{10}$$

The problem is convex maximization problem and the global optimality conditions of A. Strekalovsky [23] applied to this problem are following.

Theorem 2. *Let $(\overline{u}, \overline{r}) \in S$ satisfy $V'(\overline{r}) \neq 0$. Then $(\overline{u}, \overline{r})$ is a solution to problem (10) if and only if*

$$\langle V'(y), r - y \rangle \leq 0 \ for \ all \ y \in E_{V(\overline{r})}(V) \tag{11}$$

and $(u, r) \in S$.

where $E_c(V) = \{y \in R^k \mid V(y) = c\}$ is the level set of V at c and $V'(y)$ is the gradient of V at y.

Condition (11) can be written as

$$\sum_{i=1}^{k} y_i(r_i - y_i) \le 0, \ \forall y \in E_{V(\bar{r})}(V) = \{y \in R^k \mid \sum_{i=1}^{k} y_i^2 = \sum_{i=1}^{k} \bar{r}_i^2\}, (u, r) \in S.$$

Let D be a polyhedral set given by the following linear inequalities.

$$D = \{x \in \mathbb{R}^n \mid \langle a^i, x \rangle \le b_i, i = \overline{1, m}\}, a^i \in \mathbb{R}^n, b_i \in \mathbb{R}.$$

Assume that D is compact. Then the functions $\varphi_i(x, r)$ in (4) are computed in the following way.

$\varphi_i(x, r) = max_{\|h\| \le 1} g_i(x + rh) = max_{\|h\| \le 1}[\langle a^i, x + rh \rangle - b_i] = max_{\|h\| \le 1}[\langle a^i, x \rangle + r\langle a^i, h \rangle - b_i] = \langle a^i, x \rangle + r\|a^i\| - b_i, i = \overline{1, m}.$

Then the problem (6)–(9) has the form

$$max_{(u,r)} \ V = \frac{\pi^{n/2}}{\Gamma(\frac{n}{2} + 1)} \sum_{i=1}^{k} r_i^n \tag{12}$$

subject to:

$$\langle a^i, u^j \rangle + r_j \|a^i\| \le b_i, i = \overline{1, m}, j = \overline{1, k}. \tag{13}$$

$$\|u^i - u^j\|^2 \ge (r_i + r_j)^2, i, j = \overline{1, k}, \ i, j \in \binom{k}{2}, \ i \ne j. \tag{14}$$

$$r_1 \ge 0, r_2 \ge 0, ..., r_k \ge 0. \tag{15}$$

If we set $r_j = r, j = 1, 2, ..., k$, then problem (12)–(15) is reduced to the classical packing problem of inscribing k equal spheres into D with maximum volume:

$$max_{(u,r)} \ V = \frac{\pi^{n/2}}{\Gamma(\frac{n}{2} + 1)} kr^n \tag{16}$$

subject to:

$$\langle a^i, u^j \rangle + r\|a^i\| \le b_i, i = \overline{1, m}, j = \overline{1, k}. \tag{17}$$

$$\|u^i - u^j\|^2 \ge 4r^2, i, j = \overline{1, k}, \ i, j \in \binom{k}{2}, \ i \ne j. \tag{18}$$

$$r \ge 0. \tag{19}$$

If D is a box set, $D = \{x \in \mathbb{R}^n \mid \alpha_i \le x_i \le \beta_i, i = \overline{1, n}\}$ then problem (16)-(19) is equivalent to the problem of inscribing k equal spheres into a box set which is formulated as:

$$max_{(u,r)} \ V = \frac{\pi^{n/2}}{\Gamma(\frac{n}{2} + 1)} kr^n \tag{20}$$

subject to:

$$u_i^j + r \le \beta_i, i = \overline{1, n}, j = \overline{1, k}. \tag{21}$$

$$-u_i^j + r \leq -\alpha_i, i = \overline{1,n}, j = \overline{1,k}. \tag{22}$$

$$\|u^i - u^j\|^2 \geq 4r^2, i, j = \overline{1,k}, \ i, j \in \binom{k}{2}, \ i \neq j. \tag{23}$$

$$r \geq 0. \tag{24}$$

If we set $n = 2, m = 3$ and $k = 3$ in problem (12)–(15), then problem $SP(2,3,3)$ becomes Malfatti's problem which was first formulated in 1803 [18]. Indeed, the problem has the form:

$$max_{(u,r)} \ V = \pi \sum_{i=1}^{3} r_i^2 \tag{25}$$

subject to:

$$\langle a^i, u^j \rangle + r_j \|a^i\| \leq b_i, i, j = 1, 2, 3. \tag{26}$$

$$\|u^i - u^j\|^2 \geq (r_i + r_j)^2, i, j = 1, 2, 3, i \neq j. \tag{27}$$

$$r_1 \geq 0, r_2 \geq 0, r_3 \geq 0. \tag{28}$$

Malfatti's problem as well its high dimensional case was solved numerically in [5,7]. Note that problem $SP(n, m, k)$ can be easily reduced to D.C. programming with d.c.constraints so that one can apply D.C programming approach developed in [24]. Computational algorithms for solving the problem $SP(n, m, k)$ based on D.C programming has been discussed in [13].

3 Multi-objective Optimization Approach to Sphere Packing Problem

We consider the problem $SP(n, m, k)$ which is the extension of Malfatti's problem with k disks. Let us now consider this problem from a point of view of game theory. Suppose k players are involved in the game. Assume that k players who correspond to each disk have to maximize their area simultaneously. One can denote by $y = (u, r) \in \mathbb{R}^{n+1}$. In this case, we introduce $l = \frac{k(k-1)}{2}$ and $g(y) = (g_1(y), ..., g_{m \cdot k}(y))^T$, $h(y) = (h_1(y), ..., h_l(y))^T$ and $s(y) = (s_1(y), ..., s_k(y))^T$ as the left side vector-functions of the constraints (7), (8) and (9) respectively. In other words,

$$\begin{cases} g_1(y) = \langle a^i, u^i \rangle + r_j \|a^i\| - b_i \\ \quad \cdots \cdots \\ g_{m \cdot k}(y) = \langle a^m, u^k \rangle + r_k \|a^m\| - b_m \\ h_1(y) = (r_1 + r_2)^2 - \|u^1 - u^2\|, \\ \quad \cdots \cdots \\ h_l(y) = (r_{k-1} + r_k)^2 - \|u^{k-1} - u^k\| \\ s_j(y) = -r_j, \ j = 1, ..., k. \end{cases}$$

Denote by S the feasible set of the problem:

$$S = \{y \in \mathbb{R}^{n+1} \| g(y) \underset{\mathbb{R}^{m \cdot k}}{\leq} 0, h(y) \underset{\mathbb{R}^l}{\leq} 0, s(y) \underset{\mathbb{R}^k}{\leq} 0\}. \tag{29}$$

Let us consider multi-objective optimization problem [9] based on Malfatti's problem in the following:

$$
\begin{cases}
\max_{y \in S} f_1(y) = \pi r_1^2 \\
\quad \cdots \cdots \\
\max_{y \in S} f_k(y) = \pi r_k^2
\end{cases}
\tag{30}
$$

Denote the problem (30) by $MOSP(n, m, k)$.

Definition 1. $\bar{y} \in S$ *is called a Pareto optimal point of* $MOSP(n, m, k)$ *if there is no* $y \in S$ *with*

$$
f_i(y) \geq f_i(\bar{y}) \text{ for all } i \in \{1, ..., k\}
$$

and $f(y) \neq f(\bar{y})$, $f = (f_1, ..., f_k)$.

The Pareto optimal concept is the main optimality notion used in multi-objective optimization. The main approach for determination of Pareto optimal point is the weighted sum approach. Therefore, one can introduce appropriate weights $\alpha_1 > 0, ..., \alpha_k > 0$, $\sum_{i=1}^{k} \alpha_i = 1$, so we can formulate the corresponding scalarized optimization problem for problem (30):

$$
\max_{y \in S} F(y) = \sum_{i=1}^{k} \alpha_i f_i(y)
\tag{31}
$$

Problem (31) can be also considered as the convex maximization problem over nonconvex constraint set. A relationship between Pareto optimal point and solution of the scalarized problem is given by the following well-known result.

Proposition 1. *A solution* $y^\star$ *to problem (31) is a Pareto optimal point of problem* $MOSP(n, m, k)$.

Proof. Let $y^\star$ be a solution to problem (31) but is not a Pareto optimal point of problem (30). This means that there exist a point $\tilde{y} \in S$ such that $f_i(\tilde{y}) \geq f_i(y^\star)$ for all $i \in \{1, ..., k\}$ with $f_k(\tilde{y}) > f_k(y^\star)$, $\exists j \in \{1, ..., k\}$.

$$
F(y^\star) = \sum_{i=1}^{k} \alpha_i f_i(y^\star) \leq \sum_{i=1}^{k} \alpha_i f_i(\tilde{y}) < \sum_{i \neq j}^{k} \alpha_i f_i(\tilde{y}) + f_j(\tilde{y}) = F(\tilde{y})
\tag{32}
$$

This contradicts that $y^\star$ is a solution to problem (31).

4   Game Theory Approach to Sphere Packing Problem

Generalized Nash equilibrium problem(GNEP) is a N players non-cooperative game where each player's strategy set depends on rival players' strategies. The GNEP was introduced in 1952 by Debreu [1]. Consider the generalized Nash equilibrium problem where player k ($k=1,...,N$) controls $x_k \in \mathbb{R}^{n_k}$ and tries to solve the following optimization problem

$$P_k(x_{-k}) \quad \max_{x_k} f_k(x_k, x_{-k})$$

$$s.t. \quad g^k(x_k, x_{-k}) \leq 0$$

with given $f_k : \mathbb{R}^n \to \mathbb{R}$ and $g^k : \mathbb{R}^n \to \mathbb{R}^{m_k}$. Here, $n := n_1 + \cdots + n_k$ denotes the total number of variables, $m := m_1 + \cdots + m_N$ will be the total number of constraints. Each player k controls his strategy vector

$$x_k := (x_1^k, ..., x_{n_k}^k)^T \in \mathbb{R}^{n_k}$$

of n_k decision variables. The vector

$$x := (x_1, ..., x_N)^T \in \mathbb{R}^n$$

contains the $n = \sum_{k=1}^{N} n_k$ decision variables of all players. To emphasize the $k-$th palyer's variables within x, one can write (x_k, x_{-k}) instead of x, where

$$x_{-k} := (x_{k'})_{k'=1, k' \neq k}^{N} \in \mathbb{R}^{n-k}.$$

A vector $x := (x_1, ..., x_N)^T$ is called feasible for the GNEP if it is satisfies the constraints $g^k(x) \leq 0$ for all players $k = 1, ..., N$. A feasible point $\bar{x}$ is a solution of the GNEP if, for all players $k = 1, ..., N$, we have

$$f_k(\bar{x}_k, \bar{x}_{-k}) \geq f_k(x_k, \bar{x}_{-k}), \quad \forall x_k : \quad g^k(x_k, x_{-k}) \leq 0.$$

i.e., if, for all players k, $\bar{x}_k$ is the solution of the kth player's problem when the other players set their variables to $\bar{x}_{-k}$. In other words, the problem $P_k(x_{-k})$ are N parallel and parametric optimization problems. These problems are extremely difficult for solving due to nonconvexities. In practice, convex class is more popular than the class nonconvex of the GNEP. That is why, compared with convex class of the GNEPs, the study of theory and computational methods of nonconvex GNEPs is still in its infancy.

Now we extend Malfatti's problem from a view point of game theory. Assume that three players who correspond to each disk have to maximize their area simultaneously in a given triangle. First, we introduce the following variables:

$$x_1 = (x_1^1, x_2^1, x_3^1), \quad u = (x_1^1, x_2^1).$$

$$x_2 = (x_1^2, x_2^2, x_3^2), \quad v = (x_1^2, x_2^2).$$

$$x_3 = (x_1^3, x_2^3, x_3^3), \quad p = (x_1^3, x_2^3).$$

Then Malfatti's problem can be rewritten as:

$$\max G = \pi((x_3^1)^2 + (x_3^2)^2 + (x_3^3)^2) \tag{33}$$

$$\langle a^i, u \rangle + x_3^1 \|a^i\| \le b_i, \ i = 1, 2, 3. \tag{34}$$

$$\langle a^i, v \rangle + x_3^2 \|a^i\| \le b_i, \ i = 1, 2, 3. \tag{35}$$

$$\langle a^i, p \rangle + x_3^3 \|a^i\| \le b_i, \ i = 1, 2, 3. \tag{36}$$

$$(x_3^1 + x_3^2)^2 - (x_1^2 - x_1^1)^2 - (x_2^2 - x_2^1)^2 \le 0. \tag{37}$$

$$(x_3^1 + x_3^3)^2 - (x_1^3 - x_1^1)^2 - (x_2^3 - x_2^1)^2 \le 0. \tag{38}$$

$$(x_3^2 + x_3^3)^2 - (x_1^3 - x_1^2)^2 - (x_2^3 - x_2^2)^2 \le 0. \tag{39}$$

$$x_3^1 \ge 0, \ x_3^2 \ge 0, \ x_3^3 \ge 0. \tag{40}$$

Then three players generalized Nash equilibrium problem based on Malfatti's problem is formulated as follows:

$$(P_1) \qquad \max_{x_1} f_1(x) = \pi(x_3^1)^2, \ u = (x_1^1, x_2^1)$$

$$X_1(x_{-1}) = \{x_1 \in \mathbb{R}^3 \mid \ g_i^1(x) = \langle a^i, u \rangle + x_3^1 \|a^i\| - b_i \le 0, \ \ i = 1, 2, 3,$$

$$g_4^1(x) = (x_3^1 + x_3^2)^2 - (x_1^2 - x_1^1)^2 - (x_2^2 - x_2^1)^2 \le 0,$$

$$g_5^1(x) = (x_3^1 + x_3^3)^2 - (x_1^3 - x_1^1)^2 - (x_2^3 - x_2^1)^2 \le 0,$$

$$g_6^1(x) = -x_3^1 \le 0\}.$$

$$(P_2) \qquad \max_{x_2} f_2(x) = \pi(x_3^2)^2, \ v = (x_1^2, x_2^2)$$

$$X_2(x_{-2}) = \{x_2 \in \mathbb{R}^3 \mid \ g_i^2(x) = \langle a^i, v \rangle + x_3^2 \|a^i\| - b_i \le 0, \ \ i = 1, 2, 3,$$

$$g_4^2(x) = (x_3^1 + x_3^2)^2 - (x_1^2 - x_1^1)^2 - (x_2^2 - x_2^1)^2 \le 0,$$

$$g_5^2(x) = (x_3^2 + x_3^3)^2 - (x_1^3 - x_1^2)^2 - (x_2^3 - x_2^2)^2 \le 0,$$

$$g_6^2(x) = -x_3^2 \le 0\}.$$

$$(P_3) \qquad \max_{x_3} f_3(x) = \pi(x_3^3)^2, \ p = (x_1^3, x_2^3)$$

$$X_3(x_{-3}) = \{x_3 \in \mathbb{R}^3 \mid \ g_i^3(x) = \langle a^i, p \rangle + x_3^3 \|a^i\| - b_i < 0, \ \ i = 1, 2, 3,$$

$$g_4^3(x) = (x_3^1 + x_3^3)^2 - (x_1^3 - x_1^1)^2 - (x_2^3 - x_2^1)^2 \le 0,$$

$$g_5^3(x) = (x_3^2 + x_3^3)^2 - (x_1^3 - x_1^2)^2 - (x_2^3 - x_2^2)^2 \le 0,$$

$$g_6^3(x) = -x_3^3 \le 0\}.$$

Now the constraints (34)–(36) will be shared constraints for the problem (P_1) – (P_3). Therefore, we can note some relations between constraints in the following:

$$g_4^1(x) = g_4^2(x).$$

$$g_5^1(x) = g_4^3(x).$$

$$g_5^2(x) = g_5^3(x).$$

The problem $(P_1)-(P_3)$ is a generalized Nash equilibrium problem with nonconvex shared constraints. This problem is extremely difficult for solving numerically. It is possible only to define a Nash equilibrium of this game in the ordinary sense, which is based on a global maximum of each player's problem. However, from the viewpoint of computation, it is difficult to find an equilibrium that guarantee global maxima of all players' problems. Therefore, it may be reasonable to try to find a stationary Nash equilibrium, which consists of stationary points of all players' problems. Here, a stationary point means a point that satisfies the first-order optimality condition for each player's problem. We denote by

$$L^k(x, \lambda^k) := -f_k(x) + \sum_{i=1}^{5} \lambda_i^k g_i^k(x) + \mu_k g_6^k(x), \quad k = 1, 2, 3.$$

the Lagrangian of player k. Hence we can write down concatenating KKT conditions of problem $(P_1) - (P_3)$ as follows:

$$-\nabla_{x_1} f_1(x) + \sum_{i=1}^{5} \lambda_i^1 \nabla_{x_1} g_i^1(x) + \mu_1 \nabla_{x_1} g_6^1(x) = 0. \tag{41}$$

$$\lambda_i^1 g_i^1(x) = 0, \ \mu_1 g_6^1(x) = 0 \ \lambda_i^1 \geq 0, \ g_i^1(x) \leq 0, \ \mu_1 \geq 0, \ g_6^1(x) \leq 0, \ \forall i = 1, ..., 5. \tag{42}$$

$$-\nabla_{x_2} f_2(x) + \sum_{i=1}^{5} \lambda_i^2 \nabla_{x_2} g_i^2(x) + \mu_2 \nabla_{x_2} g_6^2(x) = 0. \tag{43}$$

$$\lambda_i^2 g_i^2(x) = 0, \ \mu_2 g_6^2(x) = 0, \ \lambda_i^2 \geq 0, \ g_i^2(x) \leq 0, \ \mu_2 \geq 0, \ g_6^2(x) \leq 0, \ \forall i = 1, ..., 5. \tag{44}$$

$$-\nabla_{x_3} f_3(x) + \sum_{i=1}^{5} \lambda_i^3 \nabla_{x_3} g_i^3(x) + \mu_3 \nabla_{x_3} g_6^3(x) = 0. \tag{45}$$

$$\lambda_i^3 g_i^3(x) = 0, \ \mu_3 g_6^3(x) = 0, \ \lambda_i^3 \geq 0, \ g_i^3(x) \leq 0, \ \mu_3 \geq 0, \ g_6^3(x) \leq 0, \ \forall i = 1, ..., 5. \tag{46}$$

There is the following relationship between the generalized Nash equilibrium problem $(P_1) - (P_3)$ and Malfatti's problem (12)–(19).

Proposition 2. *([11]) A generalized Nash equilibrium of problem $(P_1) - (P_3)$ satisfies the optimality conditions for Malfatti's problem.*

5   Optimal Control Sphere Packing

Let us consider a problem of maximizing a total volume of k non-overlapping spheres inscribed in $D \subset \mathbb{R}^n$ centers of which move along trajectories given by differential equations. These trajectories are:

$$\begin{cases} \dot{x}^j = f^j(x^j, t), & x^j \in \mathbb{R}^n. \\ x^j(0) = x_0^j, & j = 1, 2, \ldots, k. \end{cases} \tag{47}$$

$$x^j = (x_1^j, x_2^j, \ldots, x_n^j), \quad j = 1, 2, \ldots, k,$$

where $f^j : \mathbb{R}^{n+1} \to \mathbb{R}$ are differentiable functions, $f^j = (f_1^j, f_2^j, \ldots, f_k^j)$.

Denote by $u_1(t), \ldots, u_k(t)$ radii of spheres at moment t.

Assume that $u_j : \mathbb{R} \to \mathbb{R}, j = 1, 2, \ldots, k$ are continuously differentiable functions.

Let $B(x^0(t), u^0(t))$ be a ball with a center $x^0(t) \in \mathbb{R}^n$ and radius $u^0(t)$ at moment t.

$$B(x^0, u^0) = \{x \in \mathbb{R}^n \mid \|x - x^0\| \le u^0\}.$$

Then condition $B(x^0, u^0) \subset D$ becomes

$$\langle a^q, x^0(t) \rangle + u^0(t)\|a^q\| \le b_q, q = 1, 2, \ldots, m. \tag{48}$$

Non-overlapping conditions for spheres $B(x^i, u_i), i = 1, 2, \ldots, k$ are

$$\|x^i - x^j\|^2 \ge (u_i + u_j)^2, \quad i, j = 1, 2, \ldots, k, \quad i < j. \tag{49}$$

Volume of spheres inscribed in D at the moment t is

$$V(t) = \frac{\pi^{n/2}}{\Gamma(\frac{n}{2} + 1)} \sum_{j=1}^{k} u_j^n(t),$$

$$u(t) = (u_1(t), \ldots, u_k(t)) \in \mathbb{R}^n, \quad t \in (-\infty, +\infty).$$

Then under the above assumptions, the volume maximization problem is formulated as follows:

$$\max_{(u,T,t_0)} J = \max_{(u,T,t_0)} \frac{\pi^{n/2}}{\Gamma(\frac{n}{2} + 1)} \sum_{j=1}^{k} u_j^n(T) \tag{50}$$

subject to constraints:

$$\langle a^q, x^j(t) \rangle + u_j(t)\|a^q\| \le b_q. \tag{51}$$

$$t_0 \le t \le T, \quad q = 1, 2, \ldots, m, \quad j = 1, 2, \ldots, k.$$

$$\begin{cases} \dot{x}^j = f^j(x^j, t), \\ x^j(t_0) = x_0^j, \quad j = 1, 2, \ldots, k, \quad t_0 \le t \le T. \end{cases}$$

$$\|x^i(t) - x^j(t)\|^2 \ge (u_i(t) + u_j(t))^2, \quad i, j = 1, 2, \ldots, k, \quad i < j. \tag{52}$$

$$u_1(t) \ge 0, u_2(t) \ge 0, \ldots, u_k(t) \ge 0.$$

$$\begin{cases} \langle a^q, x^j(T) \rangle \le b_q \\ \langle a^q, x_0^j(t_0) \rangle \le b_q, \quad q = 1, 2, \ldots, m, \quad j = 1, 2, \ldots, k. \end{cases} \tag{53}$$

Problem (50)–(53) can be written in the form:

$$\max_{(u,T,t_0)} J = \frac{\pi^{n/2}}{\Gamma(\frac{n}{2} + 1)} \sum_{j=1}^{k} u_j^n(T) \tag{54}$$

$$\langle a^q, x_0^j + \int_{t_0}^t f^j(x^j, \tau)\mathrm{d}\tau \rangle + u_j(t)\|a^q\| \le b_q. \tag{55}$$

$$t_0 \le t \le T, \quad q = 1, 2, \ldots, m, \quad j = 1, 2, \ldots, k.$$

$$\left\| \int_{t_0}^t [f^j(x^j, \tau) - f^i(x^i, \tau)] \, \mathrm{d}\tau + (x_0^j - x_0^i) \right\|^2 \ge (u_i(t) + u_j(t))^2. \tag{56}$$

$$i, j = 1, 2, \ldots, k, \quad i < j.$$

$$\begin{cases} \langle a^q, x_0^j + \int_{t_0}^T f^j(x^j, \tau)\mathrm{d}\tau \rangle \le b_q \\ \langle a^q, x_0^j \rangle \le b_q, \quad q = 1, 2, \ldots, m, \quad j = 1, 2, \ldots, k. \end{cases} \tag{57}$$

Now we can write the discrete optimal control sphere packing problem

$$\max \ J = \max_{(u, t_p)} \frac{\pi^{n/2}}{\Gamma(\frac{n}{2} + 1)} \sum_{j=1}^k u_j^n(t_p), \quad p \in \{0, 1, \ldots, N\}.$$

$$\langle a^q, x^j(t_{p+1}) \rangle + u_j(t_{p+1})\|a^q\| \le b_q.$$

$$q = 1, 2, \ldots, m, \quad j = 1, 2, \ldots, k, \quad p = 0, 1, \ldots, N - 1.$$

$$\begin{cases} x^j(t_{p+1}) = x^j(t_p) + f^j(x^j(t_p)), \quad j = 1, 2, \ldots, k. \\ x^j(t_0) = x_0^j, \quad p = 0, 1, \ldots, N - 1. \end{cases}$$

$$\left\| x^i(t_p) - x^j(t_p) \right\|^2 \ge (u_i(t_p) + u_j(t_p))^2.$$

$$p = 0, 1, \ldots, N - 1, \quad i, j = 1, 2, \ldots, k, \quad i < j.$$

$$\begin{cases} \langle a^q, x^j(N) \rangle \le b_q, \quad q = 1, 2, \ldots, m, \\ \langle a^q, x_0^j \rangle \le b_q, \quad j = 1, 2, \ldots, k. \end{cases}$$

Denote by $u_j(t_p) = u_{jp}$, $x^j(t_{p+1}) = x_{p+1}^j$, $u_j(t_{p+1}) = u_{jp+1}$, $x^j(t_p) = x_p^j$, $f^j(x^j(t_p)) = f^j(x_p^j)$, $u_i(t_p) = u_{ip}$, $x^i(N) = x_{iN}, i, j = 1, 2, \ldots, k; \quad i < j$. Then the problem is equivalent to

$$\max_{(u, t_0, N)} \ J = \frac{\pi^{n/2}}{\Gamma(\frac{n}{2} + 1)} \sum_{j=1}^k u_{jN}^n$$

$$\langle a^q, x_0^j + \int_{t_0}^{t_p} f^j(x^j, \tau)\mathrm{d}\tau \rangle + u_{jp}\|a^q\| \le b_q.$$

$$p = 0, 1, \ldots, N - 1, \quad j = 1, 2, \ldots, k, \quad q = 1, 2, \ldots, m.$$

$$\begin{cases} x_{p+1}^j - x_p^j - f^j(x_p^j) = 0, \quad j = 1, 2, \ldots, k, \\ x^j(t_0) = x_0^j = x_{j0}, \quad p = 0, 1, \ldots, N - 1. \end{cases}$$

$$\|x_p^i - x_p^j\|^2 \ge (u_{ip} + u_{jp})^2, \quad i, j = 1, 2, \ldots, k, \quad i < j.$$

$$\begin{cases} \langle a^q, x_N^j \rangle - b_q \leq 0, & q = 1, 2, \ldots, m. \\ \langle a^i, x_0^j \rangle \leq b_q, & j = 1, 2, \ldots, k. \end{cases}$$

Since Malfatti's problem is a particular case of the general sphere packing problem for $k = 3, m = 3, n = 2$, then Malfatti's optimal control problem can be formulated as follows.

$$\max_{(u, T, t_0)} \; J = \pi(u_1^2(T) + u_2^2(T) + u_3^2(T))$$

subject to constraints:

$$\langle a^1, x^1(t) \rangle + u_1(t) \| a^1 \| \leq b_1$$
$$\langle a^2, x^2(t) \rangle + u_2(t) \| a^2 \| \leq b_2$$
$$\langle a^3, x^3(t) \rangle + u_3(t) \| a^3 \| \leq b_3$$
$$t_0 \leq t \leq T < +\infty.$$
$$\begin{cases} \frac{dx^1}{dt} = f^1(x^1, t) \\ x^1(t_0) = x_0^1. \end{cases}$$
$$\begin{cases} \frac{dx^2}{dt} = f^2(x^2, t) \\ x^2(t_0) = x_0^2. \end{cases}$$
$$\begin{cases} \frac{dx^3}{dt} = f^3(x^3, t) \\ x^3(t_0) = x_0^3. \end{cases}$$
$$\| x^1(t) - x^2(t) \|^2 \geq (u_1(t) + u_2(t))^2$$
$$\| x^1(t) - x^3(t) \|^2 \geq (u_1(t) + u_3(t))^2$$
$$\| x^2(t) - x^3(t) \|^2 \geq (u_3(t) + u_2(t))^2.$$

$$\langle a^1, x^1(T) \rangle \leq b_1$$
$$\langle a^2, x^2(T) \rangle \leq b_2$$
$$\langle a^3, x^3(T) \rangle \leq b_3$$
$$\langle a^1, x_0^1(T) \rangle \leq b_1$$
$$\langle a^2, x_0^2(T) \rangle \leq b_2$$
$$\langle a^3, x_0^3(T) \rangle \leq b_3$$
$$u_1(t) \geq 0, u_2(t) \geq 0, u_3(t) \geq 0,$$

where

$$f^1 = (f_1^1(x^1, t), f_2^1(x^1, t)),$$
$$f^2 = (f_1^2(x^2, t), f_2^2(x^2, t)),$$
$$f^3 = (f_1^3(x^3, t), f_2^3(x^3, t)),$$
$$x^1(t) = (x_1^1(t), x_2^1(t)),$$
$$x^2(t) = (x_1^2(t), x_2^2(t)),$$
$$x^3(t) = (x_1^3(t), x_2^3(t)),$$
$$-\infty < t_0 \leq t \leq T < +\infty.$$

6 Applications

6.1 Mathematical Model of Flotation Process [10]

Metal recovery of copper of the flotation process depends on its technological parameters and it can be characterized by a linear regression function using a statistical data of flotation process for a given period.

$$f = \sum_{j=1}^{n} a_j x_j + a_0,$$

where $x_j, j = 1, 2, ..., n$ are variables.

Let γ be a given level of copper recovery.

We define the set of safe system perturbation as:

$$K(a, \gamma) = \left\{ x \in \mathrm{IR}^n \,\middle|\, \sum_{j=1}^{n} a_j x_j + a_0 \geq \gamma, \right.$$

$$\left. x_j^{min} \leq x_j \leq x_j^{max}, j = 1, 2, .., n \right\}.$$

Assume that $K(a, \gamma) \neq \emptyset$.

This set defines the best operating conditions for a given level of γ. The wider this set is, the higher survival of the system. In order to improve the survival of the system, we need to find a sphere with the maximum radius inscribed in $K(a, \gamma)$ by solving the following linear programming problem [10]:

$$\max r \tag{58}$$

$$\sum_{j=1}^{n} a_j x_j - r \sqrt{\sum_{j=1}^{n} a_j^2} \geq \gamma - a_0. \tag{59}$$

$$-x_j + r \leq -x_j^{min}, j = 1, 2, .., n. \tag{60}$$

$$x_j + r \leq x_j^{max}, j = 1, 2, .., n. \tag{61}$$

$$x_j \geq 0, j = 1, 2, .., n. \tag{62}$$

$$r \geq 0. \tag{63}$$

Numerical Experiments. In our experiments, we used the following technological parameters of 9 variables.

x_1: addition of collector agent VK-901 (in grams per ton),
x_2: consumption of foaming agent MIBK (in grams per ton),
x_3: content of $74\,\mu$m grain class in the hydro cyclone overflow (in percentage
of mass),
x_4: monflot-03(in grams per ton),
x_5: total copper grade in the rougher(in percentage of mass),
x_6: total content of oxidized copper in the feed (in percentage of mass),
x_7: total content of primary copper in the feed (in percentage of mass),
x_8: $\pm$ 74mkm grinding level percentage,
x_9: total copper grade in the feed, in percentage of mass,
f: metal copper recovery.

A linear regression model constructed on a real data for the period (June
2019) of Erdenet Mining Corporation of Mongolia in coded variables in $[-1,1]$
was:

$$f = 0.86 - 0.004x_1 + 0.006x_2 - 0.012x_3 + 0.005x_4 + 0.011x_5 - 0.014x_6$$
$$+ 0.002x_7 + 0.006x_8 - 0.0005x_9$$

with the determination coefficient $R^2 = 0.75$ which is the proportion of the
variance in the dependent variables that is predictable from the independent
variables. In other words, 75% of the flotation process can be explained by the
above linear equation. Increase of variable x_j in one unit effects value of f by a_j
unit.

Then problems (58)–(63) reduces to the following problem

$$\max \; r \tag{64}$$
$$- 0.004x_1 + 0.006x_2 - 0.012x_3 + 0.005x_4 + 0.011x_5 \tag{65}$$
$$-0.014x_6 + 0.002x_7 + 0.006x_8 - 0.0005x_9 - 0.86385r \geq \gamma - 0.86.$$
$$-x_j + r \leq 1, j = 1, 2, .., 9. \tag{66}$$
$$x_j + r \leq 1, j = 1, 2, .., 9. \tag{67}$$
$$x_j \geq 0, j = 1, 2, .., 9. \tag{68}$$
$$r \geq 0. \tag{69}$$

Since problems (64)–(69) are required to be solved for each period of shift at
the industry, then the best performance of the flotation is found by the following
algorithm ST.

Algorithm ST

Step 1. Set $k := 0$ and choose a value of $\gamma_0 = 0.85$ which is a current level
of copper recovery and x^0 is its corresponding technological variable, and
$\delta = 0.001$.
Step 2. Solve problem (64)–(69) for $\gamma := \gamma_k$.

Step 3. If the problem has a no solution then γ_k is the best performance of the copper recovery and x^k is an optimal solution.

Step 4. If the problem has a solution (r^k, x^k) then set $\gamma_{k+1} := \gamma_k + \delta, k := k + 1$,

return to **Step 2**.

After converting the coded variables into real variables, we get optimal solutions to problem (64)–(69) computed on Matlab as:

$x_1^* = 4.25, x_2^* = 21.9735, x_3^* = 10.25, x_4^* = 9.9837, x_5^* = 0.5996, x_6^* = 2.51,$
$x_7^* = 79.9389, x_8^* = 64.9858, x_9^* = 15.00,$ and $r^* = 0.004.$

The maximum value of $\gamma^* = 0.893$. It means that we can increase copper recovery up to 89.3%.

Note that for $\gamma = 0.894$ problem (64)–(69) has a no solution.

Optimal range of original variables to ensure the level γ^* of copper recovery is the following: $4.2486 \leq x_1 \leq 4.2514, 21.9722 \leq x_2 \leq 21.9749, 10.2486 \leq x_3 \leq 10.2514, 9.9824 \leq x_4 \leq 9.9851, 0.5982 \leq x_5 \leq 0.6009, 2.5086 \leq x_6 \leq 2.5114, 79.9376 \leq x_7 \leq 79.9403, 64.9844 \leq x_8 \leq 64.9871, 14.9986 \leq x_9 \leq 15.0014.$

6.2 Profit Analysis [25]

The total profit of a company for a multi-product case can be written as:

$$\pi = \sum_{j=1}^{n} p_j x_j - \sum_{j=1}^{n} c_j x_j - F. \tag{70}$$

where π-total profit, x_j-quantity of with product sold, c_j-variable cost per unit of j-th product, F-total fixed cost, $p_j > c_j$.

It is important for a company to operate profitable, in other words

$$\pi \geq 0,$$

or the set D defined by

$$D(p, x, c, F) = \{(p, x, c, F) \in R^{3n+1} | \sum_{j=1}^{n} (p_j - c_j)x_j \geq F, x_j^{min} \leq x_j \leq x_j^{max},$$

$$p_j^{min} \leq x_j \leq p_j^{max}, \ c_j^{min} \leq x_j \leq c_j^{max}, \ F_j^{min} \leq x_j \leq F_j^{max}, \ j = 1, \ldots, n\}$$

is nonempty.

We introduce the set of profitable volumes as follows:

$$D_x = \left\{ x \in R^n \ | \ \sum_{j=1}^{n} (p_j - c_j)x_j \geq F, \ x_j^{min} \leq x_j \leq x_j^{max}, \ j = 1, \ldots, n \right\}. \tag{71}$$

where x_j^{min}, x_j^{max}-minimum and maximum capacity volumes for j-th product, $j = 1, \ldots, n$ and $\|x\|$ denotes the norm of the vector x in R^n.

Denote by $B(x^0, r^0)$ a sphere with a center $x^0 \in R^n$ and radius $r^0 \in R$, $r^0 > 0$,

$$B(x^0, r^0) = \{x \in R^n \mid \|x - x^0\| \leq r^0\}. \tag{72}$$

It is easy to see that any point $y \in B$ can be presented as

$$y = x^0 + \alpha \frac{h}{\|h\|} \tag{73}$$

for any $h \in R^n$ and $0 \leq \alpha \leq r^0$.

If $B(x^0, r^0) \subset D_x$ then any point of $B(x^0, r^0)$ satisfies profitable condition (71).

According to sphere packing theory [8], we must find a sphere $B(x^0, r^0)$ with the maximum radius r^0 such that $B(x^0, r^0) \subset D_x$.

For this purpose, we recall the following assertion in [8] which is useful for company's profit analysis.

Theorem 3. $B(x^0, r^0) \subset D_x$ *if and only if* (x^0, r^0) *is a solution to the following linear programming problem*

$$r \to max \tag{74}$$

$$\sum_{j=1}^{n} (p_j - c_j)x_j - r \sqrt{\sum_{j=1}^{n} (p_j - c_j)^2} \geq F. \tag{75}$$

$$-x_j + r \leq -x^{min}, j = 1, \ldots, n. \tag{76}$$

$$x_j + r \leq x_j^{max}, j = 1, \ldots, n. \tag{77}$$

If we take any $h = (h_1, h_2, \ldots, h_n) \in R$ then

$$x^h = (x_1, \ldots, x_n) = \left(x_1^0 + r^0 \frac{h_1}{\|h\|}, \ldots, x_i^0 + r^0 \frac{h_i}{\|h\|}, \ldots, x_n^0 + r^0 \frac{h_n}{\|h\|} \right)$$

$$\bar{x}^h = (\bar{x}_1, \ldots, \bar{x}_n) = \left(x_1^0 - r^0 \frac{h_1}{\|h\|}, \ldots, x_i^0 - r^0 \frac{h_i}{\|h\|}, \ldots, x_n^0 - r^0 \frac{h_n}{\|h\|} \right)$$

$$\|h\| = \sqrt{h_1^2 + \ldots, h_n^2},$$

and $x^h, \bar{x}^h \in B$, consequently $x^h, \bar{x}^h \in D_x$ satisfy profitable condition (71). It means that in order to ensure company's profit, the company must keep its volumes in the following intervals:

$$x_j^0 - r^0 \frac{h_j}{\|h\|} \leq x_j \leq x_j^0 + r^0 \frac{h_j}{\|h\|}, \ j = 1, \ldots, n.$$

7   Conclusion

We survey recent advances of the sphere packing problems as well as their applications in mining and business. We examine the general sphere packing problem which is to pack non-overlapping spheres into a convex set with the maximum volume from a view point of theory of global optimization. We show that the classical circle packing problems, the design centering and Malfatti's problems are particular cases of the general sphere packing problem. Also, for the first time, a new optimal control problem has been formulated for sphere packing problem.

References

1. Debreu, G.: A social equilibrium existence theorem. Proc. National Acad. Sci. United States of America **38**, 886–893 (1952)
2. Dowsland, K.A.: Optimising the palletisation of cylinders in cases. OR Spectr. **13**, 204–212 (1991)
3. Drezner, Z., Erkut, E.: Solving the continuous p-dispersion problem using nonlinear programming. J. Oper. Res. Soc. **46**, 516–520 (1995)
4. Drezner, Z.: DISCON: a new method for the layout problem. Oper. Res. **28**, 1375–1384 (1980)
5. Enkhbat, R.: Global optimization approach to Malfatti's problem. J. Global Optim. **65**, 33–39 (2016)
6. Enkhbat, R.: An algorithm for maximizing a convex function over a simple set. J. Global Optim. **8**, 379–391 (1996)
7. Enkhbat, R., Barkova, M., Strekalovsky, A.: Solving Malfatti's high dimensional problem by global optimization. Numer. Algebra Control Optim. **6**(2), 153–160 (2016)
8. Enkhbat, R.: Convex maximization formulation of general sphere packing problem. Bulle Irkutsk State Univ. Ser. Math. **31**, 142–149 (2020)
9. Enkhbat, R., Battur, G.: Multi-objective optimization approach to Malfatti's problem. Contrib. Game Theory Manag. **14**, 82–90 (2021)
10. Enkhbat, R., Tungalag, N., Enkhbayar, J., Battogtokh, O., Enkhtuvshin, L.: Application of survival theory in mining industry. Numer. Algebra Control Optim. **11**(3), 443–448 (2021)
11. Enkhbat, R., Battur, G.: Generalized Nash equilibrium based on Malfatti's Problem. Numer. Algebra Control Optim. **11**(2), 209–220 (2021)
12. Enkhbat, R., Davaadulam, J.: Maximizing the sum of radii of balls inscribed in a polyhedral set. Bull. Irkutsk State Univ. Ser. Math. **28**, 138–145 (2019)
13. Enkhbat, R., Barkova, M., Batbileg, S.: D.C.programming approach to Malfatti's problem. Math. Ser. Math. Inf. Buryat State Univ. **4**, 72–83 (2018)
14. Erkut, E.: The discrete p-dispersion problem. Eur. J. Oper. Res. **46**, 48–60 (1990)
15. Fraser, H.J., George, J.A.: Integrated container loading software for pulp and paper industry. Eur. J. Oper. Res. **77**, 466–474 (1994)
16. Hifi, M., Hallah, R.M.: Approximate algorithms for constrained circular cutting problems. Comput. Oper. Res. **31**, 675–694 (2004)
17. Kirkpatrick, S., Gelatt, C.D., Vecchi, M.P.: Optimization by simulated annealing. Science **220**, 671–680 (1983)

18. Malfatti, G.: Memoria sopra una problema stereotomico. Memoria di Matematica e di Fisica della Societa italiana della Scienze **10**(1), 235–244 (1803)

19. Martin, C.: How many robots can talk at the same time? Department of Mathematics. The Royal Institute of Technology, Stockholm (2004). https://www.math.kth.se/optsyst/seminar/martinIII.html

20. Munkhdalai, D., Enkhbat, R.: A Circle packing problem and its connection to Malfatti's problem. In: Pardalos, P.M., Migdalas, A. (eds.) Open Problems in Optimization and Data Analysis. SOIA, vol. 141, pp. 225–248. Springer, Cham (2018). https://doi.org/10.1007/978-3-319-99142-9_12

21. Nguyen, V.H., Strodiot, J.J.: Computing a global optimal solution to a design centering problem. Math. Program. **53**, 111–123 (1992)

22. Paris, R.B., Kaminski, D.: Asymptotics and Mellin-Barnes integrals. Cambridge University Press, (2001)

23. Strekalovsky, A.: On the global extrema problem. Soviet Math. Doklad **292**(5), 1062–1066 (1987). (In Russian)

24. Strekalovsky, A.: On local search in d.c. optimization problems. Appli. Math. Comput. **255**, 73–83 (2015)

25. Tungalag, N., Enkhbat, R.: Application of sphere packing theory in cost volume profit analysis. Mongolian Math. J. **23**, 1–6 (2021)

26. Vidigal, L.M., Director, S.W.: A design centering algorithm for nonconvex regions of acceptability. IEEE Trans. Comput. Aided Des. Integr. Circ. Syst. CAD **1**, 13–24 (1982)

27. Wu, Q.J., Bourland, J.D.: Morphology-Guided radiosurgery treatment planning and optimization for multiple isocenters. Med. Phys. **26**, 2151–2160 (1992)

28. Wu, A.: Physics and dosimetry of the gamma knife. Neurosurg. Clin. N. Am. **3**, 35–50 (1992)

29. Equation 5.19.4, NIST Digital Library of Mathematical Functions (2013). https://dlmf.nist.gov/5.19E4

Integer Programming
and Combinatorial Optimization

Genetic Algorithm for the Variable Sized Vector Bin-Packing Problem with the Limited Number of Bins

Pavel Borisovsky[1]([⊠])[iD] and Evgeniya Fedotova[1,2]

[1] Sobolev Institute of Mathematics SB RAS, Novosibirsk, Russia
pborisovsky@ofim.oscsbras.ru
[2] Dostoevsky Omsk State University, Omsk, Russia

Abstract. In this paper, we consider a generalization of the well-known bin packing problem, in which several types of bins are given and the number of bins of each type is limited. Unlike the classic bin packing problem, this variant is not well represented in the literature, although it has a considerable practical value. For solving this problem, a genetic algorithm is proposed. It is based on a new representation scheme that uses first fit decreasing algorithm for decoding genotypes to solutions. The computational evaluation on the test instances have shown a competitive performance of the proposed approach comparing to the heuristic algorithms previously known from the literature and Gurobi solver.

Keywords: Variable sized bin packing · Limited number of bins · Genetic algorithm

1 Introduction

The bin packing problem (BPP) have been studied since the 1930s and remains one of the most popular applied discrete optimizaton problems. Its classic formulation is as follows: given a set of items, each of which has a given weight, and an unlimited number of identical bins of a fixed capacity, it is required to pack all items in the minimum number of bins so that in any bin the total weight of items does not exceed its capacity. The solution approaches include the greedy-like First Fit (FF) and First Fit Decreasing (FFD) algorithms that show good performance in practice and are widely used as subroutines in other algorithms. The exact approaches are mostly based on column generation, branch and bound, and arc-flow modeling techniques [4,10,18]. A survey of metaheuristic algorithms can be found in [15].

A natural generalization of BPP is its multidimensional (vector) formulation [5], where each item weight and bin capacity are represented by p-dimensional vectors, and a feasible packing assumes that the sum of weights of items in a bin does not exceed the bin capacity componentwise. As an example,

The research of the first author was funded in accordance with the state task of the IM SB RAS, project FWNF-2022-0020.

Y. Kochetov et al. (Eds.): MOTOR 2022, CCIS 1661, pp. 55–67, 2022.
https://doi.org/10.1007/978-3-031-16224-4_3

one may consider items with the given weights and volumes and bins with the certain limits on the total values of these parameters. The kind of problem has a number of applications in transportation and warehousing. It also especially useful in distributed computing systems, in which certain tasks must be assigned to a set of computers in a cloud, and each task is characterized by many parameters, such as network bandwidth, CPU, RAM, and disk space requirements. Note that the vector formulation essentially differs from another well know multidimensional geometric packing problems, in which p-dimensional rectangular items are packed in a set of p-dimensional rectangular containers (see, e.g., [2]).

In this paper, we are dealing with the further generalization of BPP known as the variable sized (heterogeneous) BPP, in which several types of bins are present, for each type the capacity and cost parameters are known, and the total packing cost is to be minimized. It seems that this formulation first appeared in [9] and since then is extensively studied. As before, column generation [7] and branch and bound [12,14] approaches are used for exact solution. A rather competitive genetic algorithm and a set partitioning heuristic are given in [11]. Quite an advanced variable neighbourhood search that showed an outstanding experimental performance is presented in [13].

We will also assume that the number of bins of each type is limited. This restriction suits better for real practical situations and helps to model more general cases. Consider for example an enterprise, which owns a certain number of different containers, and when the total needs exceed the own capacity, additional containers must be rented at the market. This can be easily modeled by introducing the set of bin types with given capacities, number of available bins, and zero costs, and another set of types with the market costs and possibly unlimited number of bins.

Despite the considerable practical value, this kind of BPP is very rarely represented in the literature. In [6] and [7] the problem with individual bins each one having its own cost and capacity is studied. In [16] the vector BPP and the variable sized BPP are considered separately and solved by the path enumeration model and the branch-cut-and-price algorithm. We are not aware of any studies, devoted to the variable sized vector BPP.

In this paper, we propose a genetic algorithm with a new encoding that uses FFD for decoding genotypes to solutions. This idea allows to exploit good performance of FFD without excluding optimal solutions from the search space. An experimental evaluation and comparison with the previously known approaches is performed.

The rest of the paper is organized as follows. Section 2 introduces the problem formulation and presents the corresponding mathematical model. Section 3 describes the proposed genetic algorithm. Section 4 discusses the results obtained in numerical experiments and Sect. 5 concludes this study.

2   Problem Formulation

In this section, we give the formal description of the problem and its Integer Linear Programming (ILP) model that can be solved by general-purpose optimization software, e.g., CPLEX or Gurobi. Introduce the following notation.

n is the number of items for packing;
p is the number of dimensions of item weights and bin capacities;
D_j is a p-dimensional vector representing the weight of item $j = 1, \ldots, n$;
m is the number of bin types;
W_i is a p-dimensional vector representing the capacity of bin type $i = 1, \ldots, m$;
C_i is a cost of using a bin of type i, $i = 1, \ldots, m$;
L_i is the limit on the number of bins of type i, $i = 1, \ldots, m$.

The problem asks to assign all items to available bins so that the bin capacities are not violated and the total cost is minimized. The ILP model uses the following decision variables:

x_{ijk} is a binary variable that states if item j is placed in k-th bin of type i;
y_{ik} is a binary variable that states if k-th bin of type i is used.

The model is formulated as follows.

$$\text{Minimize } C(x, y) = \sum_{i=1}^{m} \sum_{k=1}^{L_i} C_i y_{ik}; \tag{1}$$

subject to

$$\sum_{i=1}^{m} \sum_{k=1}^{L_i} x_{ijk} = 1, \ j = \overline{1, n}; \tag{2}$$

$$\sum_{j=1}^{n} D_{jq} x_{ijk} \leq W_{iq} y_{ik}, \ i = \overline{1, m}, \ k = \overline{1, L_i}, \ q = \overline{1, p}; \tag{3}$$

$$x_{ijk}, y_{ik} \in \{0, 1\}. \tag{4}$$

Objective function (1) is a total cost of used bins. Constraints (2) state that each item must be placed in some bin, and (3) bound the load of each bin by its capacity. Note that the numbers of available bins are not present explicitly in the model, but are used in the definition of the variables.

Clearly, the problem is NP-hard even in the sense of finding at least a feasible solution. Indeed, otherwise a classic one-dimensional BPP could be solved applying a binary search for minimal value of L in the interval of some polynomially solvable lower and upper bounds.

3   Genetic Algorithm

The genetic algorithm (GA) is a well-known random search optimization algorithm that simulates an evolution of a population according to the principle of survival of fittest individuals. The GA starts from constructing an initial population of individuals, that correspond to solutions of the given optimization problem. Then, at each iteration, two individuals are selected from the population using a probabilistic selection operator. The crossover and mutation operators are applied to them to build new solutions. After that, new solutions are added to the population, and the two worst individuals are removed. The process is repeated until some stopping criterion is met, which can be a given limit on the number of iterations or the running time.

A formal outline of the described GA is given below.

General scheme of the genetic algorithm.

1. Randomly build the initial population.
2. Until the stop criterion is met:
   2.1. Select two parent individuals p_1 and p_2 from the population.
   2.2. Using the crossover operator, construct two children c_1 and c_2.
   2.3. Apply the mutation operator to c_1 and c_2.
   2.4. Apply the repair and improvement operators to c_1 and c_2.
   2.5. Remove the two worst individuals from the population and put c_1 and c_2 in their place.

3.1   Solutions Representation

The encoding is an especially important issue for a GA in the case of BPP. Some simple representations are known from the literature, e.g., a vector $(r_1, r_2, ..., r_n)$, in which r_j is the index of the bin containing item j, or a permutation of items $(j_1, j_2, ..., j_n)$ that defines the order, in which the bins are loaded. One of the earliest papers, [8], discusses the high redundancy of these encodings, i.e. a large set of genotypes may correspond to one actual solution. A better approach called a grouping representation was proposed, which stores a solution in its natural form as a partitioning of items into sets corresponding to their placement to bins (no encoding). A further development of a GA with such a representation was done in [17] where a special crossover operator was introduced.

A rather competitive encoding for the VSBPP was proposed in [12]. The solution is represented by a sequence of items, and the decoding consists in cutting the sequence into several parts, so that each part corresponds to one bin. The problem of cutting the sequence can be solved optimally in polynomial time using the shortest path approach. This scheme showed good experimental results, but unfortunately, it cannot be straightforwardly extended to the case of the limited number of bins.

In this paper, we propose a new encoding that uses FFD for decoding genotypes to solutions. This idea allows to exploit good performance of FFD without excluding optimal solutions from the search space. To represent a packing, two n-dimensional vectors are used: vector t and vector s. Each coordinate j of these vectors corresponds to the j-th item ($j = 1, ..., n$). Each component t_j stores an index of the bin type, in which item j is packed. Vector s defines some deviations from the plain FFD solution; the details will be described below. When decoding a genotype, first the items are distributed to subsets corresponding to the bin types defined by t. Then in each subset, a packing is built with the modified FFD algorithm that uses additional information from vector s. Recall that FFD is a heuristic that looks over the items in the decreasing order of their weights and tries to place each current item to the first suitable bin. Assume that all the items are sorted by decreasing of their total relative weights, defined as follows:

$$\overline{D}_j = \sum_{q=1}^{p} D_{jq}/\overline{W_q}, \ j = 1, ..., n,$$

where $\overline{W_q}$ is a q-th coordinate of the average bin capacity: $\overline{W_q} = \sum_{i=1}^{m} W_{iq}/m$.

We consider two FFD-decoders. The first one, which will be referenced to as *Plain-FFD decoder* simply applies FFD algorithm to the subsets of items assigned to the same type of bins according to vector t. In this case, vector s is not used and in the simple case of one bin type ($m = 1$), the whole genetic algorithm is degraded to the simple FFD. We do not expect an outstanding performance of this scheme, and it is used as a baseline for estimating the effect of another encoding.

The second decoder scheme called *Shift-FFD decoder* uses vector s to define possible shifts of each item to the previous or the subsequent bins relative to their positions in the FFD solution. This decoder works as follows:

1. For each bin type, collect the items assigned to it according to vector t and run FFD algorithm. As a result, each item j is given a *position* l_j, which is an index of the bin, where this item is put.
2. Start from the beginning (assume that all bins are empty) and consider the items j in the previously defined order (decreasing by $\overline{D}_j$) and selecting only such items for which $s_j \neq 0$. Try to place such items in positions $l_j + s_j$, if placing the current item causes violation of the capacity then try to allocate it to the next position and so on.
3. The rest of the items (with $s_j = 0$) are allocated by the FFD algorithm. If for some bin type, the Shift-FFD produced worse packing than the simple FFD, then the FFD solution is restored and returned as a result for this bin type.

To illustrate the described decoder, consider the following example. Let the problem have one dimension and one type of bin with capacity $W = 13$ and the cost is equal to one (classic BPP problem). The weights of items are given by the vector $D = (7, 5, 4, 4, 2, 2, 2)$, which are sorted in descending order. Since we have only one type of bin, the vector t is redundant.

The Plain-FFD algorithm finds the solution shown in Fig. 1 on the left and it defines the initial positions of items (1,1,2,2,2,2,3). Recall that a position means the index of bin containing the given item. The value of the objective function in this case is 3. The optimal solution for this example can be seen in Fig. 1 on the right, and the optimal number of bins is 2. It easy to see that this solution can be obtained if the item with $D = 5$ is shifted one position to the right and the other items are placed by the Plain-FFD algorithm. It means that the vector $s = (0, 1, 0, 0, 0, 0, 0)$ being passed to the Shift-FFD decoder represents the optimal solution.

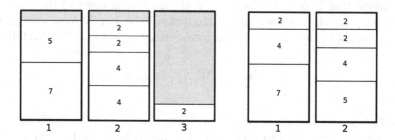

Fig. 1. The solution found by the FFD algorithm and the optimal solution

Due to the limits on the number of bins, the FFD decoders can not guarantee obtaining feasible solutions. If some items are not allocated after the decoder, an additional *repair procedure* is applied. It simply traverses over the list of unallocated items, and checks each used bin of any possible type for enough capacity for this item. If no such bins are found, new bins of some randomly chosen type are added if it is allowed. Finally, the unallocated items are placed to randomly selected bins and the penalty for exceeding the capacities is introduced.

In addition, in attempt to improve the solution a similar *improvement procedure* is applied to the last bin of each type. Namely, for each bin type the items in the last bin are identified and moved to other bins of any types if enough free place is found.

For a given individual, its fitness is evaluated as the total cost of used bins, the penalty of exceeding the capacity constraints and an additional term for encouraging dense packing (see, e.g. [17]). Denote the total computed penalty by P, the load of k-th bin of type i in the q-th coordinate by S_{ikq} and assuming that any solution obtained by the decoder can be represented by variables x and y as of Sect. 2, the fitness function (for minimization) can be expressed by

$$F(g) = C(x, y) + \alpha P - \beta \sum_{i=1}^{m} \sum_{k=1}^{L_i} \sum_{q=1}^{p} (S_{ikq}/W_{iq})^2.$$

The penalty coefficient α is chosen large enough so that any feasible solutions would have better fitness that any infeasible one. On the other hand, β should be rather small, so that the last term could not exceed the smallest cost of one bin.

The genotypes for the initial population are built randomly. For each item j, the value t_j is chosen uniformly among the bin types with enough capacity for this item. The components of vector s are set so that the solution would not be very far from the one obtained by FFD, i.e. value $s_j = 0$ should have quite a large probability, and for the other values the probabilities are represented by a decreasing sequence. A simple way to implement this is to use the geometric distribution. Note that s_j may admit both positive and negative values, so eventually the random variable is given by $s_j = \mathcal{H}(\mathcal{G}-1)$, where $\mathcal{G}$ is a geometrically distributed random variable, and $\mathcal{H}$ takes values 1 and -1 with equal probabilities. For example, if the parameter of the geometric distribution is 0.5 then s_j is as follows

Value	...	-2	-1	0	1	2	...
Probability	...	0.0625	0.125	0.5	0.125	0.0625	...

For selection of the parent genotypes at step 2.1 of the GA, the well-known S-tournament operator is used. It chooses S genotypes from the current population and returns the best one as the first parent. Then the second parent is chosen the same way.

3.2   Crossover and Mutation Operators

In our GA we use the classic one-point and two-points crossing operators and several mutations. In the one-point crossover, an index in the parent genotypes string called a *cut point* is randomly selected and all the elements before the cut point are simply copied to the offspring, the elements after the cut point are exchanged, i.e. the elements of the first parent are copied to the second offspring and vice versa. In our case, the cut point is chosen between 1 and n and by the j-th element of the genotype we mean two numbers (t_j, s_j) where $j = 1, ..., n$. In the two-point crossover, two cut points are chosen randomly, the middle parts are exchanged and the left and right parts are copied to the offspring. In the GA, at Step 2.2 one of the two crossovers is chosen with probability $1/3$ and if no one is applied (with probability $1/3$) the parent genotypes are simply copied to the offspring.

Different mutation operators are used to introduce small random changes in vectors t and s. The first mutation moves some item to another bin type. It traverses over all items and for each $j = 1, .., n$ with probability $1/n$ the value of t_j is changes to a randomly chosen bin type among the ones suitable for item j (i.e. the capacities of the bin type allow storing the item j). The second mutation selects randomly a bin type and each item assigned to it is moved to another bin type with probability 0.5.

The third mutation chooses randomly two items j and l assigned to different bin types, i.e. $t_j \neq t_l$, but both types t_j and t_l are suitable for j and l. Then the bin types are exchanged: item j is assigned to t_l and vice versa. Note that in case of a single bin type these three mutation operators make no any effect.

The forth mutation traverses over items and for each $j = 1, ..., n$ with probability $1/n$ changes s_j to a randomly generated value $\mathcal{H}(\mathcal{G} - 1)$ as described in Sect. 3.1. The last mutation increments or decrements s_j by one for each item j with probability $1/n$. In the GA, at Step 2.3 one of the mutation operators is chosen randomly with equal probabilities.

4 Experimental Evaluation

The proposed GA was implemented in C++ and run on the server with AMD EPYC 7502 CPU under Ubuntu Linux 20.04. The tunable parameters were determined in the preliminary testing. We observed that the best results are given when the population size is rather large, so it was set to 50000. The selection size $S = 20$, the parameter of the geometric distribution is 0.5.

In the experiments we use several datasets known from the literature for two particular cases of the considered problem: the two-dimensional vector packing problem (2-DVPP) from [5] and the one dimensional VSBPP with the unlimited number of bins, see [11,13,14]. Besides, on the base of the existing data we constructed two-dimensional problem instances with the limited number of bins.

The dataset of 2-DVPP instances is generated randomly and consists of ten classes with different characteristics denoted by CL_1,...,CL_10. Each class but the 10-th contains 40 instances with the number of items varying from 25 to 200. The 10-th class contains 30 instances with 24, 51, and 201 items. The detailed description can be found in [5]. As in the other papers (see, e.g., [1]) we use classes 1, 6, 7, 9, and 10 because the others are considered as rather easy and not suitable for testing. The best known upper bounds are available at http://or.dei.unibo.it/library/two-constraint-bin-packing-problem.

The VSBPP data proposed in [14] consist of six classes, denoted by A_X1, A_X2, A_X3, B_X1, B_X2, B_X3. The classes marked with "A" have three bin types, and the ones marked with "B" have five bin types. The marks X1, X2, and X3 mean different intervals of item weights. For the comparison we used the upper bounds obtained by the VRP Solver [16]. Similar instances were generated in [11] for testing the different heuristics including a genetic algorithm, and then they were used in [13] for the experiments with the VNS heuristic. According to [13], these classes are named "Set1 linear", "Set2 linear", "Set2 convex", and "Set2 concave", which reflect the kind of the cost function. In the linear case, the cost of one bin coincide with its capacity: $c_i = W_i$. The convex and concave costs are $\lceil 0.1 W_i^{3/2} \rceil$ and $\lceil 10\sqrt{W_i} \rceil$ respectively.

For each instance, the GA made ten independent runs and the best found solution was returned. In each run, the maximal number of iterations was set to one million and the running time was limited by five minutes. The results are given in Table 1. The columns of the table represent the name of the class, the number of items, and the results of the GA with the Plain-FFD decoder and the Shift-FFD decoder. Column r is a relative error computed as $(C - C^{best})/C^{best} \cdot 100\%$, where C^{best} is the optimum or the best known upper bound from the literature. Since each row corresponds to ten instances, the value of r is given

Table 1. Experimental results on the test instances known from the literature.

Class	n	Plain-FFD decoder r, %	#opt	Shift-FFD decoder r, %	#opt	Class	n	Plain-FFD decoder r, %	#opt	Shift-FFD decoder r, %	#opt
CL_1	25	1.67	9	0	10	A_X1	25	0.08	9	0.08	9
	50	3.85	5	0	10		50	0.04	9	0.08	8
	100	5.77	0	0	10		100	0.1	5	0.12	4
	200	7.65	0	1.96	0		200	0.03	6	0.03	6
CL_6	25	6	4	0	10		500	0.04	8	0.04	8
	50	4.2	1	0	10	A_X2	25	0.07	9	0	10
	100	6.1	0	1.71	3		50	0.3	1	0.34	1
	200	7.28	0	2.22	0		100	0.31	0	0.35	0
CL_7	25	3.22	7	0	10		200	0.32	0	0.27	0
	50	3.73	3	1.23	7		500	0.17	0	0.14	0
	100	1.98	2	0	10	A_X3	25	0	10	0	10
	200	1.5	0	0.87	3		50	0	10	0	10
CL_9	25	0	10	0	10		100	0	10	0	10
	50	6.19	1	0	10		200	0	10	0	10
	100	6.75	0	0	10		500	0	10	0	10
	200	7.6	0	1.95	0	B_X1	25	0.09	9	0	10
CL_10	24	13.75	0	6.25	5		50	0.11	7	0.11	7
	51	14.12	0	5.88	0		100	0.08	6	0.06	7
	201	7.01	0	4.03	0		200	0.02	8	0.01	9
Set1 linear	25	0.15	8	0.16	8		500	0.01	6	0.02	4
	50	0.11	7	0.11	7	B_X2	25	0.07	9	0.14	8
	100	0.06	7	0.08	6		50	0.23	4	0.2	4
	200	0.03	7	0.03	7		100	0.23	0	0.26	1
	500	0	9	0	9		200	0.23	0	0.23	0
Set2 linear	100	0.14	2	0.12	3		500	0.17	0	0.17	0
	200	0.08	4	0.05	6	B_X3	25	0	10	0	10
Set2 convex	100	0	10	0	10		50	0	10	0	10
	200	0	6	0.01	5		100	0	10	0	10
Set2 concave	100	0.02	5	0.02	8		200	0	10	0	10
	200	0.02	5	0.06	6		500	0	10	0	10

on average. The column named "#opt" is a number of instances, for which the optimal or the best known solution was found. The actual running time varied from 7 s for the smallest instances to five minutes for the largest ones.

Analyzing the results we may observe that the 2-DVPP instances of classes CL_1,...,CL_10 are expectedly difficult for the Plain-FFD decoder (recall that these series have only one bin type and such a variant of the GA is degraded to the simple FFD). The use of the Shift-FFD decoder gives a notable improvement even for the largest instances, for which the optimal solutions were still not obtained, but the better approximate solutions were found. We may also note that the overall results are rather competitive with the heuristics known from the literature (see [1], Table 5).

For the other series, the Plain-FFD decoder provides surprisingly good results. For all the classes the average error does not exceed half a percent. Using the Shift-FFD decoder for these problems did not help to improve the results, the differences between them are negligible. For all the instances of X3 classes the best known solutions were obtained. For instances 1_A_3_7, 1_A_4_0, and 1_A_4_4 from class A_X1 we could find better solutions than the ones provided by the solver from [16]. Comparing to the other approaches, we may conclude that out GA does not outperform rather sophisticated Branch-and-price algorithms from [3] and the VNS from [13], but is quite competitive comparing to the beam search [11] and the GA of [11].

Table 2. Experimental results on the test instances with the limited number of bins.

Items	Bin cap.	L	C	Gurobi, LB	Gurobi, UB	Plain-FFD	Shift-FFD
CL_01_200_01	(1000,1000)	34	0	23	26	25	25
	(700,700)	∞	1				
CL_06_100_01	(150,150)	24	0	19	21	25	20
	(105,105)	∞	1				
CL_06_200_01	(150,150)	55	0	35	41	40	40
	(105,105)	∞	1				
CL_07_200_01	(150,150)	56	0	34	38	36	36
	(105,105)	∞	1				
CL_09_200_01	(994,991)	36	0	20	23	23	23
	(696,696)	∞	1				
CL_10_201_01	(100,100)	47	0	30	33	33	33
	(67,67)	∞	1				

For the experiments with the vector VSBPP, we constructed two datasets with several instances on the base of 2-DVPP data. In the first dataset we model the situation, in which there are a limited number of bins with zero cost and extra bins of different capacity, the number of which has to be minimized (as it was discussed in the introduction). The problems data and the results are given in Table 2. The first column shows the 2-DVPP instance name, from which the items were taken, the next three columns describe the bin types. For the comparison we use MIP solver Gurobi 9.5 applied to formulation (1)–(4). As before, each instance was solved by the GA ten times and the best results were collected. Gurobi was tuned to use up to eight parallel threads and the time limit was set to one hour. In the table, the lower and upper bounds obtained by Gurobi within this time are given. The results show that again the two variants of decoder give much the same result, except for the case of CL_06_100_01 where the Plain-FFD decoder could find very poor solution. In all the cases, the results of the GA were the same or better than the results provided by Gurobi.

Table 3. Experimental results on the test instances constructed as a combination of two 2-DVPP instances.

Items	Bin cap.	L	UB	Gurobi, LB	Gurobi, UB	Plain-FFD	Shift-FFD
CL_07_25_01	(150,150)	9	2150	2150	2150	2150	2150
CL_10_24_01	(100,100)	8					
CL_07_50_01	(150,150)	21	4850	4750	4850	4850	4850
CL_10_51_01	(100,100)	17					
CL_01_100_01	(1000,1000)	26	32700	31640	32100	32000	31900
CL_10_201_01	(100,100)	67					
CL_07_200_01	(150,150)	80	62592	61464	–	–	62592
CL_09_200_01	(994,991)	51					
CL_06_100_01	(150,150)	42	12850	12666	–	13000	12950
CL_10_201_01	(100,100)	69					

Another dataset was constructed by combining two 2-DVPP instances from different classes. The characteristics and the results are given in Table 3. The items from both original instances are brought together as well as the bin types. The optimal or the best known upper bounds define the limits on the number of bins. For example, the instance in the first row has the bin type with the capacity vector (150,150) and the number of available bins is $L_1 = 9$ which corresponds to the optimal solution of CL_07_25_01. The second bin type has the capacity (100,100) and $L_2 = 8$. The linear cost function is used, i.e. $c_1 = 150, c_2 = 100$. A trivial upper bound can be obtained by $c_1 L_1 + c_2 L_2$ but it can be potentially improved by better reallocation of items. One exceptional case is the instance reported in the last row, in which the limits were increased, because the original values have produced too hard problem that was not solved by any algorithm. In these experiments, the number of iterations was increased to 5 millions and the time limit for Gurobi was set to three hours so that the total running times were comparable. Again we see that the GA provides the same or better results as Gurobi. In the two last cases Gurobi was not able to find any feasible solutions. The comparison of the different decoders shows that the Shift-FFD decoder provides slightly better results and for one instance the GA with the Plain-FFD decoder could not find any feasible solution.

5 Conclusions

In this paper, a new genetic algorithm for the variable sized vector bin packing problem with the limited number of bins is developed. The main part of the algorithm is an FFD based decoder used for the representation of solutions. In the experiments, we observed that the plain FFD used in the decoder in many cases can provide good results, but there could be instances, for which it works very poorly. The modification of FFD that introduces small deviations from the FFD solution greatly improves the results for such cases and in general does not exclude chances to find an optimum. Since the considered problem is new, some ideas for generation of difficult test instances were presented.

Regarding the possible future research we think that it will be worthwhile to implement and evaluate different solution representations for the GA, and test other approaches, such as variable neighborhood search, column generation, and metaheuristics that showed good performance for the similar problems. Also it would be useful to extend these studies to the temporal problem formulation, in which each item exists in a certain time interval, and consider the online variant of this problem. Also, using a high-performance parallel computing can be a promising direction for improvement of the solution quality.

Acknowledgement. The authors would like to thank Prof. Michele Monaci, Prof. Cristian Blum, and Dr. Ruslan Sadykov for the help with the test instances. An AMD EPYC based server of Sobolev Institute of Mathematics, Omsk Branch is used for computing.

References

1. Aringhieri, R., Duma, D., Grosso, A., Hosteins, P.: Simple but effective heuristics for the 2-constraint bin packing problem. J. Heuristics **24**(3), 345–357 (2018)
2. Baker, B.S., Coffman, E.G., Rivest, R.L.: Orthogonal packings in two dimensions. SIAM J. Comput. **9**(4), 846–855 (1980)
3. Baldi, M.M., Crainic, T.G., Perboli, G., Tadei, R.: Branch-and-price and beam search algorithms for the Variable Cost and Size Bin Packing Problem with optional items. Ann. Oper. Res. **222**(1), 125–141 (2012). https://doi.org/10.1007/s10479-012-1283-2
4. Brandão, F., Pedroso, J.P.: Bin packing and related problems: general arc-flow formulation with graph compression. Comput. Oper. Res. **69**, 56–67 (2016)
5. Caprara, A., Toth, P.: Lower bounds and algorithms for the 2-dimensional vector packing problem. Discret. Appl. Math. **111**(3), 231–262 (2001)
6. Correia, I., Gouveia, L., Saldanha-Da-Gama, F.: Solving the variable size bin packing problem with discretized formulations. Comput. Oper. Res. **35**, 2103–2113 (2008)
7. Crainic, T.G., Perboli, G., Rei, W., Tadei, R.: Efficient lower bounds and heuristics for the variable cost and size bin packing problem. Comput. Oper. Res. **38**, 1474–1482 (2011)
8. Falkenauer, E.: A hybrid grouping genetic algorithm for bin packing. J. Heuristics **2**(1), 5–30 (1996)

9. Friesen, D.K., Langston, M.A.: Variable sized bin packing. SIAM J. Comput. **15**(1), 222–230 (1986)

10. Gilmore, P.C., Gomory, R.E.: A linear programming approach to the cutting-stock problem. Comput. Oper. Res. **9**, 849–859 (1961)

11. Haouari, M., Serairi, M.: Heuristics for the variable sized bin-packing problem. Comput. Oper. Res. **36**(10), 2877–2884 (2009)

12. Haouari, M., Serairi, M.: Relaxations and exact solution of the variable sized bin packing problem. Comput. Optim. Appl. **48**, 345–368 (2011)

13. Hemmelmayr, V., Schmid, V., Blum, C.: Variable neighbourhood search for the variable sized bin packing problem. Comput. Oper. Res. **39**, 1097–1108 (2012)

14. Monaci, M.: Algorithms for packing and scheduling problems. Ph.D. thesis. University of Bologna, Bologna (2002)

15. Munien, C., Ezugwu, A.E.: Metaheuristic algorithms for one-dimensional bin-packing problems: a survey of recent advances and applications. J. Intell. Syst. **30**, 636–663 (2021)

16. Pessoa, A., Sadykov, R., Uchoa, E.: Solving bin packing problems using VRPSolver models. Oper. Res. Forum **2**(2), 1–25 (2021). https://doi.org/10.1007/s43069-020-00047-8

17. Quiroz-Castellanos, M., Cruz-Reyes, L., Torres-Jimenez, J., Gómez, C.S., Fraire Huacuja, H., Alvim, A.: A grouping genetic algorithm with controlled gene transmission for the bin packing problem. Comput. Oper. Res. **55**, 52–64 (2014)

18. Valério de Carvalho, J.: Exact solution of bin-packing problems using column generation and branch-and-bound. Ann. Oper. Res. **86**, 629–659 (1999)

An Approximation Algorithm for Graph Clustering with Clusters of Bounded Sizes

Victor Il'ev[1,2], Svetlana Il'eva[1], and Nikita Gorbunov[2(✉)]

[1] Dostoevsky Omsk State University, Omsk, Russia
[2] Sobolev Institute of Mathematics SB RAS, Omsk, Russia
gorbunov_nikita_v@mail.ru

Abstract. In graph clustering problems, one has to partition the set of vertices of a graph into disjoint subsets (called clusters) minimizing the number of edges between clusters and the number of missing edges within clusters. We consider a version of the problem in which cluster sizes are bounded from above by a positive integer s. This problem is NP-hard for any fixed $s \geqslant 3$. We propose a polynomial-time approximation algorithm for this version of the problem. Its performance guarantee is better than earlier known bounds for all $s \geqslant 5$.

Keywords: Graph clustering · Approximation algorithm · Performance guarantee

1 Introduction

Clustering is the problem of grouping an arbitrary set of objects so that objects in each group are more similar to each other than to those in other clusters. In other words, it is required to partition a given set of objects into some pairwise nonintersecting subsets (clusters) so that the sum of the number of similarities between the clusters and the number of missing similarities inside the clusters would be minimum.

One of the most visual formalizations of clustering is the graph clustering [16]. A version of this problem is known as the graph approximation problem. In this problem, the similarity relation between objects is given by an undirected graph whose vertices are in one-to-one correspondence to the objects and whose edges connect similar objects. The goal is to partition the set of vertices of a graph into disjoint subsets (called clusters) minimizing the number of edges between clusters and the number of missing edges within clusters. The number of clusters may be given, bounded, or undefined. The statements and interpretations of the graph approximation problem can be found in [1, 8, 9, 18, 19].

Later, the graph approximation problem was repeatedly and independently rediscovered and studied under various names (Correlation Clustering, Cluster Editing, etc. [3, 4, 17]).

Y. Kochetov et al. (Eds.): MOTOR 2022, CCIS 1661, pp. 68–75, 2022.
https://doi.org/10.1007/978-3-031-16224-4_4

This paper is organized as follows. Section 2 contains three well-known NP-hard versions of the graph clustering problem and a brief survey of basic results on computational complexity and approximability of these problems. In Sect. 3, another problem of graph clustering with some bounds on the sizes of clusters is considered. This problem is NP-hard too. Polynomially solvable cases of the problem are discussed. In Sect. 4, we propose a polynomial-time approximation algorithm with the tight performance guarantee $2(s-1)$ for the problem in which the cluster sizes are bounded from above by a given number $s \geqslant 3$. Thus, the problem of graph clustering with clusters of bounded sizes belongs to the class APX for every fixed s. Conclusion summarizes the results of the work.

2 Graph Clustering Problems

We consider only ordinary graphs, i.e., the graphs without loops and multiple edges. An ordinary graph is called a *cluster graph* if its every connected component is a complete graph [17]. Let $\mathcal{M}(V)$ be the family of all cluster graphs on the set of vertices V, let $\mathcal{M}_k(V)$ be the family of all cluster graphs on the vertex set V having exactly k connected components, and let $\mathcal{M}_{\leqslant k}(V)$ be the family of all cluster graphs on V having at most k connected components, $2 \leqslant k \leqslant |V|$.

If $G_1 = (V, E_1)$ and $G_2 = (V, E_2)$ are ordinary graphs both on the set of vertices V, then the distance $d(G_1, G_2)$ between them is defined as

$$d(G_1, G_2) = |E_1 \Delta E_2| = |E_1 \setminus E_2| + |E_2 \setminus E_1|,$$

i.e., $d(G_1, G_2)$ is the number of distinct edges in G_1 and G_2.

In the 1960s–1980s the following three graph approximation problems were under study. They can be considered as different formalizations of the graph clustering problem [8,9,11,18,19]:

Problem GC. Given an ordinary graph $G = (V, E)$, find a graph $M^* \in \mathcal{M}(V)$ such that

$$d(G, M^*) = \min_{M \in \mathcal{M}(V)} d(G, M).$$

Problem GC$_k$. Given an ordinary graph $G = (V, E)$ and an integer k, $2 \leqslant k \leqslant |V|$, find a graph $M^* \in \mathcal{M}_k(V)$ such that

$$d(G, M^*) = \min_{M \in \mathcal{M}_k(V)} d(G, M).$$

Problem GC$_{\leqslant k}$. Given an ordinary graph $G = (V, E)$ and an integer k, $2 \leqslant k \leqslant |V|$, find a graph $M^* \in \mathcal{M}_{\leqslant k}(V)$ such that

$$d(G, M^*) = \min_{M \in \mathcal{M}_{\leqslant k}(V)} d(G, M).$$

The first theoretical results related to the graph approximation problem were obtained in the last century. In 1964, Zahn [19] studied Problem **GC** for the graphs of some special form. In 1971, Fridman [8] defined the first polynomially solvable case of Problem **GC**. He showed that Problem **GC** for a graph without triangles can be reduced to constructing maximum matching in this graph. In 1986, Křivánek and Morávek [15] showed that problem **GC** is NP-hard, but their article remained unnoticed.

In 2004, Bansal, Blum, and Chawla [3] and independently Shamir, Sharan, and Tsur [17] proved again that Problem **GC** is NP-hard. In [17] it was also proved that Problem **GC$_k$** is NP-hard for every fixed $k \geqslant 2$. In 2006, a more simple proof of this result was published by Giotis and Guruswamy [10]. In the same year, Ageev, Il'ev, Kononov, and Talevnin [1] independently proved that Problems **GC$_2$** and **GC$_{\leqslant 2}$** are NP-hard even for cubic graphs, and derived from this that both Problems **GC$_k$** and **GC$_{\leqslant k}$** are NP-hard for every fixed $k \geqslant 2$.

In [3], a simple 3-approximation algorithm for Problem **GC$_{\leqslant 2}$** was proposed. In [1], existence of a randomized polynomial-time approximation scheme for Problem **GC$_{\leqslant 2}$** was proved, and in [10], a randomized polynomial-time approximation scheme was proposed for Problem **GC$_{\leqslant k}$** (for every fixed $k \geqslant 2$). In 2008, pointing out that the complexity of the polynomial time approximation scheme of [10] deprives it the prospects of practical use, Coleman, Saunderson and Wirth [7] proposed a 2-approximation algorithm for Problem **GC$_{\leqslant 2}$**, applying a local search procedure to the feasible solutions obtained by the 3-approximation algorithm from [3]. For Problem **GC$_2$**, a 2-approximation algorithm was proposed by Il'ev, Il'eva, and Morshinin [14]. As regards to Problem **GC**, it was shown in 2005 [5] that Problem **GC** is APX-hard, and a 4-approximation algorithm was developed. In 2008, a 2.5-approximation algorithm for Problem **GC** was presented in Ailon, Charikar, and Newman [2]. Finally, in 2015, $(2.06 - \varepsilon)$-appriximation algorithm for Problem **GC** was proposed [6].

3   The Graph Clustering Problem with Clusters of Bounded Sizes

In contrast to Problems **GC$_k$** and **GC$_{\leqslant k}$**, where the restrictions on the number of clusters are imposed, we now discuss the problem of graph clustering with clusters of bounded sizes.

Let $\mathcal{M}^{\leqslant s}(V)$ be the family of all cluster graphs on V such that the size of each connected component is at most some integer s, $2 \leqslant s \leqslant |V|$. We say that a cluster graph belongs to $\mathcal{M}^s(V)$ if the size of each of its connected components is equal to s.

Problem GC$^{\leqslant s}$. Given an n-vertex graph $G = (V, E)$ and an integer s, $2 \leqslant s \leqslant n$, find $M^* \in \mathcal{M}^{\leqslant s}(V)$ such that

$$d(G, M^*) = \min_{M \in \mathcal{M}^{\leqslant s}(V)} d(G, M).$$

Problem GC^s. Given a graph $G = (V, E)$ such that $|V| = sq$, where s and q are positive integers, find $M^* \in \mathcal{M}^s(V)$ such that

$$d(G, M^*) = \min_{M \in \mathcal{M}^s(V)} d(G, M).$$

In [3], in the proof of NP-hardness of Problem **GC** without any constraints on the number and sizes of clusters, it is actually shown that Problem $\mathbf{GC}^{\leqslant 3}$ is NP-hard. In [12], the following is proved:

Theorem 1. [12] *Problems $\mathbf{GC}^{\leqslant s}$ and $\mathbf{GC}^s$ are NP-hard for every fixed $s \geqslant 3$.*

In [12], the cases when the optimal solutions to Problems $\mathbf{GC}^{\leqslant s}$ and $\mathbf{GC}^s$ can be found in polynomial time are also considered:

Theorem 2. [12] *Problems $\mathbf{GC}^{\leqslant 2}$ and $\mathbf{GC}^2$ are polynomially solvable. Problem $\mathbf{GC}^{\leqslant 3}$ on graphs without triangles is polynomially solvable.*

In [13], it is shown that the latter result can be generalized to the case of an arbitrary s.

Theorem 3. [13] *Problem $\mathbf{GC}^{\leqslant s}$ on graphs without triangles is polynomially solvable for all $s \geqslant 2$.*

4   An Approximation Algorithm for Problem $\mathbf{GC}^{\leqslant s}$

In [13], for Problem $\mathbf{GC}^{\leqslant s}$ ($s \geqslant 3$) a polynomial-time approximation algorithm with the performance guarantee $\lfloor \frac{(s-1)^2}{2} \rfloor + 1$ is proposed. In this section, we offer a polynomial-time approximation algorithm with the performance guarantee $2(s-1)$ for this problem.

Algorithm A
Input: An arbitrary graph $G = (V, E)$, an integer s, $3 \leqslant s \leqslant |V|$.
Output: A cluster graph $M = (V_M, E_M) \in \mathcal{M}^{\leqslant s}$ – approximate solution to problem $\mathbf{GC}^{\leqslant s}$.
Iteration 0. $V_M \leftarrow \emptyset$, $E_M \leftarrow \emptyset$, $V' \leftarrow V$, $E' \leftarrow E$, $G' \leftarrow (V', E')$. Go to iteration 1.
Iteration i ($i \geqslant 1$). Constructing i-th cluster K^i.
    Step 1. If $V' = \emptyset$, then **END**.
Else select an arbitrary vertex $v \in V'$, $K^i \leftarrow \{v\}$.
    Step 2. If $|K^i| < s$, then go to step 3, else go to step 5.
    Step 3. If there exists $u \in V'$ such that $uv \in E'$ for all $v \in K^i$, then go to step 4, else go to step 5.
    Step 4. $K^i \leftarrow K^i \cup \{u\}$, go to step 2.
    Step 5. $V_M \leftarrow V_M \cup K^i$, $E_M \leftarrow E_M \cup \{uv \mid u, v \in K^i\}$, $V' \leftarrow V' \setminus K^i$, $E' \leftarrow E' \setminus \{uv \mid \{u, v\} \cap K^i \neq \emptyset\}$. Go to iteration $i + 1$.

Comments.
    Step 0. V' is the set of undistributed by clusters vertices of the graph G, G' is the subgraph of G induced by the set V'.
    Step 1. Start constructing a new cluster K^i of the graph M.
    Steps 2 − 4. Increasing the cluster.
    Step 5. Put in M the clique induced by the set K^i and delete it from the graph G' with all edges incident to its vertices.

Remark 1. Time complexity of Algorithm A is $\mathcal{O}(sn^2)$, where $n = |V|$.

Indeed, the number of iterations is less or equal to n. Steps 3 and 5 are the most time-consuming, each of them requires $\mathcal{O}(sn)$ operations.

Theorem 4. *Let $G = (V, E_G)$ be an arbitrary graph, s be an integer, $3 \leqslant s \leqslant |V|$. Then*

$$\frac{d(G, M)}{d(G, M^*)} \leqslant 2(s - 1), \tag{1}$$

where $M^* = (V, E_{M^*})$ is an optimal solution to problem $\mathbf{GC}^{\leqslant \mathbf{s}}$ on the graph G, $M = (V, E_M)$ is the cluster graph constructed by Algorithm A.

Proof. 1) By the definition, $d(G, M^*) = |E_G \backslash E_{M^*}| + |E_{M^*} \backslash E_G|$. Let $E_G \backslash E_{M^*} = E_0 \cup E_1$, where
E_0 is the set of edges of the graph G that are not placed neither in M^*, nor in M by Algorithm A;
E_1 is the set of edges of the graph G that are not placed in M^*, but are placed in M by Algorithm A.
Denote $E_2 = E_{M^*} \backslash E_G$. Then $d(G, M^*) = |E_0| + |E_1| + |E_2|$.
2) By constructing, the graph M is a subgraph of the graph G, therefore $d(G, M) = |E_G \backslash E_M|$. Write the difference $E_G \backslash E_M$ in the following form: $E_G \backslash E_M = E_0 \cup \tilde{E}$, where $\tilde{E}$ is the set of edges of G not included in M, but placed in M^*. Edges of the set $\tilde{E}$ are edges of the graph G between clusters (cliques) of the graph M, hence any edge of the set $\tilde{E}$ has the form $e = uv$, where $v \in K^i$, $u \in K^j$ for some i, j, $i < j$ (see step 5).
3) Show that any edge of $\tilde{E}$ either is adjacent to an edge of the set E_1 in the graph G, or is adjacent to an edge of the set E_2 in the graph M^*.
Consider an arbitrary edge $e = uv$, where $v \in K^i$, $u \in K^j$, $i < j$. Since $i < j$, hence $u \in V'$ by constructing the cluster K^i (on step 1 of iteration i). At the same time, the vertex u isn't included in the cluster K^i. It is possible only in two cases:
   a) $|K^i| = s$. Then $|K^i \cup \{u\}| = s+1$, therefore there exists at least one vertex $w \in K^i$ such that w and v are in different clusters of the graph M^*. And since K^i is the vertex set of a clique of the graph G, hence vertices v and w are adjacent in G. Therefore, $vw \in E_1$.
   b) $|K^i| < s$, but there exists $w \in K^i$ such that $uw \notin E$. If w is placed in the same cluster of the graph M^* as u, then $uw \in E_2$. Otherwise, if w and u are in different clusters of the graph M^*, then vertices w and v are also in different clusters of M^*. As in case a) we get that $vw \in E_1$.
So, it is shown that any edge from $\tilde{E}$ either is adjacent to an edge of the set E_1 in G, or is adjacent to an edge of the set E_2 in M^*. Hence $\tilde{E} = \tilde{E}_1 \cup \tilde{E}_2$, where

$$\tilde{E}_1 = \{uv \in \tilde{E} \mid \exists w \text{ such that } uw \in E_1 \text{ or } vw \in E_1\},$$

$$\tilde{E}_2 = \{uv \in \tilde{E} \mid \exists w \text{ such that } uw \in E_2 \text{ or } vw \in E_2\}.$$

(it is possible that $\tilde{E}_1 \cap \tilde{E}_2 \neq \emptyset$).

4) Note, that for every edge $uv \in E_1$ there exist at most $2(s-1)$ edges of $\tilde{E}_1$ incident with u or v in G, and for every edge $uv \in E_2$ there exist at most $2(s-2)$ edges of $\tilde{E}_2$ incident with u or v in G. Hence

$$d(G,M) \leqslant |E_0| + |\tilde{E}_1| + |\tilde{E}_2| \leqslant |E_0| + 2(s-1)|E_1| + 2(s-2)|E_2|$$
$$\leqslant |E_0| + 2(s-1)(|E_1| + |E_2|).$$

Therefore,

$$\frac{d(G,M)}{d(G,M^*)} \leqslant \frac{|E_0| + 2(s-1)(|E_1| + |E_2|)}{|E_0| + |E_1| + |E_2|} \leqslant \frac{2(s-1)(|E_1| + |E_2|)}{|E_1| + |E_2|} = 2(s-1).$$

Theorem 4 is proved.

Corollary 1. Problem $\mathbf{GC}^{\leqslant s}$ belongs to the class APX for every fixed $s \geqslant 3$.

By the following assertion, bound (1) is tight for any value of s.

Remark 2. For every $s \geqslant 3$, there exists a graph G_s such that

$$d(G_s, M) = 2(s-1)d(G, M^*). \tag{2}$$

The graph G_s has $2s$ vertices and consists of the two cliques K_s connected by one edge (bridge). Figure 1 shows an example for the case $s = 4$.

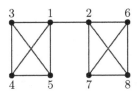

Fig. 1. The graph G_4

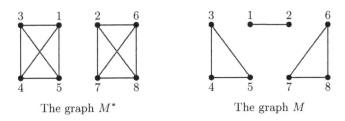

The graph M^* The graph M

Fig. 2. The graphs M^* and M

As we see in Fig. 2, $d(G_4, M^*) = 1$, $d(G_4, M) = 6$, and equality (2) holds.

Clearly, $d(G_s, M^*) = 1$. In the worst case the cluster graph M for G_s found by Algorithm A consists of a clique K_2 and two cliques K_{s-1}, and hence $d(G_s, M) = 2(s-1)$. Thus, we obtain (2).

Remark 3. In the cases $s = 3$ and $s = 4$ the previously known $(\lfloor \frac{(s-1)^2}{2} \rfloor + 1)$-approximation algorithm from [13] guarantees better performance than $2(s - 1)$-approximation Algorithm A. But for all $s \geqslant 5$ the performance guarantee of Algorithm A is better.

5 Conclusion

A version of the graph clustering problem is considered. In this version sizes of all clusters don't exceed a given positive integer s. This problem is NP-hard for every fixed $s \geqslant 3$. A new polynomial-time approximation algorithm is presented and a bound on worst-case behaviour of this algorithm is obtained.

Acknowledgement. The research of the first author was funded in accordance with the state task of the IM SB RAS, project FWNF-2022-0020.

References

1. Ageev, A.A., Il'ev, V.P., Kononov, A.V., Talevnin, A.S.: Computational complexity of the graph approximation problem. Diskretnyi Analiz i Issledovanie Operatsii. Ser. 1. **13**(1), 3–11 (2006). (in Russian). English transl. J. Appl. Ind. Math. **1**(1), 1–8 (2007)
2. Ailon, N., Charikar, M., Newman, A.: Aggregating inconsistent information: ranking and clustering. J. ACM **55**(5), 1–27 (2008)
3. Bansal, N., Blum, A., Chawla, S.: Correlation clustering. Mach. Learn. **56**, 89–113 (2004)
4. Ben-Dor, A., Shamir, R., Yakhimi, Z.: Clustering gene expression patterns. J. Comput. Biol. **6**(3–4), 281–297 (1999)
5. Charikar, M., Guruswami, V., Wirth, A.: Clustering with qualitative information. J. Comput. Syst. Sci. **71**(3), 360–383 (2005)
6. Chawla, S., Makarychev, K., Schramm, T., Yaroslavtsev, G.: Near optimal LP algorithm for correlation clustering on complete and complete k-partite graphs. In: STOC 2015 Symposium on Theory of Computing. ACM, New York, pp. 219–228 (2015)
7. Coleman, T., Saunderson, J., Wirth, A.: A local-search 2-approximation for 2-correlation-clustering. In: Halperin, D., Mehlhorn, K. (eds.) ESA 2008. LNCS, vol. 5193, pp. 308–319. Springer, Heidelberg (2008). https://doi.org/10.1007/978-3-540-87744-8_26
8. Fridman, G.Š.: A graph approximation problem. Upravlyaemye Sistemy. Izd. Inst. Mat., Novosibirsk **8**, 73–75 (1971). (in Russian)
9. Fridman, G.Š.: Investigation of a classifying problem on graphs. Methods of Modelling and Data Processing (Nauka, Novosibirsk), pp. 147–177 (1976). (in Russian)
10. Giotis, I., Guruswami, V.: Correlation clustering with a fixed number of clusters. Theory Comput. **2**(1), 249–266 (2006)
11. Il'ev, V.P., Fridman, G.Š.: On the problem of approximation by graphs with a fixed number of components. Dokl. Akad. Nauk SSSR. **264**(3), 533–538 (1982). (in Russian). English transl. Sov. Math. Dokl. **25**(3), 666–670 (1982)

12. Il'ev, V.P., Navrotskaya, A.A.: Computational complexity of the problem of approximation by graphs with connected components of bounded size. Prikl. Diskretn. Mat. **3**(13), 80–84 (2011). (in Russian)
13. Il'ev, V.P., Il'eva, S.D., Navrotskaya, A.A.: Graph clustering with a constraint on cluster sizes. Diskretn. Anal. Issled. Oper. **23**(3), 50–20 (2016). (in Russian). English transl. J. Appl. Indust. Math. **10**(3), 341–348 (2016)
14. Il'ev, V., Il'eva, S., Morshinin, A.: A 2-approximation algorithm for the graph 2-clustering problem. In: Khachay, M., Kochetov, Y., Pardalos, P. (eds.) MOTOR 2019. LNCS, vol. 11548, pp. 295–308. Springer, Cham (2019). https://doi.org/10.1007/978-3-030-22629-9_21
15. Křivánek, M., Morávek, J.: NP-hard problems in hierarchical-tree clustering. Acta Informatica **23**, 311–323 (1986)
16. Schaeffer, S.E.: Graph clustering. Comput. Sci. Rev. **1**(1), 27–64 (2005)
17. Shamir, R., Sharan, R., Tsur, D.: Cluster graph modification problems. Discrete Appl. Math. **144**(1–2), 173–182 (2004)
18. Tomescu, I.: La reduction minimale d'un graphe à une reunion de cliques. Discrete Math. **10**(1–2), 173–179 (1974)
19. Zahn, C.T.: Approximating symmetric relations by equivalence relations. J. Soc. Ind. Appl. Math. **12**(4), 840–847 (1964)

Approximation Non-list Scheduling Algorithms for Multiprocessor System

Natalia Grigoreva[(✉)][iD]

St. Petersburg State University, St. Petersburg, Russia
n.s.grig@gmail.com

Abstract. The multiprocessor scheduling problem is defined as follows: jobs have to be executed on several parallel identical processors. Each job has a positive processing time. At most one job can be processed at a time, but all jobs may be simultaneously delivered. We study the case where precedence constrains exist between jobs and preemption on processors is not allowed. The objective is to minimize the time, by which all jobs are done. The problem is NP-hard in the strong sense. The best-known approximation algorithm is the critical path algorithm, which generates the list no delay schedules. We define an IIT (inserted idle time) schedule as a feasible schedule, in which a processor is kept idle at a time when it could begin processing a job. The paper proposes a 2-1/m approximation inserted idle time algorithm for the multiprocessor scheduling. To illustrate the efficiency of our approach, we compared two algorithms on randomly generated sets of jobs.

Keywords: Parallel identical processors · Critical path · Makespan · Inserted idle time · Approximation algorithm

1 Introduction

We consider the scheduling problem of a job set $V = \{v_1, v_2, \ldots, v_n\}$ on parallel identical processors to minimize the total processing time (makespan). The precedence constraints $v_i \prec v_j$ between jobs are given, that is, the job v_i should be finished before starting job v_j. Each job v_i has a positive processing time $t(v_i)$.

Resources consist of m identical parallel processors, which execute the set of jobs V. At any given time, the processor can execute only one job and each job requires one processor. Preemption on processors is not allowed. To generate a schedule means to define a start time of its execution $\tau(v_i)$ for each job v_i and a processor $num(v_i)$ which performs job v_i. A schedule is feasible, if the precedence constraints are satisfied

$$\tau(v_i) + t(v_i) \leq \tau(v_j), \quad v_i \prec v_j.$$

Y. Kochetov et al. (Eds.): MOTOR 2022, CCIS 1661, pp. 76–88, 2022.
https://doi.org/10.1007/978-3-031-16224-4_5

The goal is to find a feasible schedule minimizing the maximum completion time of all jobs (makespan)

$$C_{\max} = \max\{\tau(v_i) + t(v_i) \mid v_i \in V\}.$$

We denote this problem by $P|prec|C_{\max}$, following the three-field notation introduced in [1]. The problem is NP-hard in the strong sense when $2 \le m < n$, and we are interested in approximation algorithms for the problem.

Almost all problems where precedence relations exist between jobs and preemption is not allowed are NP-hard, except two problems. The first problem, where processing time for all jobs is equal to 1 and the job graph is an in-tree, is $P|in - tree, p_i = 1|C_{\max}$. The algorithm which generates the optimal schedule was proposed in [8].

The second problem is $P2|prec, p_i = 1|C_{\max}$, where the number of processors is equal to 2, processing time for all jobs is equal to 1, but the job graph has arbitrary precedence relations [9].

We consider the general scheduling problem, where processing time for jobs is different, precedence relations and the number of parallel processors are arbitrary.

Most researchers have developed heuristic algorithms, which use a list scheduling technique. Whenever a processor is free the list-scheduling algorithm selects a job with the largest priority from a set of ready jobs.

The critical path (CP) algorithm gives the highest priority to the job at the head of the longest path of jobs in the precedence graph. The LNS (Largest number of successors first) gives the highest priority to the job which has the largest number of successors.

The paper [11] proposes the CP/MISF (critical path/most immediate successors first) method, an improved version of the CP-method, which has so far been regarded as the most effective heuristic algorithm for solving this type of scheduling problems.

In the CP/MISF method level $l(v_i)$ of job v_i is the length of the longest path from the exit node to job v_i, but there are two jobs which have the same level, then the job with the largest number of immediately successive jobs is assigned the highest priority.

List-scheduling algorithms are fast and usually generate good approximation schedules. But list-scheduling algorithms generate non-delay schedules, these algorithms do not allow processors to be idle if there is a ready job, even if the priority of the job is low. The optimal schedule may not be non-delay, if there are unrelated machines in parallel and preemptions are not allowed. The list schedules do not allow such downtime, then in these cases the list schedules cannot generate the optimal schedule.

We are interested in non-list algorithms that have the same computational complexity and the same guaranteed accuracy estimation as list algorithms. Then solving the problem twice with a list and non-list algorithm, we get the best solution.

We proposed non-list approximation algorithms for $P|r_j, q_j|C_{max}$ and $1|r_j, q_j|C_{max}$ and investigated their worst-case performance.

We proposed an approximation IIT algorithm for $P|r_j, q_j|C_{max}$ problem, proved that $C_{max}(S) - C_{opt} < t_{max}(2m - 1)/m$, and this bound is tight, where $C_{max}(S)$ is the objective function of the approximation IIT schedule, and C_{opt} is the makespan of the optimal schedule. [5]

The problem $P|r_j, q_j|C_{max}$ is a generalization of the single-machine scheduling problem with release and delivery times $1|r_j, q_j|C_{max}$, it is NP-hard too [12].

In [4] we proposed the approximation algorithm ICA which creates two permutations, one by the Schrage method, and the second by the algorithm with inserted idle time. The construction of each permutation requires $O(n \log n)$ operations. We proved that the worst-case performance ratio of the algorithm ICA is equal to $3/2$ and the bound of $3/2$ is tight. For the single machine scheduling problem, we proved that the combined algorithm has better guaranteed accuracy than either algorithm alone.

The main idea of greedy algorithms for solving these problems is the choice the highest priority job at each step, before the execution of which the processor could be idle.

For problem $P|prec|C_{max}$, list algorithms are $2 - 1/m$ approximation algorithms. Moreover, this assessment does not depend on the method of determining the job priorities.

The algorithm CP/IIT allows processors to idle while waiting for a more priority job. Additional conditions are checked, under which it is profitable to perform a higher priority job, but the downtime is not too large.

The goal of this paper is to propose new approximation algorithms for $P|prec|C_{max}$ scheduling problem and investigate the worst-case performance of the CP/IIT no-list scheduling algorithm. We prove that the no-list scheduling algorithm has the same relative accuracy as the list algorithm. Then by combining two approximation algorithms we propose the combined algorithm which generates two schedules and selects the best one.

First, in Sect. 2, we propose an approximation CP/IIT (critical path/inserted idle time) algorithm and investigate the worst-case performance of the CP/IIT algorithm. To confirm the effectiveness of our approach, we tested our algorithms on randomly generated job graphs. In Sect. 3 we present the results of the algorithm testing. Summary of this paper is in Sect. 4.

2   Approximation Algorithm CP/IIT

To solve the problem $P|prec|C_{max}$, the following algorithm is proposed.

We assume that the precedence constraints are given by the acyclic graph $G = (V^*, E)$, where jobs are represented by nodes $V^* = V \cup j_- \cup j_+$ and arcs E define the partial order. There are two dummy nodes: the entry node j_- and the exit node j_+, and each node lies on some path from the entry node j_- to the exit node j_+. In the graph $G = (V^*, E)$, for each node v_i, we consider the path from the entry node j_- to v_i and from v_i to the exit node j_+.

The length $L(P)$ of a path $P = (i_1, i_2, \ldots, i_k)$ in the graph $G = (V, E)$ is the sum of the processing time of the nodes of this path $L(P) = \sum_{j=1}^{k} t(i_j)$.

Let $u(v_i)$ be the length of the maximum path from j_- to the node v_i (without $t(v_i)$), then $u(v_i)$ is the time of the earliest job v_i start (or t-level, top level) and $w(v_i)$ be the length of a maximum path from v_i to j_+ (including $t(v_i)$), then $w(v_i)$ is the minimum time from the start of the job v_i to the total completion of all jobs (b-level, bottom level).

The value $u(j_+)$ is called the length of the critical path t_{cp} in the graph $G = (V^*, E)$ and determines the minimum possible execution time of all jobs and, accordingly, a lower bound for the makespan of the optimal schedule.

Definition 1. *We call a job v_i a ready job if all its predecessors have been included in the schedule S.*

This definition differs from the definition which uses in the list algorithms, where the job is called the ready job, if all its predecessors have been finished.

For the schedule to be valid, it is necessary for each job $v_i \in V^*$ to find $\tau(v_i)$, such that

$$u(v_i) \leq \tau(v_i).$$

First, we find the t-level $u(v_i)$ and the b-level $w(v_i)$ of all jobs V.

At each step of the method, we add one job to the constructed schedule, define the start time $\tau(v_i)$ and the processor, performing this job.

We introduce some designations: let S_k be a partial schedule, which includes k jobs, let $time[l]$ for $l \in 1 : m$ be the latest time when processor l is busy, U_k is the set of ready jobs on the step k, $\tau(v_i)$ is the start time of the job v_i, $r(v_i)$ is the idle time of the processor before the start of job v_i in the schedule S_k.

First, we describe the procedure $MAP(p, q, k)$ which sets job p to processor q. **Procedure $MAP(p, q, k)$ schedules job p to processor q and adds the job p to S_k.**

1. Define the start time of job p $\tau(p) = \max\{time[q], u(p)\}$.
2. Define the idle time of processor q before the start of job p

$$r(p) = \tau(p) - time[q].$$

3. $nom(p) = q$ —fix processor q, which computes job p.
4. Delete job p from the set of ready jobs U_k.
5. Add job p to S_{k-1}, $S_k = S_{k-1} \cup p$.
6. Add all jobs, such that all predecessors of these jobs have already been set in the schedule S_k to the set of ready jobs U_k and create U_{k+1}.
7. Define a new shutdown time of processor q : $time[q] = \tau(p) + t(p)$.
8. If $\tau(p) > u(p)$, then put $u(p) = \tau(p)$ and recalculate the maximum paths $u(v_i)$ (t-levels) from the node p to all other nodes v_i of the graph G.

The algorithm CP/IIT is a greedy algorithm, but is not a list algorithm.

Algorithm CP/IIT

1. Initialization
   (a) Find the t-level of all jobs V. Set $t_{cp} = u(j_+)$.
   (b) Find the b-level of all jobs V.
   (c) Set $time[i] := 0$, $\forall i \in 1 : m$, $S_0 = \emptyset$.
   (d) Add all jobs, which have not any predecessors in graph G, to U_1.
2. For $k = 1$ to n do
   (a) Find a processor l_0 such that $time[l_0] = \min\{time[i] \,|\, i \in 1 : m\}$.
   (b) Select job v_0 with the largest priority from the set of ready jobs U_k such that
   $$w(v_0) = \max\{w(v_i) \,|\, v_i \in U_k\}.$$
   (c) If $u(v_0) \leq time[l_0]$, then map job v_0 on the processor l_0 by procedure $Map(v_0, l_0, k)$, and go to 2.
   (d) If $u(v_0) > time[l_0]$, then
       i. for j=1 to m do
       ii. if $time[j] = u(v_0)$ then map job v_0 on processor j by procedure $Map(v_0, j, k)$, and go to 2.
   (e) If there is job v^* such that
   $$w(v^*) = \max\{w(v_i) \,|\, v_i \in U_k \backslash \{v_0\}, \max\{time[l_0], u(v_i)\} + t(v_i) \leq u(v_0)\},$$
   then map job v^* on processor l_0 by procedure $Map(v^*, l_0, k)$, and go to 2.
   (f) Map job v_0 on the processor l_0 by procedure $Map(v_0, l_0, k)$.
3. If $k = n$, then schedule $S_A = S_n$ is generated.
   Find the makespan $C_{\max}(S_A) = \max\{\tau(v_i) + t(v_i) \mid v_i \in V\}$.

Computational complexity of the CP/IIT algorithm is $O(n^2 m)$.

Step 2 for the CP (critical path) algorithm

1. For $k = 1$ to n do
   (a) Find a processor l_0 such that $time[l_0] = \min\{time[i] \,|\, i \in 1 : m\}$.
   (b) Select job v_0, such that $u(v_0) \leq time[l_0]$ with maximum priority from the set of ready jobs U_k .
   $$w(v_0) = \max\{w(v_i) \,|\, v_i \in U_k, u(v_i) \leq time[l_0]\}.$$
   (c) If there are not any jobs, such that $u(v_0) \leq time[l_0]$, then $time[l_0] = \min\{u(v_i) \,|\, v_i \notin S_k\}$. Go to (b).
   (d) Map job v_0 on processor l_0 by procedure $Map(v_0, l_0, k)$.

Computational complexity of the CP algorithm is $O(n^2 m)$.

Lemma 1. *Let S_A be the schedule constructed by algorithm CP/IIT. Let I_1, ..., I_l be intervals where there is an idle processor and $L(I_j)$ is the length of I_j. Let t_{cp} be the critical path in the graph $G = (V^*, E)$.*
   Then
   $$\sum_{j=0}^{l} L(I_j) \leq t_{cp}.$$

Proof. Let the schedule S_A have length $C_{\max}(S_A)$. Let $I_1, \ldots, I_l$ be interval where there is an idle processor, and $B_1, B_2, \ldots, B_k$ are intervals where all processors are busy. Let us define a partition of the interval $(0, C_{\max}]$ into two subsets IS and BS as follows: let BS be the union of disjoint half-open intervals in which all processors are busy. Then $BS = \bigcup_{j=1}^{k} B_j$ is the union of the half-open intervals B_j. Then $IS = \bigcup_{j=1}^{l} I_j$.

The algorithm CP/IIT has the following property: let $beg(I_j)$, $end(I_j)$ be the start and the end of the idle interval I_j, respectively $I_j = (beg(I_j), end(I_j)]$. At the time $end(I_j)$ m jobs start executing and at least one ends.

We build several chains of jobs, each of these chains covers several idle intervals. We denote the chain $j_1 \prec j_2 \prec, \ldots, \prec j_k$ as $P(j_1, j_k)$ and $L(P)$ is the length of the of jobs chain, $L(P) = \sum_{j_i \in P} t(j_i)$.

Let us describe the construction a job chain, starting from the time $end(I_j)$.

Consider the time moment $end(I_l) = \max\{t \mid t \in IS\}$. This is the last moment of the idle of processors in the schedule S_A. There is a job i_e, which ends at time $end(I_l)$.

The current chain $P = (i_e)$ consists of job i_e.

At each step we add a node to the chain P as a result we have constructed the chain $P(i_b, i_e)$. Then we consider the start time $\tau(i_b)$ of job i_b in the schedule S_A.

If $\tau(i_b) \leq beg(I_1)$, then we have completed the construction of the chain $P(i_b, i_e)$. The chain of jobs $P(i_b, i_e)$ begins with job i_b and ends with i_e and has the following property

$$t_{cp} \geq L(P(i_b, i_e)) \geq \sum_{j \in 1:l} L(I_j).$$

Otherwise, there are two possibilities: $\tau(i_b) \in BS$ and $\tau(i_b) > beg(I_1)$ or $\tau(i_b) \in IS$.

If $\tau(i_b) \in IS$ then there is a job $i^* \prec i_b$ such that $\tau(i^*) + t(i^*) = \tau(i_b)$. We add job i^* to chain $P = (i_b, i_e)$, set $i_b = i^*$ and continue to build the chain P.

If $\tau(i_b) \in BS$ and $\tau(i_b) > beg(I_1)$ then we find the time

$$t^* = \max\{t \in IS \mid t \leq \tau(i_b)\}.$$

Let $t^* = end(J_p)$ be the end of some period of inactivity J_p. We have to consider two cases.

Case 1. If there is a job i^*, which ends at the time $end(J_p)$ and $i^* \prec i_b$ then we add the job i^* to the chain P, set $i_b = i^*$, and continue construction of $P = (i_b, i_e)$.

Case 2. If all jobs which end at the time $end(J_p)$ are independent with i_b, then we finish the construction of chain $P_d = P(i_b, i_e)$. We introduce the set $J(P_d)$ which is the set of indexes of the idle processor intervals covered by the chain $P_d(i_b, i_e)$.

Then

$$J(P_d(i_b, i_e)) = \{j \in 1 : l \mid beg(I_j) \geq \tau(i_b), end(I_j) \leq \tau(i_e) + t(i_e)\}.$$

The chain of jobs $P_d(i_b, i_e)$ has the following property

$$L(P_d(i_b, i_e)) \geq \sum_{j \in J(P_d)} L(I_j) = \sum_{j=p+1}^{l} L(I_j)$$

We begin to construct a new chain.

We find two jobs : job v_1 which ends at the time $end(J_p)$ and job v_2 which starts at the time $end(J_p)$ and $v_1 \prec v_2$. We construct a chain of jobs P_{d-1}, starting with the job v_1 by the same way as we built the chain P_d, starting from i_e.

We fix job v_2 and denote it as $v(P_{d-1})$, it is clear that $i_e \prec v(P_{d-1})$.

We repeat this procedure until the event $\tau(i_b) \in BS$ and $\tau(i_b) \leq beg(I_1)$ occurs. As a result, we have constructed chains $P_1, P_2, \ldots P_d$, for each of which we have

$$L(P_k(i_b, i_e)) \geq \sum_{j \in J(P_k)} L(I_j).$$

$$\sum_{i=1}^{d} L(P_i) \geq \sum_{j=1}^{l} L(I_j).$$

We consider two chains $P_{d-1} = (j_b, j_e)$ and $P_d = (i_b, i_e)$, $v(P_{d-1}) = v_2$,

In this case, according to the properties of the algorithm CP/IIT, it is true, that

$$w(v(P_{d-1})) \geq w(i_b).$$

Let $P_{\max}(v, j_+)$ be the maximum path from v to node j_+ in the graph G, then $w(v) = L(P_{\max}(v, j_+))$.

By definition of b-levels, it is true that

$$L(P_{\max}(v(P_{d-1}), j_+)) \geq L(P_{\max}(i_b, j_+)) \geq L(P_d).$$

Then

$$L(P_{\max}(v(P_{d-1}), j_+)) \geq \sum_{j \in J(P_d)} L(I_j). \tag{1}$$

For the chain P_{d-1}
it is true that,

$$L(P_{d-1}) \geq \sum_{j \in J(P_{d-1})} L(I_j), \tag{2}$$

we add two inequalities (1) and (2) and get

$$L(P_{d-1}) + L(P_{\max}(v(P_{d-1}), j_+)) \geq \sum_{j \in J(P_{d-1} \cup P_d)} L(I_j).$$

We know that $j_e \prec v(P_{d-1})$ then

$$L(P_{\max}(j_b, j_+)) \geq L(P_{d-1}) + L(P_{\max}(v_{d-1}, j_+)),$$

therefore

$$L(P_{\max}(j_b, j_+)) \geq \sum_{j \in J(P_{d-1} \cup P_d)} L(I_j).$$

Further, we will similarly consider the chains P_{d-2} and P_{d-1}. If the chain $P_1(v_b, v_e)$ begins with v_b, then as a result, we get

$$L(P_{\max}(v_b, j_+)) \geq \sum_{j=1}^{l} L(I_j).$$

Hence

$$t_{cp} \geq L(P_{\max}(v_b, j_+)) \geq \sum_{j=1}^{l} L(I_j).$$

The lemma is proved.

Theorem 1. *Let $C_{\max}(S_A)$ be the makespan of the schedule S_A, m be the number of processors, C_{opt} be the makespan of the optimal schedule, then $C_{\max}(S_A)/C_{opt} \leq 2 - 1/m$ and this bound is tight.*

Proof. Let $T = \sum_{i=1}^{n} t_i$. Then $C_{opt} \geq T/m$. Let I be the total idle time of processors. Then

$$I/(m-1) \leq \sum_{j=1}^{p} L(I_j) \leq t_{cp}.$$

By definitions I and T we have

$$C_{\max}(S_A) = (I + T)/m \leq C_{opt}(m-1)/m + C_{opt},$$

hense

$$C_{\max}(S_A)/C_{opt} \leq 2 - 1/m$$

Consider the example that proves, that this bound is tight.

Example 1. Consider the $m^2 + m + 1$ jobs and the m machines instance (Fig. 1, $m = 3$).

There are two sets of jobs A and V. $A = \{a_1, a_2, \ldots, a_m\}$ and $V = V_1 \cup V_2 \cup, \ldots, V_{m-1}$, where $V_j = \{v_1^j, v_2^j, \ldots, v_m^j\}$. For all jobs $v_i^j \in V_j$ the processing time $t(v_i^j) = m + (j+1)\varepsilon$, for all jobs $a_i \in A$ the processing time $t(a_i) = m + (i-1) * \varepsilon$,

$$a_j \prec v_i^j, \quad \forall v_i^j \in V_j, \quad j \in 1 : m-1,$$

$$V_j \prec V_{j+1}, \quad u \prec v, \quad \forall u \in V_j, \forall v \in V_{j+1}, \quad j \in 1 : m-2.$$

The makespan of the CP/IIT schedule is equal $C_{\max}(S_A) = \sum_{i=1}^{m} t(a_i) = \sum_{j=1}^{m-1} t(v_1^j) = \sum_{i=1}^{m}(m + (i-1)\varepsilon) = \sum_{j=1}^{m-1}(m + (j+1)\varepsilon) = m * m + \varepsilon((m-1)m)/2 + m(m-1) + \varepsilon(m(m+1)/2 - 1) = m(2m-1) + \varepsilon(m^2 - 1)$. The optimal makespan C_{opt} is equal to $m(m-1) + m + \varepsilon(m^2 - 2) = m^2 + \varepsilon(m^2 - 2)$.

The schedule S_A constructed by the algorithm CP/IIT for case $m = 3$ is shown in Table 1, the optimal schedule S_{opt} is shown in Table 2.

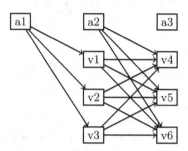

Fig. 1. Graph $G(V, E)$ Example 1.

Table 1. Schedule S_A makespan $C_{\max}(S_A) = m(2m-1) + \varepsilon(m^2 - 1)$.

t	m	$m + 2 * \varepsilon$	$m + \varepsilon$	$m + 3 * \varepsilon$	$m + 2 * \varepsilon$
$P1$	a_1	v_1	a_2	v_4	a_3
$P2$	idle	v_2	idle	v_5	idle
$P3$	idle	v_3	idle	v_6	idle

Table 2. Optimal schedule $C_{opt} = m^2 + \varepsilon(m^2 - 2)$

t	$m + 2 * \varepsilon$	$m + 2 * \varepsilon$	$m + 3 * \varepsilon$
$P1$	a_1	v_1	v_4
$P2$	a_2	v_2	v_5
$P3$	a_3	v_3	v_6

Example 2. Consider the example at Fig. 2 (for case $m = 3$). There are jobs $a, v_1, v_2, ..., v_m, w$, such that

$$a \prec v_i, \quad v_i \prec w \quad \forall v_i \in V.$$

There are $m-1$ independent jobs u_i. For all jobs v_j the processing time $t(v_j) = 1$, the processing time $t(a) = \varepsilon$, for all jobs u_i the processing time $t(u_i) = m - 1$ and $t(w) = m - 1$. These jobs are executed on m processors. Any list scheduling algorithm generates the schedule with makespan $C_{\max} = 2m - 1$ [1]. The CP/IIT algorithm generates the optimal schedule with makespan $C_{\max} = m + \varepsilon$.

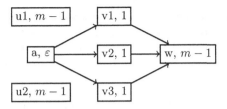

Fig. 2. Graph $G(V, E)$ Example 2.

We propose the combined algorithm A that builds two schedules (one by the CP algorithm, the other by the algorithm CP/IIT) and selects the best solution. The algorithms CP and CP/IIT are in a certain sense opposite: if the algorithm CP generates a schedule with a large error, the algorithm CP/IIT works well and vice versa.

3 Computational Experiments

To illustrate the effectiveness of our algorithm we tested it on randomly generated graphs. We used graphs from Standard Task Graph Set, which is available at http://www.kasahara.elec.waseda.ac.jp/schedule/.

But we changed the method of generating the processing time of jobs. The processing time of jobs has been chosen randomly from the interval [1:50] for 50% jobs and the processing time has been chosen randomly from the interval [1000 : 1250] for others 50% jobs.

The low bound LB of the optimal schedule length is

$$LB = \max\{t_{cp}, \lceil \sum_{i=1}^{n} t(u_i)/m \rceil\},$$

where t_{cp} is the length of critical path in the job graph.

To examine the effectiveness of the CP/IIT algorithm and the A algorithm, we tested 500 randomly generated instances with the number of jobs ranging from 100 to 300.

The results of the experiment are shown in Tables 3, 4 and 5. In the Table 3, we vary the number of processors from 2 to 12 for the series of 100 tests. We select the tests with the number of jobs $n = 100$. We compare the relative errors of approximate solutions, obtained by the CP/IIT algorithm, the CP algorithm and by the combined algorithm A.

We compare the length T_{ICP} of the approximation solution (obtained by the CP/IIT algorithm) and the length T_{CP} of the approximation solution obtained by the CP algorithm.

Comparative results are presented in Table 3. The first column of this table contains the name of test series, in the second one m is a number of processors. The third one contains the average relative error of an approximation solution

$R(ICP) = T_{ICP}/LB$ and the fourth one contains the average relative error of an approximation solution $R(CP) = T_{CP}/LB$ and the last one contains the average relative error $R(A) = T_{best}/LB$, where $T_{best} = \min\{T_{ICP}, T_{CP}\}$ is the length of the best schedule.

Table 3. Performance of algorithms according to the variation of m.

Series	m	$R(ICP)$	$R(CP)$	$R(A)$
s1	2	1.0200	1.0090	1.0090
s1	4	1.0102	1.0104	1.0102
s1	6	1.1058	1.1034	1.1027
s1	8	1.0674	1.0705	1.0460
s1	10	1.0297	1.0298	1.0256
s1	12	1.0000	1.0041	1.0000
Average		1.0388	1.0378	1,0322

We can see that the maximum deviation from the lower bound takes place when the number of processors is $m = 6$ and $m = 8$. Table 3 shows that for all m, the relative error decreases for the combined algorithm A, and for $m = 8$ the improvement is maximum.

Table 4 shows data for the third series of 10 tests. We select $n = 100$ and $m = 8$. In the first column there is the file name, in the second column there is the lower bound of the objective function. The third and fourth columns contain the length of the schedules, constructed by the CP/IIT and the CP algorithms, respectively. In the fifth, the sixth and the last columns there are the relative errors of approximate solutions, constructed by the CP/IIT, the CP and the A algorithms, respectively.

We see that there is a slight difference in the values of the objective function in 8 tests. In tests r32 and r36, we see the difference of 20%, and in the test r32, the best algorithm is the CP algorithm, in the test r36, the best algorithm is the CP/IIT algorithm. As a result, the A algorithm have improved the solution by 20% in these two tests.

In Table 5, we select 26 tests from 180 where the algorithm CP/IIT generates better schedules than the CP algorithm. We see that the improvement in the relative estimate ranges from 2 to 20%. The number of tests where the algorithm CP/IIT generates the best schedule is 14 %.

In each group, about 80% of the tests have the same value of the objective function. In 10% of the tests the critical path algorithm CP generates the best solutions, and in 10% of the tests, the critical path algorithm with idle times CP/IIT generates the best solutions.

Table 4. Performance of algorithms for instances from $s3$, $n = 100$, $m = 8$.

Test	LB	$T(ICP)$	$T(CP)$	$R(ICP)$	$R(CP)$	$R(A)$
r30	4589	5652	5650	1.232	1.231	1.231
r31	5652	5652	5884	1.000	1.040	1.000
r32	4652	5644	4681	**1.213**	**1.006**	1.006
r33	5675	5675	5676	1.000	1.000	1.000
r34	5656	5658	5656	1.000	1.000	1.000
r35	5638	5652	5638	1.002	1.000	1.000
r36	5738	5738	6907	**1.000**	**1.204**	1.000
r37	5704	5728	5704	1.004	1.000	1.000
r38	4660	5695	5697	1.222	1.223	1.222
r39	5668	5668	5668	1.000	1.000	1.000

Table 5. Data for the series of tests $n = 100$, $m = 8$.

Test	LB	$T(ICP)Z$	$T(CP)$	$R(ICP)$	$R(CP)$	$R(A)$
r20	7920	7920	9103	1.000	1.232	1.000
r21	5695	5695	5735	1.000	1.040	1.000
r33	5675	5675	5676	1.000	1.000	1.000
r36	5738	5738	6907	**1.000**	**1.204**	1.000
r40	5666	5666	6840	1.000	1.207	1.000
r49	4680	4698	4906	1.004	1.048	1.004
r52	5655	5655	5869	1.000	1.037	1.000
r101	5729	6900	6904	1.204	1.207	1.204
r108	4680	4698	4906	1.004	1.048	1.004
r109	4659	5652	5659	1.213	1.214	1.213
r113	5737	6934	6937	1.208	1.209	1.208
r101	5727	5736	5739	1.001	1.002	1.001
r108	4656	5745	5747	1.233	1.234	1.223
r109	5700	5700	5702	1.000	1.000	1.000
r113	5737	6934	6937	1.208	1.209	1.208
r115	5727	5736	5739	1.001	1.002	1.001
r118	4656	5745	5747	1.233	1.234	1.233
r131	5700	5700	5702	1.000	1.000	1.000
r142	4661	4745	4746	1.018	1.018	1.018
r146	5505	6692	6694	1.215	1.216	1.215
r165	5701	5701	5712	1.000	1.001	1.000
r167	5716	5732	5739	1.002	1.003	1.002
r168	6933	6933	6935	1.000	1.000	1.000
r170	5751	6916	6925	1.202	1.204	1.202
r172	5713	5720	5751	1.001	1.006	1.001
r175	5685	5691	6860	1.001	1.207	1.001

4   Conclusions

In this paper we investigated algorithms for $P|prec|C_{\max}$ problem. We proposed the algorithm CP/IIT and proved that no-list scheduling algorithm CP/IIT is a $(2 - 1/m)$-approximation algorithm. The number of tests where the algorithm CP/IIT generates the best schedule than CP algorithm is 14 %. Then by combining two approximation algorithms we propose the combined algorithm A, which generates two schedules and selects the best one. Computational experiments have shown that the combined algorithm works better than each of the algorithms separately.

References

1. Graham, R.L., Lawler, E.L., Lenstra, J.K., Rinnooy Kan, A.H.G.: Optimization and approximation in deterministic sequencing and scheduling: a survey. Ann. Disc. Math. **5**(10), 287–326 (1979)
2. Lenstra, J.K., Rinnooy Kan, A.H.G., Brucker P.: Complexity of machine scheduling problems. Ann. Disc. Math. **1**, 343–362 (1977)
3. Baker, K.R.: Introduction to Sequencing and Scheduling. Wiley, New York (1974)
4. Grigoreva, N.S.: Worst-case analysis of an approximation algorithm for single machine scheduling problem. In: Proceedings of the 16th Conference on Computer Science and Intelligence Systems, Annals of Computer Science and Information System, vol. 25, pp. 221–225 (2021)
5. Grigoreva, N.S.: Multiprocessor scheduling problem with release and delivery times. In: Ganzha, M., Maciaszek, L., Paprzycki, M. (eds.) Proceedings of the 2020 Federated Conference on Computer Science and Information Systems, Annals of Computer Science and Information Systems, vol. 21, pp. 263–270 (2020)
6. Gusfield, D.: Bounds for Naive multiple machine scheduling with release times and deadlines. J. Algorithms **5**, 1–6 (1984)
7. Carlier, J.: Scheduling jobs with release dates and tails on identical machines to minimize the Makespan. Eur. J. Oper. Res. **29**, 298–306 (1987)
8. Hu, T.C.: Parallel sequencing and assembly line problem. Oper. Res. **9**, 841–848 (1961)
9. Coffman, E.G., Graham, R.L.: Optimal scheduling for two processor systems. Acta. Informatica **1**, 200–213 (1972)
10. Kanet, J.J., Sridharan, V.: Scheduling with inserted idle time: problem taxonomy and literature review. Oper. Res. **48**(1), 99–110 (2000)
11. Kasahara, H., Narita, S.: Practical multiprocessor scheduling algorithms for efficient parallel processing. IEEE Tran. Comp. **4**(11), 22–33 (1984)
12. Ullman, J.: NP-complete scheduling problems. J. Comp. Sys. Sci. **171**, 394–394 (1975)

Approximate Algorithms for Some Maximin Clustering Problems

Vladimir Khandeev[1]([⊠]) [ID] and Sergey Neshchadim[2] [ID]

[1] Sobolev Institute of Mathematics, 4 Koptyug Avenue, 630090 Novosibirsk, Russia
khandeev@math.nsc.ru
[2] Novosibirsk State University, 2 Pirogova Street, 630090 Novosibirsk, Russia
s.neshchadim@g.nsu.ru

Abstract. In this paper we consider three problems of searching for disjoint subsets among a finite set of points in an Euclidean space. In all three problems, it is required to maximize the size of the minimal cluster so that each intra-cluster scatter of points relative to the cluster center does not exceed a predetermined threshold. In the first problem, the centers of the clusters are given as input. In the second problem, the centers are unknown, but belong to the initial set. In the last problem, the center of each cluster is defined as the arithmetic mean of all its elements. It is known that all three problems are NP-hard even in the one-dimensional case. The main result of the work is constant-factor approximation algorithms for all three problems. For the first two problems, 1/2-approximate algorithms for an arbitrary space dimension are constructed. Also we propose a 1/2-approximate algorithm for the one-dimensional case of the third problem.

Keywords: Euclidean space · Clustering · Max-min problem · NP-hardness · Bounded scatter · Approximation algorithm

1 Introduction

The subject of this study is three problems which model the following applied one: find disjoint subsets (clusters) in a collection of objects so that each subset consists of objects similar in the sense of a certain criterion. The applications for which this problem is typical include data analysis, data mining, statistical learning, pattern recognition and machine learning [2–4, 7]. The main aim of this paper is construction of approximate algorithms for the considered problems.

In the considered problems, it is required to find two clusters with the maximum value of the minimum cluster size so that in each cluster, the total intra-cluster scatter of points relative to the center of the cluster does not exceed a predetermined threshold. The difference between these three problems is in the way the cluster center is determined.

The considered problems statements are as follows.

© The Author(s), under exclusive license to Springer Nature Switzerland AG 2022
Y. Kochetov et al. (Eds.): MOTOR 2022, CCIS 1661, pp. 89–103, 2022.
https://doi.org/10.1007/978-3-031-16224-4_6

Problem 1. *Given* an N-element set $\mathcal{Y} = \{y_1, \ldots, y_N\}$ of points in Euclidean space $\mathbb{R}^d$, points z_1, z_2, and a real number $A \in \mathbb{R}_+$.

Find non-empty disjoint subsets $\mathcal{C}_1, \mathcal{C}_2 \subset \mathcal{Y}$ such that the minimal size of a subset is maximal. In other words,

$$\min(|\mathcal{C}_1|, |\mathcal{C}_2|) \to \max, \tag{1}$$

where

$$F_1(\mathcal{C}_i) = F(\mathcal{C}_i, z_i) := \sum_{y \in \mathcal{C}_i} \|y - z_i\|_2 \leq A, \; i = 1, 2, \tag{2}$$

$\|v\|_2 = \sqrt{v_1^2 + \ldots + v_d^2}$ for each $v = (v_1, \ldots, v_d) \in \mathbb{R}^d$.

Problem 2. *Given* an N-element set $\mathcal{Y} = \{y_1, \ldots, y_N\} \subset \mathbb{R}^d$ and a real number $A \in \mathbb{R}_+$.

Find non-empty disjoint subsets $\mathcal{C}_1, \mathcal{C}_2 \subset \mathcal{Y}$ and points $u_1, u_2 \in \mathcal{Y}$ such that (1) holds and $\mathcal{C}_1, \mathcal{C}_2$ satisfy

$$F_2(\mathcal{C}_i, u_i) = F(\mathcal{C}_i, u_i) \leq A, \; i = 1, 2. \tag{3}$$

Problem 3. *Given* an N-element set $\mathcal{Y} = \{y_1, \ldots, y_N\} \subset \mathbb{R}^d$ and a real number $A \in \mathbb{R}_+$.

Find non-empty disjoint subsets $\mathcal{C}_1, \mathcal{C}_2 \subset \mathcal{Y}$ such that (1) holds and $\mathcal{C}_1, \mathcal{C}_2$ satisfy

$$F_3(\mathcal{C}_i) = F(\mathcal{C}_i, \bar{y}(\mathcal{C}_i)) \leq A, \; i = 1, 2, \tag{4}$$

where $\bar{y}(\mathcal{C}_i) = \frac{1}{|\mathcal{C}_i|} \sum\limits_{y \in \mathcal{C}_i} y$, $i = 1, 2$, are the centroids (geometric centers) of the clusters $\mathcal{C}_i$.

The problems have the following interpretation. There is a set of objects in which there are two groups (two clusters) of similar objects. Each object is defined by a vector of real characteristics. The similarity of objects is set by the restriction on the cluster scatter which is defined by the sum of the distances from objects to the reference "center". In each of the problems, the centers are defined in their own way: i) as objects with specified characteristics; ii) as unknown objects from the original set; iii) as objects with characteristics equal to the arithmetic mean of the characteristics of objects from the cluster. It is required to find two clusters of similar objects that satisfy the scatter constraint so that the minimum cluster capacity is maximized.

Previously it was proved [11] that all three problems are NP-hard even in the one-dimensional case. In addition, a similar property holds [8] for analogues of the first two problems, in which the squares of norms are summed in (2), (3) and (4). At the same time, there are no known algorithms with guaranteed estimates for both Problems 1–3 and problems with squared norms.

Some other clustering problems in which the cardinalities of the clusters are maximized can be found, for example, in [1,5,9,10]. Like Problems 1–3, these problems model the search for homogeneous subsets, which is typical for data editing [6] and data cleaning [12]. However, it is worth noting that despite the similarity of formulations, the algorithms known for problems from [1,5,9,10] are not applicable either to Problems 1–3 or to their quadratic analogues.

The main result of the paper is $\frac{1}{2}$-approximate algorithms for Problems 1–2 and for the special case of Problem 3.

The paper has the following structure. Section 2 describes the formulation of the problem which generalizes Problem 1–3. There is also an approach that allows us to construct a 1/2-approximate algorithm for the generalized problem. In Sect. 3, this approach is applied to Problem 1. The next section shows how to get an approximate algorithm for Problem 2 from the algorithm for Problem 1, and then using a generalized approach, an algorithm is constructed with the same accuracy, but with better running time. Finally, Sect. 5 shows the existence of a 1/2-approximate algorithm for the one-dimensional case of Problem 3.

2   Generalized Problem

Instead of considering Problems 1–3 individually, we will consider a generalized problem, for which we will present an algorithm and show that it is $\frac{1}{2}$-approximate. This approach will allow us to reduce amount of proofs and make paper clearer.

We need to give some additional definition before generalized problem formulation. Let's call the d-dimensional scatter function an arbitrary function $\mathcal{F} : \mathcal{P}(\mathbb{R}^d) \to \mathbb{R}_+$, where $\mathcal{P}(\mathbb{R}^d)$ is the set of all finite subsets of $\mathbb{R}^d$, such that for any sets $\mathcal{C}' \subset \mathcal{C}'' \subset \mathbb{R}^d$ the inequality $F(\mathcal{C}') \leq F(\mathcal{C}'')$ holds.

Then we can formulate the following generalized two-cluster problem.

Problem CLUST2($\mathcal{Y}$, $\mathcal{F}_1$, $\mathcal{F}_2$, A). *Given* a set $\mathcal{Y} = \{y_1, \ldots, y_N\} \subset \mathbb{R}^d$, d-dimensional scatter functions $\mathcal{F}_1, \mathcal{F}_2$ and a non-negative real number $A \in \mathbb{R}$. *Find* non-empty disjoint subsets $\mathcal{C}_1, \mathcal{C}_2 \subset \mathcal{Y}$, such that their minimum cardinality is maximum:

$$\min\left(|\mathcal{C}_1|, |\mathcal{C}_2|\right) \to \max,$$

where

$$\mathcal{F}_i\left(\mathcal{C}_i\right) \leq A, \; i = 1, 2.$$

Then Problems 1–3 are special cases of the generalized problem. Indeed, Problem 1 is the same as Problem **CLUST2($\mathcal{Y}$, $F\left(\mathcal{C}, z_1\right)$, $F\left(\mathcal{C}, z_2\right)$, A)**.

Problem 2 reduces (see Sect. 4) to Problem **CLUST2($\mathcal{Y}$, F_2^M, F_2^M, A)**, where

$$F_2^M\left(\mathcal{C}\right) = \min_{u \in \mathcal{Y}} F(\mathcal{C}, u).$$

Problem 3 is the same as Problem **CLUST2($\mathcal{Y}$, $F(\mathcal{C}, \bar{y}(\mathcal{C}))$, $F(\mathcal{C}, \bar{y}(\mathcal{C}))$, A)**. Let us also formulate a generalized single-cluster problem.

Problem CLUST1($\mathcal{Y}$, $\mathcal{F}$, M). *Given* a set $\mathcal{Y} = \{y_1, \ldots, y_N\} \subset \mathbb{R}^d$, d-dimensional scatter function $\mathcal{F}$ and positive integer $M \in \mathbb{N}$. *Find* M-element subset $\mathcal{C} \subset \mathcal{Y}$ such that its scatter is minimal, i.e.,

$$\mathcal{F}\left(\mathcal{C}\right) \to \min.$$

Consider an arbitrary instance of the Problem **CLUST2**($\mathcal{Y}$, $\mathcal{F}_1$, $\mathcal{F}_2$, A). Let's assume that there is an algorithm that allows to find the optimal solution to Problems **CLUST1**($\mathcal{Y}$, $\mathcal{F}_i$, M), $i = 1, 2$, for any $M = 1, \ldots, |\mathcal{Y}|$. Then we can propose the following algorithm for Problem **CLUST2**.

Algorithm $\mathcal{A}$

Step 1. Let $M = 1$, $\mathcal{C}_1 = \emptyset$, $\mathcal{C}_2 = \emptyset$.

Step 2. Find the optimal solution $\mathcal{C}_1$ of Problem **CLUST1**($\mathcal{Y}$, $\mathcal{F}_1$, M).

If $\mathcal{F}_1(\mathcal{C}_1) > A$, then go to Step 3. Else, find the optimal solution $\mathcal{C}_2$ of Problem **CLUST1**($\mathcal{Y} \setminus \mathcal{C}_1$, $\mathcal{F}_2$, M).

If $\mathcal{F}_2(\mathcal{C}_2) > A$, then go to the next step. Else, increase M by one, save sets $\mathcal{C}_1$, $\mathcal{C}_2$ (the feasible solution of the cardinality M is constructed), and go to Step 2.

Step 3. Repeat Step 2, but first find the optimal solution $\mathcal{C}_2$ of Problem **CLUST1**($\mathcal{Y}$, $\mathcal{F}_2$, M) and then find the optimal solution $\mathcal{C}_1$ of Problem **CLUST1** ($\mathcal{Y} \setminus \mathcal{C}_2$, $\mathcal{F}_1$, M).

Step 4. Take $M - 1$ as the maximum cluster size and last saved sets $\mathcal{C}_1$ and $\mathcal{C}_2$ as the solution to the problem.

The next proposition and theorem establish the quality of the solution constructed by the algorithm $\mathcal{A}$.

Proposition 1. *Consider an arbitrary instance CLUST2($\mathcal{Y}$, $\mathcal{F}_1$, $\mathcal{F}_2$, A) of generalized two-cluster problem, where $\mathcal{Y} \subset \mathbb{R}^d$; $\mathcal{F}_1$, $\mathcal{F}_2$ are the d-dimensional scatter functions; $A \in \mathbb{R}_+$. Let $\mathcal{C}_1^*$, $\mathcal{C}_2^*$ be an arbitrary feasible solution to the problem under consideration. Then algorithm $\mathcal{A}$ applied to this problem constructs a solution of at least $\lfloor \frac{\min(|\mathcal{C}_1^*|, |\mathcal{C}_2^*|)}{2} \rfloor$ cardinality.*

Proof. Let $M^* = \min(|\mathcal{C}_1^*|, |\mathcal{C}_2^*|)$. We will show that the result of algorithm $\mathcal{A}$ is sets of at least $\lfloor \frac{M^*}{2} \rfloor$ cardinality. Let's assume that this is not the case. Let M^L be the value of M with which the algorithm $\mathcal{A}$ has finished (i.e., the sets found by algorithm $\mathcal{A}$ have cardinality $M^L - 1$).

Consider the last iteration of algorithm $\mathcal{A}$, i.e. the iteration where $M = M^L$. Let $\mathcal{C}_1$ be the optimal solution to Problem **CLUST1**($\mathcal{Y}$, $\mathcal{F}_1$, M) constructed at Step 2. Let's show that its scatter $\mathcal{F}_1(\mathcal{C}_1)$ does not exceed A. Indeed, let $\mathcal{C}_1^{**}$ be an arbitrary M^L-element subset of $\mathcal{C}_1^*$. Then

$$\mathcal{F}_1(\mathcal{C}_1^{**}) \leq \mathcal{F}_1(\mathcal{C}_1^*) \leq A, \tag{5}$$

where the first inequality follows from monotonicity $\mathcal{F}_1$ and the second holds because $\mathcal{C}_1^*$, $\mathcal{C}_2^*$ is the feasible solution to Problem **CLUST2**($\mathcal{Y}$, $\mathcal{F}_1$, $\mathcal{F}_2$, A). In addition, since $\mathcal{C}_1$ is the optimal solution to the Problem **CLUST1**($\mathcal{Y}$, $\mathcal{F}_1$, M^L), then

$$\mathcal{F}_1(\mathcal{C}_1) \leq \mathcal{F}_1(\mathcal{C}_1^{**}). \tag{6}$$

Combining (5) and (6), we get $\mathcal{F}_1(\mathcal{C}_1) \leq A$. Therefore, further at Step 2, the set $\mathcal{C}_2$ will be found.

Let $\mathcal{C}_2$ be the optimal solution to the Problem **CLUST1**($\mathcal{Y} \setminus \mathcal{C}_1$, $\mathcal{F}_2$, M^L) constructed at Step 2. Now we are going to show that its scatter $\mathcal{F}_2(\mathcal{C}_2)$ does not exceed A. Consider the set $\mathcal{C}_2^* \setminus \mathcal{C}_1$. At first note that $\mathcal{C}_2^* \setminus \mathcal{C}_1 \subset \mathcal{Y} \setminus \mathcal{C}_1$. The cardinality of this set is at least $|\mathcal{C}_2^*| - |\mathcal{C}_1| \geq \lfloor \frac{M^*}{2} \rfloor \geq M^L$, therefore, its arbitrary M^L-element subset $\mathcal{C}_2^{**}$ is the feasible solution to Problem **CLUST1**$(\mathcal{Y} \setminus \mathcal{C}_1, \mathcal{F}_2, M^L)$ and has scatter less or equal than A. Then, similarly to the reasoning for $\mathcal{C}_1$, we get inequalities $\mathcal{F}_2(\mathcal{C}_2) \leq \mathcal{F}_2(\mathcal{C}_2^{**}) \leq \mathcal{F}_2(\mathcal{C}_2^*) \leq A$.

Therefore, Step 2 ends with an increasing of M and hence M^L is not the value of M at the end of algorithm $\mathcal{A}$. We got a contradiction with our assumption. Thus, algorithm $\mathcal{A}$ constructs feasible solution to Problem **CLUST2**$(\mathcal{Y}, \mathcal{F}_1, \mathcal{F}_2, A)$ with sets of cardinality of at least $\lfloor \frac{M^*}{2} \rfloor$. $\square$

Theorem 1. *Algorithm $\mathcal{A}$ is a $\frac{1}{2}$-approximate algorithm for Problem **CLUST2**.*

Proof. Consider an arbitrary instance of Problem **CLUST2**: $\mathcal{Y} \subset \mathbb{R}^d$; $\mathcal{F}_1$, $\mathcal{F}_2$ are the d-dimensional scatter functions; $A \in \mathbb{R}_+$. Let $\mathcal{C}_1^*$, $\mathcal{C}_2^* \subset \mathcal{Y}$ be the optimal solution to Problem **CLUST2** and $M^* = \min(|\mathcal{C}_1^*|, |\mathcal{C}_2^*|)$. Since $\mathcal{C}_1^*$, $\mathcal{C}_2^*$ is a feasible solution to Problem **CLUST2**, then following inequality holds.

$$\max(\mathcal{F}_1(\mathcal{C}_1^*), \mathcal{F}_2(\mathcal{C}_2^*)) \leq A$$

It follows from the previous statement that algorithm $\mathcal{A}$ constructs a feasible solution to Problem **CLUST2** with sets of cardinality of at least $\lfloor \frac{M^*}{2} \rfloor$. If M^* is even, then algorithm $\mathcal{A}$ constructs a $\frac{1}{2}$-approximate solution to Problem **CLUST1** (since $\lfloor \frac{M^*}{2} \rfloor = \frac{M^*}{2}$). Thus, it remains to consider the case when M^* is odd, i.e. $M^* = 2K^* + 1$.

In the previous statement, we proved that the algorithm will reach the step at which $M - \lfloor \frac{M^*}{2} \rfloor + 1 = K^* + 1$. Suppose that at this step the algorithm didn't construct a feasible solution with sets of cardinality of $K^* + 1$. Let $\mathcal{C}_1^{**}$ be the set constructed at Step 2. Its scatter is not greater than A ($\mathcal{F}_1(\mathcal{C}_1^{**}) \leq A$), since any $(K^* + 1)$-element subset of $\mathcal{C}_1^*$ is a feasible solution to Problem **CLUST1**$(\mathcal{Y}, \mathcal{F}_1, K^* + 1)$ and has a scatter of no more than A. Then, let $\mathcal{C}_2^{**}$ be the second set constructed at Step 2. From our assumption that no feasible solution was found at the iteration at which $M = K^* + 1$, it follows that the pair $\mathcal{C}_1^{**}$ and $\mathcal{C}_2^{**}$ is not a feasible solution. Using that, we will show that $\mathcal{C}_1^{**}$ is a subset of $\mathcal{C}_2^*$.

Suppose the opposite is true: there is an element from $\mathcal{C}_1^{**}$ that is not contained in $\mathcal{C}_2^*$.

Then the set $\mathcal{C}_2^* \setminus \mathcal{C}_1^{**}$ contains at least $K^* + 1$ elements and its scatter does not exceed A. However, since $\mathcal{C}_2^* \setminus \mathcal{C}_1^{**} \subset \mathcal{Y} \setminus \mathcal{C}_1^{**}$, then any $K^* + 1$-element subset of $\mathcal{C}_2^* \setminus \mathcal{C}_1^{**}$ is a feasible solution to Problem **CLUST1**$(\mathcal{Y} \setminus \mathcal{C}_1^{**}, \mathcal{F}_2, K^* + 1)$ and has a scatter of no more than A. Therefore, $\mathcal{C}_2^{**}$ has a scatter of no more than A, which contradicts the fact that pair $\mathcal{C}_1^{**}$ and $\mathcal{C}_2^{**}$ is not a feasible solution to Problem **CLUST2**. Hence, $\mathcal{C}_1^{**} \subset \mathcal{C}_2^*$.

Since no solution was found at Step 2, the $\mathcal{A}$ algorithm will proceed to Step 3.

By analogy with Step 2, the constructed set $\mathcal{C}_2^{***}$ has a scatter not exceeding A, therefore the second set will be constructed further. Let $\mathcal{C}_1^{***}$ and $\mathcal{C}_2^{***}$ be the sets of cardinality $K^* + 1$ which are constructed at Step 3. According to our assumption, a pair $\mathcal{C}_1^{***}$ and $\mathcal{C}_2^{***}$ of sets is not a feasible solution. Then, by analogy with the reasoning for Step 2, we get that $\mathcal{C}_2^{***} \subset \mathcal{C}_1^*$.

The sets C_1^{**} and C_2^{***} don't intersect because they are subsets of the sets C_1^* and C_2^* that do not intersect by assumption. Hence, $C_2^{***} \subset \mathcal{Y} \setminus C_1^{**}$ and Problem **CLUST1**$(\mathcal{Y} \setminus C_1^{**}, \mathcal{F}_1, K^* + 1)$ has a feasible solution with a scatter of no more than A and a scatter of C_2^{**} does not exceed A. As a result, we get that C_2^{**} satisfies the restriction on the scatter and also have the cardinality of $K^* + 1$. It follows from this that at Step 2, a feasible solution for Problem **CLUST2** is constructed with the sets of $K^* + 1$ cardinality, which contradicts our assumption.

Therefore, at the step with $M = K^* + 1$, Algorithm $\mathcal{A}$ constructs a feasible solution to Problem **CLUST2** with the cardinalities of $K^* + 1$. Thus, the result of the algorithm is a feasible solution with the cardinalities of sets of at least $K^* + 1$ and $\mathcal{A}$ is a $\frac{1}{2}$-approximate algorithm to Problem **CLUST2**.   □

Proposition 2. *Suppose that at Steps 2 and 3 of algorithm $\mathcal{A}$, algorithms with $\mathcal{O}(g(N))$ and $\mathcal{O}(f(N))$ time complexity are used to solve Problems CLUST1 with $\mathcal{F}_1$ and $\mathcal{F}_2$ scatter functions respectively. Then the time complexity of algorithm $\mathcal{A}$ is $\mathcal{O}(N(g(N) + f(N)))$.*

Proof. Algorithm $\mathcal{A}$ increases M no more than $\lfloor \frac{N}{2} \rfloor$ times. In the worst case, for each value of M, Problems **CLUST1**$(*, \mathcal{F}_1, *)$ and **CLUST1**$(*, \mathcal{F}_2, *)$ are solved twice and the cardinalities of the input sets do not exceed N. Thus, the time complexity of algorithm $\mathcal{A}$ can be estimated as $\mathcal{O}(N(g(N) + f(N)))$.   □

Remark 1. If algorithm $\mathcal{A}$ uses polynomial algorithms to solve **CLUST1** problems, then algorithm $\mathcal{A}$ is also polynomial.

3   Algorithm for Problem with Fixed Centers

As mentioned earlier, Problem 1 is equivalent to Problem **CLUST2**$(\mathcal{Y}, F(\mathcal{C}, z_1), F(\mathcal{C}, z_2), A)$. Thus, if we had an approximate algorithm for the **CLUST1** problem generated by Problem 1, we would be able to construct an approximate algorithm for Problem 1.

Theorem 2. *Problem CLUST1 with scatter function $F(\mathcal{C}, z)$ is solvable in $\mathcal{O}(N \log N)$ time.*

Proof. Let $\mathcal{Y} \subset \mathbb{R}^d$, $|\mathcal{Y}| = N$, $z \in \mathbb{R}^d$, $M \in \mathbb{N}$ be an arbitrary instance of Problem **CLUST1**.

Let us sort the elements of the set $\mathcal{Y}$ in non-decreasing order of the distance between the elements and the point z. Thus, we will assume that the set $\mathcal{Y}$ looks like following:

$$\mathcal{Y} = \{y_1, \ldots, y_N\},$$

where $\|y_k - z\| \le \|y_l - z\|$ for all k, l such that $k < l$.

Let us show that the set $\mathcal{C}^* = \{y_1, \ldots, y_M\}$ is an optimal solution to the problem under consideration. Consider an arbitrary feasible solution

$$\mathcal{C}' = \{y_{i_1}, \ldots, y_{i_M}\}.$$

Without loss of generality we assume that all its elements are sorted in non-decreasing of the distance between the elements and the point z. In other words, $\|y_{i_k} - z\| \leq \|y_{i_l} - z\|$ for all k, l such that $k < l$. Therefore, it is obvious that $\|y_k - z\| \leq \|y_{i_k} - z\|$ for all $k = 1, \ldots, M$. From the definition of the scatter of a set we obtain that $F(\mathcal{C}^*, z) \leq F(\mathcal{C}', z)$. Thus, $\mathcal{C}^*$ is the optimal solution. The time complexity of the algorithm is determined by sorting the set $\mathcal{Y}$, which can be performed in $\mathcal{O}(N \log N)$ operations. Thus, the total time complexity of the algorithm is $\mathcal{O}(N \log N)$. □

From Proposition 2 and Theorem 2 we get that there is an approximate algorithm for Problem 1.

Theorem 3. *There is a $\frac{1}{2}$-approximate algorithm for Problem 1 that runs in $\mathcal{O}(N^2 \log N)$ time.*

4 Algorithm for Problem with Centers from the Input Set

Now let's consider Problem 2.

Firstly we will show that from any approximate algorithm for Problem 1, we can obtain an approximate algorithm for Problem 2 with the same accuracy estimate.

Proposition 3. *If $\mathcal{A}_1'$ is an α-approximate algorithm for Problem 1 running in $\mathcal{O}(T(N))$ time, where N is the cardinality of $\mathcal{Y}$, then there is an α-approximate algorithm $\mathcal{A}_2'$ for Problem 2 running in $\mathcal{O}(N^2 T(N))$ time.*

Proof. Let $\mathcal{Y} = \{y_1, \ldots, y_N\} \subset \mathbb{R}^d$ and $\tilde{A} \in \mathbb{R}_+$ be an arbitrary instance of Problem 2. Consider the following Algorithm $\mathcal{A}_2'$.

Algorithm $\mathcal{A}_2'$.

Step 1. Let $M^{\max} = -1$, $u_1 = 0$, $u_2 = 0$, $\mathcal{C}_1 = \emptyset$, $\mathcal{C}_2 = \emptyset$, $i = 0$, $j = 0$.

Step 2. Increase i by one and set j to 0. If $i > N$, then go to Step 5.

Step 3. Increase j by one. If $j > N$, then go to Step 2.

Step 4. Using $\mathcal{A}_1'$, solve Problem 1 with $z_1 = y_i$, $z_2 = y_j$ and the same values of $\mathcal{Y}$, A as in Problem 2. Denote it's result by $\tilde{\mathcal{C}}_1$, $\tilde{\mathcal{C}}_2$. Denote minimal cardinality of resulting clusters by $\tilde{M} = \min(|\tilde{\mathcal{C}}_1|, |\tilde{\mathcal{C}}_2|)$. If $\tilde{M} > M^{\max}$, then assign $u_1 = y_i$, $u_2 = y_j$, $\mathcal{C}_1 = \tilde{\mathcal{C}}_1$, $\mathcal{C}_2 = \tilde{\mathcal{C}}_2$, $M^{\max} = \tilde{M}$.

Step 5. Return sets $\mathcal{C}_1$ and $\mathcal{C}_2$ as the result of the algorithm.

It is obvious that if algorithm $\mathcal{A}_1'$ is an α-approximate to Problem 1, then $\mathcal{A}_2'$ constructs a feasible α-approximate solution to Problem 2. Since Step 4 is repeated $O(N^2)$ times, the total time complexity of algorithm $\mathcal{A}_2'$ is $\mathcal{O}(N^2 T(N))$. □

The existence of a $\frac{1}{2}$-approximate algorithm for Problem 1 and the previous proposition imply the following corollary.

Corollary 1. *There is a $\frac{1}{2}$-approximate algorithm for Problem 2 that runs in $\mathcal{O}(N^4 \log N)$ time.*

However, the same result, but with better complexity, can be obtained using the **CLUST2** problem.

Let us show how Problem 2 reduces to Problem **CLUST2**$(\mathcal{Y}, F_2^M, F_2^M, A)$, where

$$F_2^M(\mathcal{C}) = \min_{u \in \mathcal{Y}} F(\mathcal{C}, u). \tag{7}$$

Proposition 4. *If $\mathcal{C}_1$, $\mathcal{C}_2$, u_1, u_2 is a feasible solution to Problem 2($\mathcal{Y}$, A), then $\mathcal{C}_1$, $\mathcal{C}_2$ is a feasible solution to $CLUST2(\mathcal{Y}, F_2^M, F_2^M, A)$. If $\mathcal{C}_1$, $\mathcal{C}_2$ is a feasible solution to $CLUST2(\mathcal{Y}, F_2^M, F_2^M, A)$, then $\mathcal{C}_1$, $\mathcal{C}_2$, u_1, u_2 is a feasible solution to Problem 2($\mathcal{Y}$, A), where*

$$u_i = \arg\min_{u \in \mathcal{Y}} F(\mathcal{C}_i, u), i = 1, 2. \tag{8}$$

Proof. Let's assume that $\mathcal{C}_1$, $\mathcal{C}_2$, u_1, u_2 is a feasible solution to Problem 2($\mathcal{Y}$, A). Then $\mathcal{C}_1 \cap \mathcal{C}_2 = \emptyset$, $F_2(\mathcal{C}_i, u_i) \leq A$. We will show that $\mathcal{C}_1$, $\mathcal{C}_2$ is a feasible solution to **CLUST2**$(\mathcal{Y}, F_2^M, F_2^M, A)$. From definition (7) it follows that

$$F_2^M(\mathcal{C}_i) \leq F_2(\mathcal{C}_i, u_i), i = 1, 2.$$

Therefore, scatter F_2^M of $\mathcal{C}_i$, $i = 1, 2$, does not exceed A. Sets $\mathcal{C}_i$, $i = 1, 2$, are non-intersecting due to assumption. Thus, $\mathcal{C}_1$, $\mathcal{C}_2$ is a feasible solution to **CLUST2**$(\mathcal{Y}, F_2^M, F_2^M, A)$.

Let $\mathcal{C}_1$, $\mathcal{C}_2$ be a feasible solution to **CLUST2**$(\mathcal{Y}, F_2^M, F_2^M, A)$. We define point u_i using Eq. (8). Due to the definition of u_i we get that

$$F_2(\mathcal{C}_i, u_i) = F_2^M(\mathcal{C}_i), i = 1, 2.$$

Therefore, scatter F_2 of $\mathcal{C}_i, u_i$, $i = 1, 2$, does not exceed A. Thus, $\mathcal{C}_1$, $\mathcal{C}_2$, u_1, u_2 is a feasible solution to Problem 2($\mathcal{Y}$, A). □

Proposition 5. *If $\mathcal{C}_1$, $\mathcal{C}_2$, u_1, u_2 is an optimal solution to Problem 2($\mathcal{Y}$, A), then $\mathcal{C}_1$, $\mathcal{C}_2$ is an optimal solution to $CLUST2(\mathcal{Y}, F_2^M, F_2^M, A)$. If $\mathcal{C}_1$, $\mathcal{C}_2$ is an optimal solution to $CLUST2(\mathcal{Y}, F_2^M, F_2^M, A)$, then $\mathcal{C}_1$, $\mathcal{C}_2$, u_1, u_2 is an optimal solution to Problem 2($\mathcal{Y}$, A), where u_i is defined by (8).*

Proof. Let $\mathcal{C}_1$, $\mathcal{C}_2$, u_1, u_2 be an optimal solution to Problem 2($\mathcal{Y}$, A). Our goal is to show that $\mathcal{C}_1$, $\mathcal{C}_2$ is an optimal solution to **CLUST2**$(\mathcal{Y}, F_2^M, F_2^M, A)$.

Consider an arbitrary feasible solution $\hat{\mathcal{C}}_1$, $\hat{\mathcal{C}}_2$ of Problem **CLUST2**$(\mathcal{Y}, F_2^M, F_2^M, A)$. From Proposition 4 we get that there exist points $\hat{u}_1$, $\hat{u}_2$ such that $\hat{\mathcal{C}}_1$, $\hat{\mathcal{C}}_2, \hat{u}_1$, $\hat{u}_2$ is a feasible solution to Problem 2($\mathcal{Y}$, A). Using that $\mathcal{C}_1$, $\mathcal{C}_2$, u_1, u_2 is optimal we get that $\min(|\hat{\mathcal{C}}_1|, |\hat{\mathcal{C}}_2|) \leq \min(|\mathcal{C}_1|, |\mathcal{C}_2|)$ for every feasible solution $\hat{\mathcal{C}}_1$, $\hat{\mathcal{C}}_2$ of Problem **CLUST2**$(\mathcal{Y}, F_2^M, F_2^M, A)$. Thus, $\mathcal{C}_1$, $\mathcal{C}_2$ is optimal.

Let $\mathcal{C}_1$, $\mathcal{C}_2$ be an optimal solution to **CLUST2**$(\mathcal{Y}, F_2^M, F_2^M, A)$. We will show that $\mathcal{C}_1$, $\mathcal{C}_2$, u_1, u_2 is an optimal solution to Problem 2($\mathcal{Y}$, A), where u_i is defined by (8).

Consider an arbitrary feasible solution $\hat{C}_1$, $\hat{C}_2$, $\hat{u}_1$, $\hat{u}_2$ of Problem 2($\mathcal{Y}$, A). Using (4), we get that $\hat{C}_1$, $\hat{C}_2$ is a feasible solution to **CLUST2**($\mathcal{Y}, F_2^M, F_2^M, A$). Since C_1, C_2 is an optimal solution, $\min(|\hat{C}_1|, |\hat{C}_2|) \leq \min(|C_1|, |C_2|)$ for every feasible solution $\hat{C}_1$, $\hat{C}_2$, $\hat{u}_1$, $\hat{u}_2$ of Problem 2($\mathcal{Y}$, A). $\qquad\square$

Proposition 6. *If C_1, C_2, u_1, u_2 is an α-approximate solution to Problem 2($\mathcal{Y}$, A), then C_1, C_2 is an α-approximate solution to $\mathbf{CLUST2}(\mathcal{Y}, F_2^M, F_2^M, A)$. If C_1, C_2 is an α-approximate solution to $\mathbf{CLUST2}(\mathcal{Y}, F_2^M, F_2^M, A)$, then C_1, C_2, u_1, u_2 is an α-approximate solution to Problem 2($\mathcal{Y}$, A), where u_i defined by (8).*

Proof. Denote the optimal solution to Problem 2($\mathcal{Y}$, A) by C_1^*, C_2^*, u_1^*, u_2^*. From Proposition 5 it follows that C_1^*, C_2^* is an optimal solution to **CLUST2**($\mathcal{Y}$, F_2^M, F_2^M, A).

Let C_1, C_2, u_1, u_2 be an α-approximate solution to Problem 2($\mathcal{Y}$, A). It means that

$$\frac{\min(|C_1|, |C_2|)}{\min(|C_1^*|, |C_2^*|)} \geq \alpha. \tag{9}$$

Therefore C_1, C_2 is an α-approximate solution to **CLUST2**($\mathcal{Y}, F_2^M, F_2^M, A$).

Let C_1, C_2 be an α-approximate solution to **CLUST2**($\mathcal{Y}, F_2^M, F_2^M, A$), therefore (9) is satisfied. From Proposition 4 we know that C_1, C_2, u_1, u_2 is a feasible solution to Problem 2($\mathcal{Y}$, A). Thus, (9) means that C_1, C_2, u_1, u_2 is an α-approximate solution to Problem 2($\mathcal{Y}$, A). $\qquad\square$

As a result, we conclude that if an approximate solution to the **CLUST2** problem is known, then to obtain an approximate solution to Problem 2 we can keep these clusters and take such points from the original set that provide the minimum scatter to these clusters as their centers.

Similarly to the construction of the algorithm for Problem 1, we propose an algorithm for Problem **CLUST1**($\mathcal{Y}, F_2^M, M$).

Theorem 4. *Problem $\mathbf{CLUST1}$ with scatter function F_2^M can be solved in $\mathcal{O}(N^2 \log N)$.*

Proof. Let $\mathcal{Y} = \{y_1, \ldots, y_N\} \subset \mathbb{R}^d$ and $M \in \mathbb{N}$ be an arbitrary instance of **CLUST1** problem with scatter function F_2^M.

Consider the following algorithm.

Algorithm $\mathcal{A}_2$.

Step 1. Let $i = 0$, $F_{\min} = -1$, $C_{\min} = \emptyset$.

Step 2. Increase i by 1. If $i > N$ then go to Step 4.

Step 3. Solve Problem **CLUST1**($\mathcal{Y}, F(\mathcal{C}, y_i), M$) and save result to $\mathcal{C}$. If $F_{\min} < 0$ or $F(\mathcal{C}, y_i) < F_{\min}$, then assign $F_{\min} = F(\mathcal{C}, y_i)$ and $C_{\min} = \mathcal{C}$. Go to Step 2.

Step 4. Return set $C_{\min}$ and its scatter $F_{\min}$ as the result of the algorithm.

Since the **CLUST1** problem with the scatter function $F(\mathcal{C}, z)$ can be solved in $\mathcal{O}(N \log N)$ time, it is obvious that the algorithm above can be implemented with the complexity $\mathcal{O}(N^2 \log N)$.

Now let us show that algorithm $\mathcal{A}_2$ solves the **CLUST1** problem with the scatter function F_2^M optimally. Let $\mathcal{C}^A$ be the solution obtained as a result of algorithm $\mathcal{A}_2$ and its scatter be $F^{\mathcal{A}_2} = F_2^M(\mathcal{C}^A)$. Assume that this solution is not optimal, therefore, there is a set $\mathcal{C}^*$ of cardinality M such that

$$F_2^M(\mathcal{C}^*) < F^{\mathcal{A}_2}. \tag{10}$$

We define $y^* = \arg\min_{y \in \mathcal{Y}} F(\mathcal{C}^*, y)$. In other words, y^* is such a point that

$$F(\mathcal{C}^*, y^*) = F_2^M(\mathcal{C}^*).$$

There exists an index i^* such that $y^* = y_{i^*}$. Consider Step 3 of $\mathcal{A}_2$ at which $i = i^*$. At this step, the algorithm constructs an optimal solution $\mathcal{C}$ of Problem **CLUST1**$(\mathcal{Y}, F(\mathcal{C}, y^*), M)$. Since $\mathcal{C}^*$ is a feasible solution to Problem **CLUST1**$(\mathcal{Y}, F(\mathcal{C}, y^*), M)$, then $F(\mathcal{C}, y^*) \leq F(\mathcal{C}^*, y^*)$. Using (10), we get that $F(\mathcal{C}, y^*) < F^A$. At the next Step 4, the value of the variable $F_{\min}$ will be positive and not greater than $F(\mathcal{C}, y^*)$. Consequently, the result of $\mathcal{A}_2$ is a set with a scatter not greater than $F(\mathcal{C}, y^*)$. Therefore, $F^A \leq F(\mathcal{C}, y^*)$, and we ended up with two contradicting inequalities. Thus, $\mathcal{A}_2$ optimally solves the **CLUST1** problem with the scatter function F_2^M in $\mathcal{O}(N^2 \log N)$ time.   □

From Theorem 4 and Proposition 2 we get that for Problem **CLUST2** there exists a $\frac{1}{2}$-approximate algorithm with complexity $\mathcal{O}(N^3 \log N)$. Thus, we can propose the following theorem.

Theorem 5. *For Problem 2, there exists a $\frac{1}{2}$-approximate polynomial algorithm with $\mathcal{O}(N^3 \log N)$ complexity.*

Proof. Let $\mathcal{Y}, A$ be an arbitrary instance of Problem 2. Consider $\mathcal{C}_1, \mathcal{C}_2$ that is a $\frac{1}{2}$-approximate solution to Problem **CLUST2**$(\mathcal{Y}, F_2^M, F_2^M, A)$. It can be constructed in $\mathcal{O}(N^3 \log N)$. From Proposition 6 we get that if we calculate the cluster centers u_i using (8), then $\mathcal{C}_1, \mathcal{C}_2, u_1, u_2$ will be a $\frac{1}{2}$-approximate solution for Problem 1. These centers can be calculated in $\mathcal{O}(N^2)$. The resulting complexity of the algorithm is still $\mathcal{O}(N^3 \log N)$.   □

5   Algorithm for Problem with Geometric Centers

In this section, we will consider the one-dimensional case of the third problem and show that there is a polynomial approximate algorithm for it. To do this, by analogy with Problems 1 and 2, we will show that the one-dimensional case of the **CLUST1** problem corresponding to Problem 3 can be solved in polynomial time.

Proposition 7. *Suppose that $\mathcal{C} = \{y_1, \ldots, y_N\} \subset \mathbb{R}$, $m = \min_{y \in \mathcal{C}} y$ and $M = \max_{y \in \mathcal{C}} y$. Then for every point z such that $m < z < M$ at least one of the following inequalities is satisfied:*

1. $F_3(\mathcal{C} \cup \{z\} \setminus \{M\}) < F_3(\mathcal{C})$
2. $F_3(\mathcal{C} \cup \{z\} \setminus \{m\}) < F_3(\mathcal{C})$

Proof. We denote $\mathcal{C} \setminus \{m, M\}$ by $\hat{\mathcal{C}}$ (in other words, $\mathcal{C} = \{m\} \cup \hat{\mathcal{C}} \cup \{M\}$).

Consider the case when $\bar{y}(\mathcal{C}) \leq z$. Let's define $\mathcal{C}' := \mathcal{C} \cup \{z\} \setminus \{M\}$ and $\Delta = M - z > 0$. Then

$$F_3(\mathcal{C}') = \sum_{y \in \hat{\mathcal{C}}} |y - \bar{y}(\mathcal{C}')| + |m - \bar{y}(\mathcal{C}')| + |z - \bar{y}(\mathcal{C}')|. \tag{11}$$

Let's express centroid of set $\mathcal{C}'$ in terms of the centroid of the input set $\mathcal{C}$.

$$\bar{y}(\mathcal{C}') = \frac{1}{N}(\sum_{y \in \hat{\mathcal{C}}} y + m + z) = \frac{1}{N}(\sum_{y \in \hat{\mathcal{C}}} y + m + M - M + z) = \bar{y}(\mathcal{C}) - \frac{\Delta}{N}.$$

Note that $z - \bar{y}(\mathcal{C}') \geq 0$ since $z - \bar{y}(\mathcal{C}') = z - \bar{y}(\mathcal{C}) + \frac{\Delta}{N} > 0$. Express $|z - \bar{y}(\mathcal{C}')|$ in terms of $|M - \bar{y}(\mathcal{C})|$:

$$|z - \bar{y}(\mathcal{C}')| = z - \bar{y}(\mathcal{C}') = z - M + M - \bar{y}(\mathcal{C}) + \frac{\Delta}{N} = |M - \bar{y}(\mathcal{C})| - \frac{(N-1)\Delta}{N}, \tag{12}$$

where the last equality holds due to $M \geq \bar{y}(\mathcal{C})$.

Let's express $|m - \bar{y}(\mathcal{C}')|$ in terms of $|m - \bar{y}(\mathcal{C})|$. Since $\bar{y}(\mathcal{C}') > m$, the following equality holds.

$$|m - \bar{y}(\mathcal{C}')| = \bar{y}(\mathcal{C}') - m = |m - \bar{y}(\mathcal{C})| - \frac{\Delta}{N}. \tag{13}$$

Combining (12), (13) and (11) we get

$$F_3(\mathcal{C}') = \sum_{y \in \hat{\mathcal{C}}} |y - \bar{y}(\mathcal{C}) - \frac{\Delta}{N}| + |m - \bar{y}(\mathcal{C})| - \frac{\Delta}{N} + |M - \bar{y}(\mathcal{C})| - \frac{(N-1)\Delta}{N} =$$

$$= \sum_{y \in \hat{\mathcal{C}}} |y - \bar{y}(\mathcal{C}) - \frac{\Delta}{N}| + |m - \bar{y}(\mathcal{C})| + |M - \bar{y}(\mathcal{C})| - \Delta.$$

Using inequality $|y - \bar{y}(\mathcal{C}) - \frac{\Delta}{N}| \leq |y - \bar{y}(\mathcal{C})| + |\frac{\Delta}{N}|$ we get

$$F_3(\mathcal{C}') \leq \sum_{y \in \hat{\mathcal{C}}} |y - \bar{y}(\mathcal{C})| + \frac{(N-2)\Delta}{N} + |m - \bar{y}(\mathcal{C})| + |M - \bar{y}(\mathcal{C})| - \Delta$$

$$= F_3(\mathcal{C}) - \frac{2\Delta}{N} < F_3(\mathcal{C}).$$

Thus, for the case $\bar{y}(\mathcal{C}) \leq z$ the proposition is proven.

Now consider the case $\bar{y}(\mathcal{C}) > z$. Let's define $\tilde{\mathcal{C}} = \{-y : y \in \mathcal{C}\}$, $\tilde{z} = -z$, $\tilde{M} = \max_{y \in \tilde{\mathcal{C}}} y$, $\tilde{m} = \min_{y \in \tilde{\mathcal{C}}} y$. Note that $\tilde{M} = -m$ and $\tilde{m} = -M$. From our assumption that $\bar{y}(\mathcal{C}) > z$, it follows that $-\bar{y}(\mathcal{C}) < -z$. In other words, $\bar{y}(\tilde{\mathcal{C}}) < \tilde{z}$. Therefore, all conditions for the first case are satisfied for $\tilde{C}$ and $\tilde{z}$, and as a result we get the following inequality

$$F_3(\tilde{C} \cup \{\tilde{z}\} \setminus \{\tilde{M}\}) < F_3(\tilde{C}).$$

Using that $F_3(\tilde{C}) = F_3(C)$ and $F_3(\tilde{C} \cup \{\tilde{z}\} \setminus \{\tilde{M}\}) = F_3(C \cup \{z\} \setminus \{m\})$, we get $F_3(C \cup \{z\} \setminus \{m\}) < F_3(C)$. □

Theorem 6. *Problem **CLUST1** with one dimensional scatter function $F(C, \bar{y}(C))$ can be solved in $\mathcal{O}(N \log N)$ time.*

Proof. Consider an arbitrary instance of Problem **CLUST1**—$\mathcal{Y}$, $|\mathcal{Y}| = N$, and $M \in \mathbb{N}$.

Firstly sort all points of $\mathcal{Y}$ in non-decreasing order. This sorting can be performed in $\mathcal{O}(N \log N)$ operations. Therefore, the set $\mathcal{Y}$ can be represented as follows:

$$\mathcal{Y} = \{y_1, \ldots, y_N\},$$

where $y_k \leq y_l$ for all k, l such that $k < l$.

Denote by C^* the optimal solution to the considered problem. Define the values

$$m = \min_{y \in C^*} y, \quad M = \max_{y \in C^*} y.$$

Note that there exist indices i_m and i_M such that $m = y_{i_m}$, $M = y_{i_M}$. Let us show that $i_M = i_m + M - 1$ (in other words, that C^* is a set of consecutive points from the initial set $\mathcal{Y}$). Let's suppose it's not, then $i_M > i_m + M - 1$. It follows that there exists an index $i_z \in \{i_m + 1, \ldots, i_M - 1\}$ such that $y_{i_z} \notin C$. Since we have sorted the points, $m < y_{i_z} < M$. Thus, conditions of Proposition (7) are satisfied for C^* and the point $z = y_{i_z}$. Therefore, if we replace one of the points m, M in C^* by y_{i_z}, the scatter F_3 of the new M-element set will be less than the scatter of C^*, which contradicts the assumption that C^* is optimal.

Therefore, the optimal solution to **CLUST1** is one of the sets

$$\{y_i, \ldots, y_{i+M-1}\}, i = 1, \ldots, N - M + 1.$$

The following algorithm calculates the scatters of these sets and finds among them the set with the smallest scatter in $\mathcal{O}(N \log N)$ time.

Algorithm $\mathcal{A}_3$.

Step 1. Compute all sums $S[i] = \sum_{k=1}^{i} y_k$, $i = 1, \ldots, N$ using recurrent expression $S[i + 1] = S[i] + y_{i+1}$, $i = 0, \ldots, N - 1$, where $S[0] = 0$. Also we suppose that $y_{N+1} = 0$.

Step 2. Suppose $i_{cur} = 1$, $F = 0$, $F_{\min} = -1$, $i_{\min} = -1$, $i_{cent} = 1$, $\bar{y} = S[M]/M$.

Step 3. If $i_{cur} + M > N + 1$ then go to step 5.

Step 4. If $y_{i_{cent}} < \bar{y}$, then increase i_{cent} by 1 and repeat the current step.

Step 5. Calculate F using the following equation

$$F = (2(i_{cent} - i_{cur}) - M)\overline{y} + S[i_{cur} - 1] - 2S[i_{cent} - 1] + S[i_{cur} + M - 1].$$

If $F < F_{\min}$ or $F_{\min} < 0$, then $F_{\min} = F$, $i_{\min} = i_{cur}$.

Step 6. Increase i_{cur} by one. Recompute the centroid $\overline{y} = (S[i_{cur} + M - 1] - S[i_{cur} - 1])/M$. Go to step 3.

Step 7. The algorithm ends, the optimal set is $\{y_{i_{\min}}, \ldots, y_{i_{\min}+M-1}\}$ and its scatter is $F_{\min}$.

Let us prove the correctness of this algorithm.

Proposition 8. *At the time of execution of steps 3–5, the variable $\overline{y}$ contains the values of the centroid of the set $\{y_{i_{cur}}, \ldots, y_{i_{cur}+M-1}\}$.*

Proof. If $i_{cur} = 1$, then this is the first execution of Steps 3–5, and therefore the value of $\overline{y}$ was changed only at Step 2. Then $\overline{y} = S[M]/M = \overline{y}(\{y_1, \ldots, y_M\})$, i.e. the proposition for $i_{cur} = 1$ is proven.

If $i_{cur} > 1$, the last step that changed the value of $\overline{y}$ is Step 6. In the process of it's execution, the value $(S[i_{cur} + M - 1] - S[i_{cur} - 1])/M$ was written to the variable $\overline{y}$, which is the value of the centroid of the set $\{\{y_{i_{cur}}, \ldots, y_{i_{cur}+M-1}\}\}$. □

Proposition 9. *At Step 5, the variable i_{cent} contains the minimum index such that $y_{i_{cent}} \geq \overline{y}$.*

Proof. Let us prove it by induction on i_{cur}. If $i_{cur} = 1$, then only Steps 1–4 were performed before Step 5. i_{cent} was set to one at Step 2. Step 3 doesn't affect the value of i_{cent}. Step 4 stops repeating when $y_{i_{cent}}$ is no longer less than $\overline{y}$. In other words, Step 4 ends when $y_{i_N} \geq \overline{y}$. Thus, after Step 4, i_{cent} contains the smallest index of an element y_i which is not less than $\overline{y}$.

Assume that the proposition is true for $i_{cur} - 1$, $i_{cur} > 1$. Let us show that it also holds for i_{cur}.

Consider the previous execution of Step 5. By the inductive assumption, during this step i_{cent} was equal to $\arg\min_{i:y_i \geq \overline{y}} i$. After that, the value of $\overline{y}$ changed only at Step 4. Denote the centroid of the set $\{y_{i_{cur}-1}, \ldots, y_{i_{cur}+M-2}\}$ by $\overline{y}'$. Then before Step 4 i_{cent} was equal to $\arg\min_{i:y_i \geq \overline{y}'} i$. Obviously, $\overline{y} \geq \overline{y}'$, so $\arg\min_{i:y_i \geq \overline{y}} i = \arg\min_{i \geq i_{cent}:y_i \geq \overline{y}'} i$ and this value will be written to i_{cent} before transition from Step 4 to Step 5. □

Proposition 10. *At Step 5, the scatter of the set $\{y_{i_{cur}}, \ldots, y_{i_{cur}+M-1}\}$ is written to F.*

Proof. Denote $i_{cur} + M - 1$ by i_{end}. Using the scatter F_3 definition:

$$F_3(\{y_{i_{cur}}, \ldots, y_{i_{end}}\}) = \sum_{k=i_{cur}}^{i_{end}} |y_k - \overline{y}| = \left(\sum_{k=i_{cur}}^{i_{cent}-1} + \sum_{k=i_{cent}}^{i_{end}} \right)|y_k - \overline{y}| \qquad (14)$$

In the first sum of (14), $|y_k - \overline{y}| = \overline{y} - y_k$ as shown in Proposition 9. Similarly, $|y_k - \overline{y}| = y_k - \overline{y}$ in the second sum.

Let's transform these sums separately.

$$\sum_{k=i_{cur}}^{i_{cent}-1} |y_k - \overline{y}| = \sum_{k=i_{cur}}^{i_{cent}-1} (\overline{y} - y_k)$$

$$= (i_{cent} - i_{cur})\overline{y} - (S[i_{cent} - 1] - S[i_{cur} - 1]) \quad (15)$$

$$\sum_{k=i_{cent}}^{i_{cur}+M-1} |y_k - \overline{y}| = \sum_{k=i_{cur}}^{i_{cent}-1} (y_k - \overline{y})$$

$$= (S[i_{cur} + M - 1] - S[i_{cent} - 1]) - (i_{cur} + M - i_{cent})\overline{y} \quad (16)$$

Using (15) and (16), (14) can be continued.

$$F_3(\{y_{i_{cur}}, \ldots, y_{i_{cur}+M-1}\}) = (2(i_{cent} - i_{cur}) - M)\overline{y}$$
$$+ S[i_{cur} - 1] - 2S[i_{cent} - 1] + S[i_{cur} + M - 1] = F$$

Therefore, the scatter of $\{y_{i_{cur}}, \ldots, y_{i_{cur}+M-1}\}$ will be written to F at Step 5.

□

Now we continue the proof of the theorem.

The algorithm stops when $i_{cur} + M - 1 = N + 1$; before that i_{cur} cycles through $1, \ldots, N - M + 1$. For each i_{cur}, the scatter value of $\{y_{i_{cur}}, \ldots, y_{i_{cur}+M-1}\}$ is calculated. As mentioned earlier, among these sets there is an optimal solution to the considered problem **CLUST1**. The result of the algorithm is the set $\{y_i, \ldots, y_{i+M-1}\}$ with the minimum scatter, which will be the optimal solution to the problem.

Let's evaluate complexity. Step 1 runs in $\mathcal{O}(N)$ time and is performed once. Steps 2 and 7 have $\mathcal{O}(1)$ complexity and are performed once. Steps 3–6 run in $\mathcal{O}(1)$ time and are performed $\mathcal{O}(N)$ times.

The final complexity is $\mathcal{O}(N)$, but since before that we sorted the points of the original set, the final complexity is $\mathcal{O}(N \log N)$.   □

Thus, from Proposition 2 we get that for the one-dimensional Problem 3 there exists a $\frac{1}{2}$-approximate algorithm that runs in $\mathcal{O}(N^2 \log N)$.

Theorem 7. *For the one-dimensional case of Problem 3, there exists a $\frac{1}{2}$-approximate polynomial algorithm with $\mathcal{O}(N^2 \log N)$ complexity.*

6   Conclusion

In this paper, we have considered three NP-hard maximin clustering problems. We have proposed the first constant-factor approximation algorithms for the general cases of two of these problems and for the special case of the third problem. Since, in practice, the number of clusters is not limited by two very often, an important area of further research is the generalization of the considered algorithms for the case of a larger number of clusters. Another important area is the search for a constant-factor approximation algorithm for the general case of the third problem, as well as the construction of more accurate approximate algorithms for all three problems, in particular, the construction of polynomial approximate schemes.

Acknowledgments. The study presented was supported by the Russian Academy of Science (the Program of basic research), project FWNF-2022-0015.

References

1. Ageeva, A.A., Kel'manov, A.V., Pyatkin, A.V., Khamidullin, S.A., Shenmaier, V.V.: Approximation polynomial algorithm for the data editing and data cleaning problem. Pattern Recognit. Image Anal. **27**(3), 365–370 (2017). https://doi.org/10.1134/S1054661817030038
2. Aggarwal, C.C.: Data Mining: The Textbook. Springer, Switzerland (2015). https://doi.org/10.1007/978-3-319-14142-8
3. Bishop, C.M.: Pattern Recognition and Machine Learning. Springer, New York (2006)
4. Dubes, R. C., Jain, A. K.: Algorithms for Clustering Data. Prentice Hall, Hoboken (1988)
5. Eremeev, A.V., Kelmanov, A.V., Pyatkin, A.V., Ziegler, I.A.: On finding maximum cardinality subset of vectors with a constraint on normalized squared length of vectors sum. In: van der Aalst, W.M.P., et al. (eds.) AIST 2017. LNCS, vol. 10716, pp. 142–151. Springer, Cham (2018). https://doi.org/10.1007/978-3-319-73013-4_13
6. Farcomeni, A., Greco, L.: Robust Methods for Data Reduction. CRC, Boca Raton (2015)
7. James, G., Witten, D., Hastie, T., Tibshirani, R.: An Introduction to Statistical Learning. Springer, New York (2013). https://doi.org/10.1007/978-1-4614-7138-7
8. Kel'manov, A., Khandeev, V., Pyatkin, A.: NP-hardness of some max-min clustering problems. Commun. Comput. Inf. Sci. **974**, 144–154 (2018)
9. Kel'manov, A.V., Panasenko, A.V., Khandeev, V.I.: Exact algorithms of search for a cluster of the largest size in two integer 2-clustering problems. Numer. Anal. Appl. **12**(2), 105–115 (2019). https://doi.org/10.1134/S1995423919020010
10. Kel'manov, A., Pyatkin, A., Khamidullin, S., Khandeev, V., Shamardin, Y.V., Shenmaier, V.: An approximation polynomial algorithm for a problem of searching for the longest subsequence in a finite sequence of points in euclidean space. Commun. Comput. Inf. Sci. **871**, 120–130 (2018)
11. Khandeev, V., Neshchadim, S.: Max-min problems of searching for two disjoint subsets. In: Olenev, N.N., Evtushenko, Y.G., Jaćimović, M., Khachay, M., Malkova, V. (eds.) OPTIMA 2021. LNCS, vol. 13078, pp. 231–245. Springer, Cham (2021). https://doi.org/10.1007/978-3-030-91059-4_17
12. Osborne, J.W.: Best Practices in Data Cleaning: A Complete Guide to Everything You Need to Do Before and After Collecting Your Data. SAGE, London (2013)

Parameter Analysis of Variable Neighborhood Search Applied to Multiprocessor Scheduling with Communication Delays

Tatjana Jakšić-Krüger[(✉)] [iD], Tatjana Davidović[iD], and Vladisav Jelisavčić

Mathematical Institute of the Serbian Academy of Science and Arts,
Kneza Mihaila 36, 11000 Belgrade, Serbia
{tatjana,tanjad,vladisav}@mi.sanu.ac.rs

Abstract. When dealing with hard, real-life optimization problems, metaheuristic methods are considered a very powerful tool. If designed properly, they can provide high-quality solutions in reasonable running times. We are specially interested in Variable neighborhood search (VNS), a very popular metaheuristic for more than 20 years with many successful applications. Its basic form has a small number of parameters, however, each particular implementation can involve a problem-dependent set of parameters. This makes parameter analysis and performance assessment a challenging task. Contribution of this work is twofold: we develop a new variant of the VNS algorithm for the considered optimization problem and simplify the methodology for experimental analysis of metaheuristic algorithms. We conclude three stages of the parameter analysis: parameter definition, deciding the most influential parameters and analysis of their relationship. The analysis contributes to the design of VNS as a search problem in the space of its parameters. We apply the sophisticated approach that equally relies on visual as well as on the statistical and machine learning methods that have become standard practice for parameter tuning and experimental evaluation of metaheuristic algorithms. The obtained results are presented and discussed in this study.

Keywords: Stochastic algorithms · Experimental evaluation · Statistical methods · Parameter control · Parameter tuning

1 Introduction

In addition to developing a novel approach to some selected optimization problem, our aim is to also present one view into the current state of the algorithm

This research was supported by Serbian Ministry of Education, Science and Technological Development through Mathematical Institute SANU, Agreement No. 451-03-9/2021-14/200029 and by the Science Fund of Republic of Serbia, under the project "Advanced Artificial Intelligence Techniques for Analysis and Design of System Components Based on Trustworthy BlockChain Technology".

Y. Kochetov et al. (Eds.): MOTOR 2022, CCIS 1661, pp. 104–118, 2022.
https://doi.org/10.1007/978-3-031-16224-7_7

design. Ideally, we are able to detect to which degree various algorithm's parts can change the performance, to identify their interactions and to propose suitable statistical methods for this kind of analysis. When the optimality of a solution is not the imperative in the considered optimization problem, we are happy with the generation of sub-optimal solutions provided in a short amount of time. In addition, we deal frequently with the choice between the time and the quality of a solution. In practice, a *problem-oriented* heuristic algorithm often suffers from limitations such as getting stuck in local optimum, fast convergence to and plateauing at the local optimum, influence of the initial configuration, etc. Meta-heuristic methods have been developed as general methods that enable avoiding some of these shortcomings by balancing between the intensification and diversification (exploration and exploitation) of the search [19]. The main characteristic of metaheuristics is that they cannot guarantee the optimality of the generated solutions, however, in practice they provide high-quality solutions very fast.

The intention of our study is to examine the algorithm's design choices of a metaheuristic method. We are particularly interested in the VNS algorithm [12,16], single-solution method that belongs to the class of metaheuristics with the core engine supported by local search procedure. It is known as the meta-heuristic with small number of parameters, the most important one being k_{max} (Sect. 3.2, [12,16]). However, an implementation of the VNS algorithm in practice often depends on more parameters, distinguished as the model- and problem-specific [3]. In order to design an efficient VNS algorithm for the considered problem, we first need to recognize all the model-specific parts of the algorithm that may constitute possible parameters. Next, the following research questions should be addressed: (1) Which VNS parameter is the most influential? (2) How each parameter individually influences the performance? (3) Are there interactions between VNS parameters? Finally, we need to find the best combination of parameters, i.e., to identify configurations that correspond to preferable algorithm's performance of the VNS algorithm on one or more test instances.

The remainder of this paper is organized as follows. We start with motivation and related work in algorithm experiments on VNS in the next section. Description of the scheduling optimization problem and the corresponding VNS implementation are given in Sect. 3. Section 4 presents the set of parameters and the experiments that we conducted in order to identify the interconnections and parameters' most appropriate values. Section 5 concludes the paper.

2 Motivation and Related Work

The algorithm design of metaheuristic methods has been the subject of many papers in the literature [1,4–6,9,13–15]. We are particularly inspired by the work of Mc'Geoch [15] which distinguishes between algorithm design, algorithm tuning, and code tuning. Very often, to produce general and precise results, the algorithmic experiments should take place on a scale between abstraction and instantiation. In our case, the experimental goals are based on algorithm and code tuning which take place in instantiation space. Our performance indicators are selected so that we can analyse different design choices, from the numerical parameters to the modular parts of the VNS algorithm. Moreover, these indicators are matched to the investigated parameters reported in the literature [7]. As performance measurements we are using solution's quality, computational effort (platform-independent) and CPU time (platform-dependent).

Our analysis is the simplification of the statistical Design of Experiments strategy shown in [3], and it is divided into three parts: (1) variable impact, (2) modeling, and (3) tuning. The first part pertains to find the most influential parameters. In the second step we apply statistical and visualization tools to investigate possible interactions between the VNS parameters. The third part is related to the tuning process which means identifying configurations that correspond to preferable algorithm's performance. However, we do not conduct the third step in this paper and focus only on the first two stages of the parameter analysis. We start with presuming a linear model (which is often the practice) and visualize the residuals of the model. We utilize R software tools [18].

We consider Multiprocessor scheduling problem with communication delays (MSPCD) [17]. Permutation-based VNS implementation for MSPCD is proposed in [7] where the authors recognized several VNS parameters. We continue the parametric analysis of the VNS, by making the distinction between different types of parameters. We are inspired by the fact that we might be able to predict the performance of the VNS algorithm, assuming that the structure of test instances is the same throughout the benchmark set. We also hope that our results can contribute in developing the general approach to design and analysis of future VNS implementations.

The VNS algorithm is a popular method that has been applied to many optimization problems, which is why we are interested in contributing to its proper implementation design. By examining extremely reach literature on the application of VNS, we have noticed a lack of systematic approaches to algorithm design. As we could notice, the parameter tuning of VNS has always been performed by hand, i.e., by preliminary comparison and evaluation on some predefined combinations of parameters on a sub-set of (more or less) representative test instances. Often in practice, not only parameter values are determined ad-hoc, but also the set of parameters that should be analysed. This may result in missing some important parameters or some of the promising values. In addition, this approach becomes unsustainable for increasing number of algorithm's parameters. Therefore, we argue that manual analysis of stochastic algorithms with dynamic design for modular or numerical parts is not objective. The automatic tools of parameter analysis are necessary for various reasons: to help with the authors' bias, provide visualization and detect patterns that normally are not possible with human eye.

MSPCD represents an attractive combinatorial optimization problem due to its importance in modern applications not only in computer science, but in numerous other fields, such as team building and scheduling, organization of production lines, cutting and packing. Although concentrated on the particular VNS implementation as a case study, we believe that our approach can contribute to the more wide-spread utilization of the existing experimental methodology for metaheuristics.

3 Description of MSPCD and VNS Implementation

In this section we provide a short description of MSPCD, followed by the presentation of the FI-VNS and the new best improvement VNS algorithm (BI-VNS).

3.1 Problem Description

MSPCD can be defined as follows: given are n tasks (modules or jobs) that have to be scheduled on a multiprocessor system with m identical processors connected in an arbitrary way specified by the distance matrix $D_{m \times m}$. For each task, given are its processing time (duration) p_i, as well as the list of its successors and the corresponding communication costs C_{ij} (the amount of intermediate results required to complete task j) if tasks i and its successor j are to be executed on different processors. Knowing the successors of each task, it is easy to reconstruct the corresponding list of predecessors and to complete the information about precedence constraints between tasks defining the order of task execution. More precisely, a task cannot start its execution unless all of its predecessors are completed and all relevant data (defined by the communication costs) are transferred between the corresponding processors. The time required for data transfer (communication delay) depends not only on the communication costs, but also on the distance between processors executing the corresponding tasks, as well as on the hardware-related parameter defining the ratio between computation and communication speeds. Therefore, we need to decide where and when each task should be executed in order to minimize the total execution time. Formal definition of MSPCD and the corresponding mathematical formulation are presented in [8].

3.2 Variable Neighborhood Search Algorithm and Its Parameters

VNS consists of three main steps: shaking (SH), local search (LS) and neighborhood change (NC). The role of SH step is diversification, i.e., to prevent search being trapped in a local optimum, while LS step has to ensure the improvement of the current solution by visiting its neighbors (intensification) [12,16]. The main advantages of VNS are its simplicity and a small number of parameters. Basic variant of VNS [16] has a single parameter k_{max} maximal number of neighborhoods considered, i.e., number of different neighborhood types and/or maximal distance with respect to one neighborhood type. Recent implementations of VNS [12] may consider some additional parameters, however, sometimes even k_{max} may be selected in such a way to be dependent on the problem input data making VNS a parameterless method.

The general steps of VNS may be found in [12], together with various modifications. VNS is known as "a descent, *first improvement method* with randomization" [11] (p. 3981) and here we refer to this variant as the first-improvement VNS (FI-VNS). The pseudo-code of FI-VNS is given in Algorithm 1. It is an adaptation of the corresponding algorithm from [11] (p. 3979, Algorithm 7) by the inclusion of new parameters that we consider in the implementation of VNS for MSPCD. More details about the performed changes are given in Sect. 4. As

---

Algorithm 1: PSEUDO CODE OF THE FI-VNS ALGORITHM

Input: problem data, N, FI, $forward$, k_{max}, k_{step}, k_{min}, $p_{plateau}$, MAX_step

1 INITIALIZATION: $x_{bsf} = \text{INIT}(N, FI, forward)$, $STOP = FALSE$;
2 $dstep = 0$;
3 **repeat**
4   Apply k_{min} strategy;
5    $k = k_{min}$;
6   **repeat**
7       $x' = \text{SH}(x_{bsf}, N, k)$;
8       $x'' = \text{LS}(x', N, FI, forward)$;
9       $k = k + k_{step}$;
10      $dstep++$;
11   **if** $(f(x'') < f(x_{bsf}))$ **then**
12         $x_{bsf} = x''$; /* MOVE */
13         $k = k_{min}$;
14   **else if** $(f(x'') == f(x_{bsf}))$ **then**
15         $prob = rand[0,1]$;
16   **if** $prob \leq p_{plateau}$ **then**
17           $x_{bsf} = x''$;
18   **if** $dstep \leq MAX\_step$ **then**
19         $STOP = \text{TRUE}$;
20   **until** $((k > k_{max}) \,||\, STOP)$;
21 **until** $STOP$;

---

shown, a FI-VNS iteration starts from an initial solution x_{inic} and runs its step SH, LS, and NC until the best-so-far solution (the current approximation of the best solution x_{bsf}) has been improved or until VNS explores k_{max} (the maximal number of) neighbourhoods around the x_{bsf} solution. The best-so-far solution x_{bsf} represents a global knowledge exchanged between the VNS iterations. In particular, the initial x_{bsf} solution is generated by the CP+ES constructive heuristic. It utilizes critical path (CP) based permutation scheduled by Earliest start (ES) rule and improved by the local search in the initialization phase (see Algorithm 3, function INIT). In the main VNS loop, SH tries to move the search far from the current best solution. Then, once LS is finished, the newly found solution x'' is compared against the x_{bsf} solution (Algorithm 1, line 9). If the improvement is made, the VNS iteration is reset to k_{min} and SH starts from the newly discovered x_{bsf} solution. Otherwise, the value for k increases and a VNS iteration terminates if k reaches k_{max} or when the stopping criterion is satisfied. As a consequence of FI-VNS algorithm structure, the execution time might vary greatly from one iteration to another, which is our source of inspiration to utilize the *best improvement* BI-VNS strategy (see Algorithm 2) described in [11] (p. 3981, Algorithm 11). For the purpose of demonstrating differences between the two VNS variants, in the corresponding pseudo-codes we use colors to describe how global knowledge is being utilized.

Algorithm 2: PSEUDO CODE OF THE BI-VNS ALGORITHM.

Input: problem data, N, FI, $forward$, k_{max}, k_{step}, k_{min}, $p_{plateau}$, MAX_step

1 INITIALIZATION: $x_{bsf} = $ INIT$(N, FI, forward)$, $STOP = FALSE$;

2 $dstep = 0$;

3 **repeat**

4   Apply k_{min} strategy;

5     $k = k_{min}$;

6     $x_{min} = x_{bsf}$; /* current best */

7   **repeat**

8       $x' = $ SH(x_{bsf}, N, k);

9       $x'' = $ LS$(x', N, FI, forward)$;

10       $k = k + k_{step}$;

11       $dstep++$;

12   **if** $(f(x'') < f(x_{min}))$ **then**

13         $x_{min} = x''$; /* MOVE */

14   **else if** $(f(x'') == f(x_{min}))$ **then**

15         $prob = rand[0, 1]$;

16   **if** $prob \leq p_{plateau}$ **then**

17           $x_{min} = x''$;

18   **if** $(dstep \leq MAX\_step)$ **then**

19         $STOP = TRUE$;

20   **until** $(k > k_{max})$ || $STOP$;

21   **if** $(f(x_{min}) \leq f(x_{bsf}))$ **then**

22       $x_{bsf} = x_{min}$;

23 **until** $STOP$;

Unlike FI-VNS, BI-VNS algorithm does not restart the neighborhood counter after each improvement of x_{bsf}. Therefore, in Algorithm 2 we need an auxiliary variable x_{min} keeping the current best solution to be used in updating x_{bsf} when k reaches k_{max}. Working load between successive BI-VNS iterations is more balanced than in FI-VNS due to consistency in the number of explored neighbourhoods. To be able to compare performance of FI-VNS and BI-VNS, we utilize $dstep$ in Algorithm 1 and Algorithm 2 to count the number of discrete steps (i.e., the number of SH and LS executions). The counter is controlled by MAX_step which we appoint as the stopping criterion. Counting computational steps allows us more reliable experiments with regard to the reproducibility.

4   Empirical Study of the VNS Parameters

Besides the parameters related to the given definition of VNS in Sect. 3.2, each particular implementation can involve a problem-dependent set of parameters. These kind of parameters are for example solution quality measurement parameter (e.g., objective function value, fitness or utility), number and types of used neighborhoods, an improvement strategy or some other operator. To obtain an

---

Algorithm 3: PHASES WITHIN THE VNS ITERATION.

---

1 **Function** INIT(*N, FI, forward*):
2      $x = $ CP+ES(); /* constructive heuristic to generate initial solution */
3      $x' \leftarrow LS(x, N, FI, forward)$;
4   **if** $(f(x') < f(x))$ **then**
5        $\lfloor \; x = x'$;
6   **return** x ;

7 **Function** SH(*x, N, k*):
8   /* Generate feasible solution x' from k^{th} neighborhood of x */
9   **for** $(i = 1; \; i <= k; \; i + +)$ **do**
10        $x' \in N(x)$; /* at random */
11        $x = x'$
12   **return** x ;

13 **Function** LS(*x', N, FI, forward*):
14   /* Apply a local search method on x' depending on input parameters*/
15   **repeat**
16   Let $N(x) = (x_1, ..., x_p)$;
17   **if** $(\neg forward)$ **then**
18          $\lfloor$ REVERSE$(N(x))$;
19        $i \leftarrow 0$;
20        $x' \leftarrow x$;
21   **repeat**
22          $i \leftarrow i + 1$;
23   **if** $(f(x_i) < f(x))$ **then**
24            $x \leftarrow x_i$;
25   **if** FI **then**
26              $\lfloor$ **Break;**
27   **until** $i = p$;
28   **until** $f(x') \leq f(x)$;
29   **return** x';

---

efficient algorithm that provides high-quality solutions for the majority of tested instances, the appropriate values of its parameters need to be identified. Finding the best possible parameters configuration is an optimization problem itself, parameter values represent decision variables, while the objective function corresponds to the considered optimization problem. Therefore, this part of the metaheuristic design deserves special attention and it is the main subject of our paper.

Our work extends the analysis conducted in [7]. We utilize the same benchmark set and in this section we revisit the results of the previous study in order to point out the differences with the new results.

4.1   Previous Study

VNS parameters from [7] are classified as numerical and categorical variables in the following way. The categorical variables are operators: shaking rules, neighbourhood combinations, restricted neighbourhoods, the search direction and the search strategy. The numerical parameters are k_{max}, k_{step}, $p_{plateau}$. It is presumed that k_{min} equals 1. Authors have conducted the experimental study gradually, i.e. an independent study is performed for each categorical variable with the fixed configuration of other parameters. The maximal CPU time (t_{max}) is utilised as the stopping criterion.

We notice that [7] follow several steps of the experimental algorithmics as suggested in [15]. For example, choice of the data structure in which the solution is held as well as the study of modular parts of the VNS algorithm are conducted for the following operators: (1) different types of neighborhoods (Swap-1, Swap-2, Swap-3 and IntCh); (2) heuristics for initial solution generation; (3) task scheduling rule; (4) search direction (forward-backward); (5) FI- and BI- local search procedure. Additionally, tuning of the VNS parameters (maximal number of neighborhoods, neighborhood definitions, i.e. combinations and restrictions, stopping criterion) has been conducted manually. With regard to the numeric parameters in [7], the results in Tables 7 and 8 indicate high interaction between k_{max}, k_{step}, $p_{plateau}$. Therefore, the authors conclude that the scheduling process exhibits a chaotic behaviour. In the new study we address the issue of the non-linear relationship between the numerical parameters with the help of the regression model.

4.2   New Study

Differences between the previous and the new study can be summarized as follows. We implement the best improvement VNS algorithm (BI-VNS) and reintroduce the parameter k_{min}. Next, we propose new values for numerical parameters k_{max}, k_{step} and $p_{plateau}$ that were not previously studied. The four parameters (k_{min}, k_{max}, k_{step}, $p_{plateau}$) are recognized in the literature as an integral part of the VNS algorithms and are known as the *basic* VNS parameters. Several categorical parameters are introduced that, we believe, have an important impact on the performance: VNS-strategy (VNSs), LS-type, KMINS and direction (dir). Categorical parameter VNSs allows us to revisit the study of FI-VNS from [7] and compare it against the new BI-VNS. It also allows us to examine dependencies that the VNS algorithm may have with the basic VNS parameters. To control for two types of local search techniques we utilize the categorical parameter LS-type. Auxiliary categorical parameter KMINS denotes four different strategies for calculating values of k_{min}. The dir parameter is introduced in the previous study as an indicator of the neighbourhood search direction (see Fig. 4 in [7]). The overall list of VNS parameters is shown in Table 1 (numerical parameters are defined with italic and categorical with teletype font). To generate an initial solution we consider CP+ES heuristic and only one type of the neighbourhood (Swap-1). This particular algorithm's design choice is based on the report from [7].

Table 1. VNS parameters.

VNS parameters	Numerical	k_{max}	$\{1, 2, 3, 4, 5, 6, 7, 8, 9, 10\}$
		k_{step} (if $k_{max} \leq 4$)	$[1, k_{max} - 1]$
		k_{step} (if $k_{max} > 4$)	$[1, \lceil \frac{k_{max}}{2} \rceil - 1] \cup \{k_{max} - 1\}$
		$p_{plateau}$	$\{0, 0.1, 0.5, 0.9, 1\}$
	Categorical	VNS-strategy	BI-VNS (FI-VNS = 0) FI-VNS (FI-VNS = 1)
		KMINS	kmin1: $k_{min} = 1$ kminrand: $k_{min} = rand[1, k_{step}]$ kr2,kr3: mixed
		LS-type	BI-LS (FI-LS = 0) FI-LS (FI-LS = 1)
		dir	forward = 0 forward = 1

As presented in Sect. 3.2, values of VNSs produce two algorithms FI-VNS and BI-VNS, while other categorical parameters define the so-called algorithm variants. Gray cells in Table 1 mark problem-specific (i.e., closely connected to MPSCD) parameters: LS-type and dir. We compared two types of local search: the first improvement search strategy (FI-LS=1) and the best improvement local search (FI-LS=0) denoted as BI-LS. According to [7] the FI-VNS algorithm with FI-LS performs faster and better than with BI-LS. However, when we appoint MAX_step as the stopping criterion we find that BI-LS exhibits better performance within both FI- and BI-VNS algorithms (see Sect. 4.4). Parameter dir is also problem-specific, however it was reported in [7] as not significant. We show that dir is the second most important considered VNS parameter that greatly influences algorithm's performance w.r.t. d_{step}.

With KMINS our goal is to compare two main ideas of the neighbourhood search: (1) visit all the neighborhoods of the current best solution ($k_{step} = 1$), (2) skip neighborhoods ($k_{step} > 1$). We should note that the implementations of these two ideas differ w.r.t. the two main VNS search strategies and each implementation yields a different algorithm variant. Foremost, the search terminates if the better solution is found (for FI-VNS) or all k_{max} neighborhoods have been visited (for BI-VNS). In addition, in BI-VNS the variable k_{min} is defined only once, while in FI-VNS we must reinitialize k_{min} at line 13. We denote strategy (1) as kmin1 strategy, which appoints $k_{min} = 1$ at the line 4 of both algorithms, as well as at the line 13 of Algorithm 1. The second strategy, denoted as kminrand, is to change k_{min} as a function of k_{step}. In particular, in initialization and reinitialization phases of FI-VNS the value of k_{min} becomes a random number from $[1, k_{step}]$. There are two more KMINS strategies (kr2, kr3) that are possible within the FI-VNS algorithm. Algorithm variant defined by kr2 initializes variable k_{min} as a random number from $[1, k_{step}]$. However, at line 13 it resets k to 1. The variant kr3 defines $k_{min} = 1$ at line 4, and at line 13 it determines k as a random number from $[1, k_{step}]$.

Values for k_{max} are often in practice determined w.r.t. the dimension of the problem and/or other problem-related characteristics. We focused on $k_{max} \leq 10$ in order to examine the values that were not utilized in the previous study. The values for k_{step} are specified as integers from the set $[1, \lceil \frac{x}{2} \rceil - 1] \cup \{x - 1\}$, if $x \geq 4$ where $x = k_{max}$. Moreover, if $k_{max} = 1 \lor k_{max} = 2$ then $k_{step} = 1$. If $k_{max} = 4$ then the possible values for k_{step} are $\{1, 2, 3\}$, while for $k_{max} = 10$, $k_{step} \in \{1, 2, 3, 4, 9\}$.

4.3 Experimental Methodology

For a deeper understanding of the algorithm operators and numerical parameters we employ visual, statistical and machine learning tools to help us decide about the most influential parameters. The next step is the study of the interconnection of parameters and their main effects. Typical data analysis via machine learning tools may be performed over many different variables, however, visual analysis is possible for only five or at most six variables at a time. There are other methodological settings that we need to be aware of in order to procure data that helps us gain a better understanding of VNS operators.

The stochastic nature of meta-heuristics requires repeating algorithm's runs several times in order to properly estimate the *usefulness* or *utility* of the algorithm's performance. The measure of performance may refer to the solution of an optimization problem and/or computational effort such as running time, number of function evaluations, number of iterations, etc. Moreover, the definition of performance measure (PM) depends on the goal and conditions of the research. For example, PM under limited amount of computer resources is often based on central tendency estimators: the average objective function value from n independent runs, or the median [3]. We are especially interested in the *efficiency* of the algorithm, thus, we focus on determining the effort required to reach a solution of a predetermined quality (target value). Our goal is to understand under which parameter configurations and computational restrictions is possible to obtain the highest probability of producing an optimum. We argue that in our case we should employ optimal value as the target solution and base the performance on the objective function's value (i.e. solution's quality). Consequently, the goodness of performance can be related to the number of successful runs within the limited budget. A thorough review of different performance measures is provided in [3, p. 108] and in [2, p. 40].

To obtain a statistically meaningful representative of the algorithm's performance, we repeat executions n_{run} number of times. Another important aspect of the experimental methodology is the benchmark set. The problem benchmark set consists of random test instances with known optimal solution (details about problem's instance are given in [7]). We are particularly interested in problems with 50 tasks as it provides opportunity for VNS to find the optimal solution. Moreover, in order to address the goals of our study we employ one problem instance ogra50_50 to be scheduled on the hypercube of $m = 4$ processors. The performance estimator represents the number of the occurrences of the target solution within n_{run} repetitions, which we denote as opt_{count}. In particular, $opt_{count} = \frac{\#succ}{n_{run}} \cdot 100$, where $\#succ$ denotes the number of successful runs. Finally, we set $n_{run} = 100$, therefore we employ 100 different seeds.

Essential part of the experimental study of randomized algorithms is choosing the appropriate type and value of the stopping criterion (SC). The goal is to avoid *floor* effect as the solution's quality reaches stagnation. To avoid comparisons where all variants exhibit the stagnation, we should find a suitable value for MAX_step beyond which the algorithm's performance does not improve in some practical sense. In Fig. 1 convergence plots of two VNS algorithms are presented, BI-VNS:FI-LS (BI-VNS algorithm with FI-LS search) and FI-VNS:BI-LS (FI-VNS algorithm with BI-LS search). Graphics show the propagation of the average, median, minimal and maximal solutions' quality. Between different d_{step} we are able to compare descriptive statistics, thus providing details about the convergence properties. The small boxes show the best result (brackets hold the occurrence for the n_{run} repetitions). For SC analysis we first set $MAX\_step = 500$, which was large enough for the algorithms to approach the pick of their performance. We concluded that $MAX\_step = 200$ is suitable for our parameter analysis due to the fact that the algorithm variants exhibit significant differences (before reaching stagnation phase). In addition, the variance significantly decreases compared to smaller MAX_step.

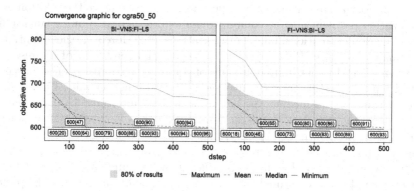

Fig. 1. Convergence plots of two VNS variants for: $p_{plato} = 0, k_{step} = 1, \texttt{forward} = 1$.

4.4   Results

With the help of graphical tools from flexplot [10], the first impressions of the parameter importance is achieved. Histograms in Fig. 2, divided among parameters dir, VNSs and LS-type, show the shape and the spread of the data distribution. Most distinguished distributions are between levels of dir. The backward direction search ($forward = 0$) produced bell-shape distribution with smaller spread compared to $foward = 1$ which has long left tail. This indicates that the VNS variant with forward search direction is much more sensitive to values of other VNS parameters. For better understanding of the reason behind such distributions, we need to conduct the parameter importance analysis followed by the regression model of their interconnections.

Estimate of the significance of each VNS parameter was determined by random forest and linear regression model (Table 2) [18]. The random forest from package *party_1.3-8* was conducted via R function *cforest* and the values in column RF of Table 2 were generated by function *varimp()* from the same package. We set *seed*(1010). The third and the fourth columns correspond to the Semi-Partial R^2 and the standardized β coefficients determined via the multivariate linear regression *lm* function[1]. In particular, Semi-Partial R^2 values are extracted from the report generated with *estimates*[2] on the result of *lm()*. The β coefficients represent normalized regression coefficients generated by R function *lm.beta()* (package *lm.beta_1.5-1*). Simple linear regression model of seven VNS parameters has identified a significant regression equation with the outcome $F(7, 3712) = 1161$, $p < 10^{-16}$, adjusted $R^2 = 0.69$.

In Table 2 the VNS parameters are ranked based on four indicators: random forest, Semi-Partial R^2, β coefficients and p-value. The higher the value of the first three indicators, larger is the importance of the parameter. The ranking also indicates the significance of proper algorithm design in order to achieve the highest performance. For example, the most influential operator is dir, followed by LS-type. VNSs has not shown to be as important as was initially expected. Numerical parameter $p_{plateau}$ showed small influence, however, the random forest

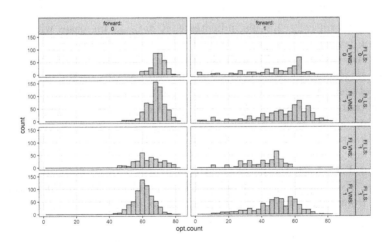

Fig. 2. Histogram of complete data for all VNS parameters.

---

[1] lm(), package *stats4*, R version 4.0.4, https://www.R-project.org/ [10].
[2] package *flexplot*.

Table 2. Size of the effect of VNS parameters

Rank	VNS parameter	Random forest	Semi-partial R^2	Standardized β coefficient	p-value
1	dir	202.93	0.333	-0.577	$< 10^{-16}$
2	k_{max}	174.70	0.291	0.547	$< 10^{-16}$
3	LS-type	33.01	0.053	-0.229	$< 10^{-16}$
4	VNS-strategy	14.46	0.002	0.056	$< 10^{-8}$
5	$p_{plateau}$	10.15	0.007	0.082	$< 10^{-16}$
6	k_{step}	9.99	0.000	-0.018	0.07
7	KMINS	1.48	0.000	-0.012	0.24

and linear regression coefficients do not agree on the ranking. Based on the p-values, parameters k_{step} and KMINS do not show statistically significant main effect on the performance.

After the parameter selection phase (screening), we are able to address the question of interactions. We do this by combining the first four ranked parameters into the model. We have produced the following significant regression equation: $opt_{count} = -55.56\text{dir} + 4.53k_{max} - 12.14\text{LS-type} - 2.36\text{VNSs} - 0.31k_{max}^2 + 5kmax \cdot \text{dir} + 16.9\text{dir} \cdot \text{LS-type} + 7.73\text{VNSs} \cdot \text{dir} + 0.77kmax \cdot \text{LS-type} - 2.36kmax \cdot \text{dir} \cdot \text{LS-type}$, with the outcome $F(10, 3709) = 2069$, $p < 10^{-16}$, adjusted $R^2 = 0.85$. The regression model is visualized in Fig. 3. The variability of data around the model in Fig. 3 is caused by the numerical VNS parameters (each point corresponds to the unique parameter configuration).

The most significant interaction term is between k_{max} and dir, thus, their configuration is intertwined. In case of the $forward = 1$ the influence of k_{max} is pronounced following the concave curve due to the negative coefficient of the second order term (k_{max}^2). This indicates that opt_{count} increases with k_{max}. In the case of $forward = 0$, the parameter k_{max} has less influence, however, the growth trend is detected. Figure 3 shows some nonlinear behavior that is not explained by the model. We suspect that the non-linearity originates from interactions between the numerical VNS parameters. The second most significant nonlinear term is between LS-type, dir and k_{max}. Thus, to achieve the highest performance we need to simultaneously configure these three parameters. However, VNSs should also be considered due to the significant interaction with dir.

We can distinguish the effects that parameters impose independently from each other. For example, the model and Fig. 3 indicate that the negative coefficient in front of LS-type suggests that BI-LS (red curve) produces optimal solutions more often than FI-LS (blue curve). $forward = 0$ indicates better performance over all categorical data and to achieve $opt_{count} \geq 80$, it is enough to employ smaller values for k_{max}. Furthermore, the plots for $forward = 0$ indicate that the stagnation phase is reached for larger k_{max} (which is the region that promises larger opt_{count}). Finally, FI-VNS and BI-VNS have demonstrated similar performance, however, we suspect that the differences increase for $k_{max} > 10$.

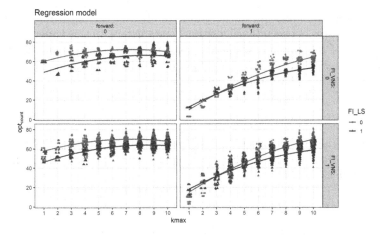

Fig. 3. Regression model of the VNS parameters (*visualize*() [10]).

To test this hypothesis, we generated additional data and graphics disclosed at https://www.mi.sanu.ac.rs/~tatjana/conf/VNS-param-analysis/. The plots show that BI-VNS demonstrated much better performance when compared to FI-VNS if combined with *forward* = 0 and FI-LS. In particular, it generated $opt_{count} > 90$ for $kmax \geq 20$.

Regardless of the small statistical influence reported in Table 1, the numerical parameters show practical impact (see Fig. 3). Data and plots may be found at the mentioned link. The results show that the parameter $p_{plateau}$ does not influence the performance in case of BI-VNS, however, it should be properly configured in case of FI-VNS. Parameter k_{step} is closely interconnected with VNS categorical parameters which means that the success of VNS variant depends on the values of k_{step} and k_{max}. Parameter KMIN has not shown an overall statistically significant influence, however, strategies kminrand and kr2 exhibited noticeable impact on the results of FI-VNS. Due to their high nonlinear relationship, to achieve the best results it is necessary to conduct automatic tuning of k_{step}, $p_{plateau}$, k_{min} via a software package such as iRace, ParamILS, SPO, etc. [4].

5   Conclusion

This paper applies a multivariate regression model, random forests and visualization tools to study the effects of the VNS parameters on one problem instance. The BI-VNS algorithm is developed and compared against FI-VNS. We promote the application of methodological experimental evaluation to metaheuristic algorithms that aids in their objective performance analysis. Three phases of the parameter analysis were followed: (a) recognition of algorithm's parameters; (b) detection of the most influential parameters; (c) data analysis of parameters' main and interaction effects. We were able to identify the most influential non-linear relationships between the algorithm's parameters. In this paper we focused on the study of categorical parameters, and pointed out the main observations about the numerical parameters. The future work is to generalize the presented methodology to work with the larger optimization problem set.

References

1. Barr, R.S., Golden, B.L., Kelly, J.P., Resende, M.G.C., Stewart Jr., W.R.: Designing and reporting on computational experiments with heuristic methods. J. Heuristics **1**(1), 9–32 (1995)
2. Barrero, D.F.: Reliability of performance measures in tree-based Genetic Programming: a study on Koza's computational effort. Ph.D. thesis (2011)
3. Bartz-Beielstein, T.: Experimental Research in Evolutionary Computation. The New Experimentalism. Natural Computing Series. Springer, Heidelberg (2006). https://doi.org/10.1007/3-540-32027-X
4. Bartz-Beielstein, T., et al.: Benchmarking in optimization: best practice and open issues (2020). arXiv
5. Beiranvand, V., Hare, W., Lucet, Y.: Best practices for comparing optimization algorithms. Optim. Eng. **18**(4), 815–848 (2017). https://doi.org/10.1007/s11081-017-9366-1
6. Czarn, A., MacNish, C., Vijayan, K., Turlach, B., Gupta, R.: Statistical exploratory analysis of genetic algorithms. IEEE Trans. Evol. Comput. **8**(4), 405–421 (2004)
7. Davidović, T., Hansen, P., Mladenović, N.: Permutation-based genetic, tabu, and variable neighborhood search heuristics for multiprocessor scheduling with communication delays. Asia-Pac. J. Oper. **22**(03), 297–326 (2005)
8. Davidović, T., Liberti, L., Maculan, N., Mladenović, N.: Towards the optimal solution of the multiprocessor scheduling problem with communication delays. In: Proceedings of MISTA 2007, pp. 128–135, Paris, France (2007)
9. Eiben, A.E., Smit, S.K.: Parameter tuning for configuring and analyzing evolutionary algorithms. Swarm Evol. Comput. **1**(1), 19–31 (2011)
10. Fife, D.: Flexplot: graphically-based data analysis. Psychol. Methods (2021)
11. Floudas, C.A., Pardalos, P.M.: Encyclopedia of Optimization. Springer, Heidelberg (2009). https://doi.org/10.1007/978-0-387-74759-0
12. Hansen, P., Mladenović, N., Todosijević, R., Hanafi, S.: Variable neighborhood search: basics and variants. EURO J. Comput. Optim. **5**(3), 423–454 (2017). https://doi.org/10.1007/s13675-016-0075-x
13. Hooker, J.N.: Needed: an empirical science of algorithms. Oper. Res. **42**(2), 201–212 (1994)
14. Kendall, G., et al.: Good laboratory practice for optimization research. J. Oper. Res. Soc. **67**(4), 676–689 (2016)
15. McGeoch, C.C.: A guide to experimental algorithmics. CUP (2012)
16. Mladenović, N., Hansen, P.: Variable neighborhood search. Comput. Oper. Res. **24**(11), 1097–1100 (1997)
17. Papadimitriou, C.H., Yannakakis, M.: Towards an architecture-independent analysis of parallel algorithms. SIAM J. Comput. **19**(2), 322–328 (1990)
18. R Core Team: R: A Language and Environment for Statistical Computing. R Foundation for Statistical Computing, Austria (2021). https://www.R-project.org/
19. Talbi, E.G.: Metaheuristics: from Design to Implementation. Wiley, New York (2009)

Competitive Location Strategies in the (r|p)-Centroid Problem on a Plane with Line Barriers

Artem Panin and Alexandr Plyasunov[✉]

Sobolev Institute of Mathematics, 4 Acad. Koptyug Avenue, Novosibirsk, Russia
apljas@math.nsc.com

Abstract. In 1982, Drezner considered the competitive facility location problem when the leader and follower each place a facility on a plane. He proposed polynomial-time algorithms for the follower and leader optimal facility location. In 2013, Davydov et al. considered a generalization of this problem when the leader has a set of $(p-1)$ facilities and wants to open another facility in the best position with the optimal response of the follower.

We examine the influence of line barriers on the optimal leader and follower strategies. The paper considers the formulations in which the number of already open facilities is fixed, and the barriers divide the plane into polygons in such a way that two different paths from one polygon to another cannot exist. We propose a polynomial-time algorithm for the Drezner problem with barriers, as well as for the problem studied by Davydov et al.

Keywords: Bilevel programming · Planar (r|p)-centroid problem · Competitive location strategies · Line barrier

1 Introduction

Planar location models have been studied over the last several decades due to their increasing importance in the current world [1–3]. These models are studied in planar continuous space, discrete space, and network environment. In continuous models, the location of objects is searched for anywhere in the plane. Therefore, an infinite number of potential object locations is considered. Discrete models limit the location of objects to a predetermined set of possible locations. In network models, the location of objects is searched for anywhere in the network (at the vertices and edges). Therefore, the number of potential object locations is also infinite. Planar competitive location problems were first studied in [4]. The choice of location and price decision of two competitors on a segment with uniformly distributed customers are considered. A large number of publications is devoted to this field of research [5–8]. A new approach to estimating the market share captured by competing facilities was proposed in [9]. The approach is based on cover location models. In [10], a hybrid genetic algorithm

Y. Kochetov et al. (Eds.): MOTOR 2022, CCIS 1661, pp. 119–131, 2022.
https://doi.org/10.1007/978-3-031-16224-4_8

with solution archive for the classical problem of competitive facility location known in the literature as the discrete (r|p)-centroid problem was developed, and a small but very informative survey of approximate algorithms for solving this problem is also given.

A number of planar location models take into account the limitations associated with obstacles or barriers [11]. These models have areas in which the placement of new facilities is impossible or the passage is prohibited. Such models allow us to assess the implications of barriers on the geometric and computational properties of planar location problems. In [12], barrier regions were first introduced to location modeling. In the barrier regions, neither placing facilities nor traveling through is allowed. In [13], the Euclidean distance single Weber location problem in the presence of barrier regions was studied. For this problem, the branch and bound algorithm BSSS (big square small square) was developed in [14].

In [15], a class of facility location problems with a barrier in which a line barrier expands across the plane, dividing it into two separate half-planes is introduced. This problem is generally non-convex and NP-hard. A solution algorithm is proposed. In [16], this problem is extended to a bi-criteria median problem. A reduction-based solution algorithm is presented. In [17] and [18], the Weber problem with barrier is studied, and how the barrier distance function can affect the location of objects is shown. The single Weber location problem with a probabilistic line barrier is introduced in [19]. A generalization to multi-Weber problem is proposed in [20] and [21]. For large-sized instances, approximate solution methods based on metaheuristics are developed. In [22], the Weber location problem in the presence of line and polyhedra barriers using the Varignon frame is studied, and an approximate local optimal solution is proposed. In [23–26], the presence of a line barrier in a rectiline-distance single-facility location problem is also studied. The problem with a non-horizontal line barrier is considered in [19]. Authors transform the problem into a horizontal barrier problem by the rotating axes. In this case, the distance between any two points is the same as for a rectiline distance, but the basic distance function is different.

In [27], the problems of location of several objects with a non-horizontal line barrier are considered. A heuristic that preserves the rectiline distance function is proposed. For the special case of a single object, an exact solution algorithm is developed. This algorithm has a polynomial-time complexity relative to the number of passages. However, its time complexity increases exponentially as the number of existing objects increases. Therefore, a fast and efficient approximation procedure is used. The procedure finds a feasible solution in polynomial computing time. For the solution of the multi-facility case, an alternate-location-allocation procedure is proposed. It reduces the original problem to solving a set of single-facility location subproblems with passages and a set of set-partitioning subproblems for allocation. In [28], this procedure is used to solve a multi-Weber location problem where traveling was possible through the barrier's extreme points. Authors applied continuous relaxation to the location subproblems and used an approximate algorithm developed by [29] to solve the location subproblems of a single object. In contrast, in [27], for the location procedure, the continuity of the location subproblems is preserved, and an exact algorithm is used to solve each subproblem.

Next, we will focus on two special cases of the (r|p)-centroid problem on the plane. In [30], Drezner considered the competitive problem of facility location on the plane when the leader and follower place one facility each. He proposed polynomial-time algorithms for the follower and leader optimal facility location. In [31], Davydov et al. considered a generalization of this problem when the leader has a set of $p-1$ facilities and wants to open another facility in the best position with the optimal response of the follower. This problem is known as $(r|X_{p-1}+1)$-centroid problem [31]. They also showed that the problem is polynomially solvable for a fixed r.

In this article, the (1|1)-centroid problem and the $(r|X_{p-1}+1)$-centroid problem are extended by the concept of line barriers, which significantly increases the complexity of these problems, but makes them more realistic and, therefore, expands their scope. We consider the Euclidean metric. It is necessary to investigate how much the behavior of companies and the complexity of calculating the optimal solution will change when splitting classical metric planes by introducing line barriers [11, 15, 16]. For these new formulations, we have developed exact polynomial-time algorithms. Given the limitations on the scope of the article, we provide a sketch of proofs in the following sections. An expanded journal version of this work is currently being prepared.

The rest of this work is organized as follows. In Sect. 2, we introduce the problem and basic definitions. In Sect. 3, we describe a polynomial-time solution algorithm for the (1|1)-centroid problem with line barriers. Section 4 presents an exact polynomial-time algorithm for the $(r|X_{p-1}+1)$-centroid problem with line barriers. Conclusions and future study directions are provided in Sect. 5.

2   Definitions, Properties, and Problem Formulation

In the classic (r|p)-centroid problem on the plane [30], the arrangement of customers (demand points) and their buying power (weights) are known. Two companies, leader and follower, want to maximize their market share (their revenue). When locating facilities (the leader places p facilities, the follower places r facilities), the leader and follower focus on the distance in a given metric (often Euclidean, as we in this article) from potential facilities to customers. The customer is serviced at the nearest facility. Revenue is determined by the total buying power of the customers served. The leader's and follower's goals are to maximize their own revenue.

Consider the situation when the set of customers is separated by a line barrier(s) with a passage (see Fig. 1).

Definition 1. *Line barrier with a passage (LBWP) is a straight line, ray, or segment with an internal point (passage) through which companies can pass when serving customers on the other side.*

LBWPs are limited (ray or segment) and unlimited (straight line). An unlimited LBWP, a straight line with a passage, divides the plane into two half-planes. To get to the other side of such a barrier, we have to use the passage. Limited LBWPs may be bypassed. To exclude this possibility, let us define the group of line barriers as follows.

Definition 2. *A set of n LBWPs is a group of line barriers with a passage (GLBWP) if and only if there exists an order such that, for all $k \leq n$, the first k LBWPs do not cross each other's interiors, and where the end(s) of the ray or segment belongs to another LBWP(s).*

Drezner noted that placing facilities outside the convex hull on the set of customers doesn't make sense [30]. In general, barriers can have multiple passages. Then several shortest paths between a pair of convex hull points can exist. This significantly complicates the search for the optimal solution. In this paper, we limited ourselves to examining single-path cases.

Definition 3 (Single-Path property). *A set of barriers with a passage has the single-path property if there exists a path between any pair of points in the plane and no two different paths with a mutually non-nested set of passages.*

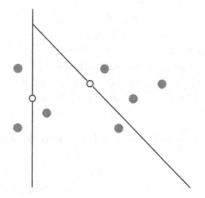

Fig. 1. An example of GLBWP

The case of several LBWPs is in Fig. 1 and Fig. 2. Consider the group of n LBWPs. The plane has $(n + 1)$ parts. Let us prove the following property.

Theorem 1 (Single-Path Property). *GLBWP has the single-path property.*

Proof. As we know, LBWP is of three kinds: lines, rays, and segments. Let's place LBWPs sequentially in order from Definition 2. Without loss of generality, we assume that the lines are placed first. Lines can only be placed parallel to each other. The set of n_l lines divides the plane into two half-planes and $(n_l - 1)$ strips. Since the passage between adjacent parts is only one, the single-path property for lines is satisfied. Consider a graph $G = (V, E)$ in which vertices (V)

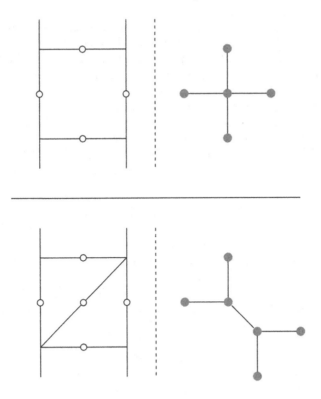

Fig. 2. Scheme for adding LBWP to the plane and graph

are parts of the plane and edges (E) are passages in LBWPs between adjacent parts. If V consists of lines, then G is a tree. The tree has no cycles. Hence, if G is a tree for all types of barriers, then GLBWP has the single-path property.

Let G be the tree on n_l lines and k rays or segments. Add one more ((k+1)-th) LBWP. The new barrier divides one part of the plane into two parts connected by a passage. It follows from Definition 2. Consequently, in graph G one of the vertices is split into two vertices connected by an edge (see Fig. 2). Then G remains the tree.

Corollary 1 (Single shortest path) For any pair of points in the plane with GLBWP, there exists a single shortest path.

Proof. The path is not the shortest when passages are repeated. Consider an arbitrary path without passage repetitions. According to Theorem 1, the path is shortest when the shortest distance to each subsequent passage.

In this study, we start with the following settings. The input data consists of coordinates of the customers, coordinates of the ends of segments, coordinates of the ends and the directing vectors of rays, parametric equations of the lines, and coordinates of the passages. The tree G defining the mutual arrangement of barriers and passages in the plane is given. Each customer is assigned a part of the plane in which he is placed. We define the following two problems.

The Follower problem. It is required to maximize revenue from customer service by placing r facilities when the location of p facilities of the leader is known.

The Leader problem. It is required to maximize revenue from customer service when placing p-th facility when $(p-1)$ facilities of the leader are already placed and at the optimal decision of the Follower problem.

The Leader problem was first introduced and investigated in [31] where it was called the problem of $(r|X_{p-1}+1)$-centroid.

3   (1|1)-Centroid Problem with Line Barriers

In this section, we study the single facility problem where $p = r = 1$. Let us first consider the single LBWP case. We denote two half-planes by P_1 and P_2 and the passage by O. If the leader's facility is in P_1, the follower can place the facility next to the passage to serve P_2 customers. Therefore, to solve the Follower problem, we can shift all customers from P_2 to O and remove LBWP. Thus, we obtain the problem studied by Drezner [30]. For the optimal location without barriers, the follower places the facility near the leader's facility to maximize its share of the market. We draw a perpendicular bisector to the segment connecting the leader and follower facilities. The perpendicular bisector divides the market between the players (companies). Then, to find the optimal solution, we need to build a straight line through the facility of the leader, separating the maximum market share. It is enough to check all lines passing through customers [30].

The scheme for finding an optimal solution to the Follower problem with GLBWP is slightly different. Again, we turn to the graph representation of the problem. Let the facility of the leader be located in some part of the plane that divides the tree G into subtrees. What do we mean? We consider the adjacent edges (passages) with the part of the plane with the facility of the leader. If the follower's facility is on the same side of the corresponding LBWP as the leader's facility, the customer on the other side of LBWP is served by the one who is closer to LBWP (the edge). Consequently, the idea of Drezner's algorithm where we place the follower's facility next to the leader's facility still applies. It is only necessary to collect customers in the passages of the adjacent part of the plane corresponding to the facility of the leader.

To do this, we divide G into subtrees. First, we delete the vertex of the tree G with the leader's facility and its adjacent edges. Then search in-depth collects all vertices and the corresponding customers of each subtree into one in linear time. Place all collected customers in the appropriate passages and use Drezner's algorithm [30].

Now we move on to the Leader problem. We consider an arbitrary subset of customers of the Leader problem without barriers. If the leader placed the facility in the convex hull of this subset, then the follower will not be able to serve all customers at once [30]. Consequently, the follower will serve customers whose convex hull does not contain the leader's facility and whose total buying power is maximum. Therefore, the leader should place his facility so as not to give away the most profitable convex hulls. A convex hull is a polyhedron. Then for the optimum, it is enough to look through the segments connecting pairs of customers [30].

To solve the Leader problem with GLBWP, we consider the following sub-problems. Each subproblem is defined by a vertex of the tree G (a part of the plane). We suppose that the leader places his facility in the appropriate part of the plane. We collect all customers in the adjacent passages as described earlier. After that, we apply Drezner's algorithm [30]. The optimal solution to the Leader problem is the best solution to the subproblems.

Solution algorithm scheme:

- Step 1. View each part of the plane in sequence. Take the unviewed part of the plane;
- Step 2. For each part of the plane, collect the external customers in the corresponding passages;
- Step 3. Remove barriers and solve Drezner's problem;
- Step 4. Update the best found solution and return to Step 1.

Since Drezner's algorithms find an optimum in polynomial time, the Leader and Follower problems with GLBWP are solvable in polynomial time. Consequently, we have proved the following statement.

Theorem 2. *The Leader and Follower problems with GLBWP and $p = r = 1$ are solvable in polynomial time.*

4 $(r|X_{p-1} + 1)$-Centroid Problem with Line Barriers

First, we consider the trivial case where $p = 1$. In this case, the follower only needs two facilities to clamp the leader's facility from two sides. Therefore, the leader must place his facility in the most profitable customer. The remaining customers are served by the follower.

In the case of placement of two or more facilities, the follower's goal remains the same to maximize its share of the market. The follower should intercept as many customers as possible (with maximum buying power). What does it mean? In the absence of barriers, we can represent the Follower problem as follows [31]. At each customer, draw a circle with the radius equals to the distance to the serving facility of the leader (see Fig. 3). The follower intercepts the customer if and only if he places the facility inside the circle.

In the statement with GLBWP, we rebuild the circles as follows. We use passages to continue circles to the opposite side of barriers. Consider an arbitrary customer near LBWP. Denote the distance from the customer to the service facility by r_1 and the distance from the customer to the passage by r_2. If $r_2 \geq r_1$, then the circle does not cross the passage. Otherwise, on the opposite side of

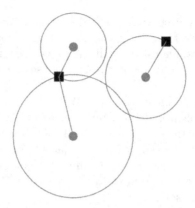

Fig. 3. Scheme of the Follower problem

LBWP, construct a semicircle with the center in the passage and radius equals $r_1 - r_2$ (see Fig. 4). If the semicircle crosses another LBWP, then continue the semicircle through LBWP in the same way. And so on.

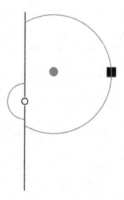

Fig. 4. Scheme of the Follower problem with GLBWP.

In [31], a polynomial algorithm for a fixed number of follower's facilities is proposed to solve the Follower problem without barriers. It is sufficient to consider each circle separately, pairs and triples to determine the intersection regions of the circles for placing facilities in them [31,32]. Such regions are $\phi(m) = O(m^3)$, where m is the number of customers. Then the number of solutions is $C^r_{\phi(m)}$. If r is a constant, it is a polynomial number.

GLBWP replaces each circle with many parts of circles of different radii (see Fig. 4). The number of such parts does not exceed $n + 1$. So for each part of the plane, we have to consider the intersection of the circles (and their extensions across the passage) separately. Consequently, the number of possible locations of facilities is at most $C^r_{\phi(m)}(n + 1)$.

Consider the Leader problem. It requires another facility to be placed. Let $r-1$ facilities have already been placed. Construct the corresponding circles for the customers and serving facilities. If the new leader's facility is located outside the circles, then the Follower problem does not change. Consequently, the leader's revenue also remains unchanged. Placing the facility inside one or more circles changes the service facility for the corresponding customers. Therefore, the radii and intersections of the circles are changed.

The algorithm for solving the problem without barriers is proposed in [31]. It has polynomial running time under fixed r. The algorithm is based on determining the regions of the possible placement of the leader's facility in the circles. The main idea of the algorithm is to change the intersection of circles. Take the set A of k intersecting circles. To remove the intersection, it is necessary to reduce the radii of at least one circle. But each circle from A can intersect with other circles. Consequently, in changing the radius of a circle from A, we must take into account intersections with circles outside A.

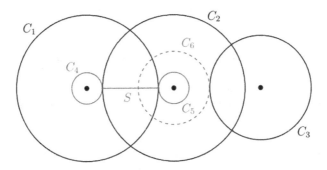

Fig. 5. Scheme of the Follower problem with GLBWP.

Consider the following example (see Fig. 5). There are three customers and three corresponding circles C_1, C_2, C_3. Place the facility in the circles C_4, C_5, or in the segment S to remove the intersection of the circles C_1 and C_2. Placing the facility in C_4 changes the radius of the circle C_1. The segment S allows us to change the radii of both circles C_1 and C_2. C_5 refers to the radius of the circle C_2. But changing the radius of C_2 may eliminate the region of intersection with C_3. This can happen if and only if the facility is placed in the circle C_6. Therefore, we have to consider two cases separately: 1) When we place the facility in $C_4 \cup C_5 \cup S \setminus C_6$; 2) When we place the facility in $C_4 \cup C_5 \cup S \cap C_6$. Paper [31] shows that there are a polynomial number of such regions to place the facility.

The Leader problem with GLBWP has the following specifics. We have to consider not circles but parts of circles in particular parts of the plane. For the Leader problem with GLBWP, we consider each part of the plane separately. To find an optimal placement in a part of the plane we have to take into account the circles (or parts of circles) that are located in the part under consideration as well as the circles (or parts of circles) that intersect with the circles (or parts of

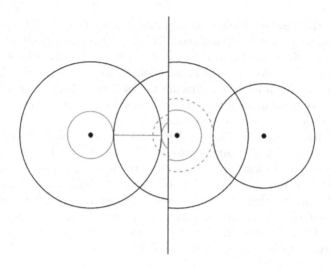

Fig. 6. Scheme of the Follower problem with GLBWP.

circles) of the clients whose parts of circles are in the part of the plane. Consider Fig. 6 with the instance from Fig. 5 supplemented by LBWP. If we are going to place the facility on the left side of the plane, then we should continue the circles C_5 and C_6 through the barrier and take it into account when determining the regions to place the facility. Now we can remove everything outside of the part of the plane, continue the parts of the circles to whole circles and apply the algorithm from [31]. The regions obtained by the algorithm outside LBWP are not interesting. Such regions are considered at another part of the plane outside LBWP.

Solution algorithm scheme:

- Step 1. View each part of the plane in sequence. Take an unviewed part of the plane;
- Step 2. For each part of the plane, take the circles (or parts of circles) that are located in the part of the plane and the circles (or parts of circles) that intersect with the circles (or parts of circles) of the customers whose parts of circles are in the part of the plane. Complete the part of the plane with circles as in Fig. 6. Remove everything outside the part of the plane, continue the parts of the circles to whole circles, and apply the algorithm from [31] to find regions to place the facility.
- Step 3. For each facility location, return the removed barriers and circles, recalculate the radii of the circles taking into account the new facility, and solve the Follower problem;
- Step 4. Update the best found solution and return to Step 1.

For each part of the plane, we are looking for the location of the facility separately. Therefore, the running time of the algorithm [31] increases in $(n + 1)m^2$ times and remains polynomial when r is fixed. Consequently, we have proved the following statement.

Theorem 3. *The Leader and Follower problems with GLBWP are solvable in polynomial time.*

5′ Conclusion

We have examined the influence of line barriers on the optimal leader and follower strategies in the (1|1)-centroid problem [30] and the $(r|X_{p-1} + 1)$-centroid problem [31]. We have considered the formulations that require opening a new leader facility. In the formulations, the barriers divide the plane into polygons in such a way that two different paths from one polygon to another cannot exist. We proposed a polynomial-time algorithm for the Drezner problem with barriers, as well as for the problem studied by Davydov et al.

Facility location problems on the plane with barriers are an interesting area of research from a practical and theoretical point of view. Studies of bilevel formulations, even with the simplest barriers, are almost non-existent in the current literature. Therefore, we plan to continue studying the (r|p)-centroid problem with different types of barriers and metrics. This problem is well suited for studying the effect of barriers on the computational properties of bilevel formulations for the following reasons. For the (r|p)-centroid problem, exact and approximate algorithms have been developed, and its connections with the polynomial hierarchy of complexity classes have been studied.

Acknowledgements. This work has been supported by the grant of the Russian Science Foundation, RSF-ANR 21-41-09017.

References

1. Eiselt, H.A., Marianov, V. (eds.): Foundations of Location Analysis. ISORMS, vol. 155. Springer, New York (2011). https://doi.org/10.1007/978-1-4419-7572-0
2. Ashtiani, M.: Competitive location: a state-of-art review. Int. J. Ind. Eng. Comput. **7**(1), 1–18 (2016)
3. Mallozzi, L., D'Amato, E., Pardalos, P.M. (eds.): Spatial Interaction Models. SOIA, vol. 118. Springer, Cham (2017). https://doi.org/10.1007/978-3-319-52654-6
4. Hotelling, H.: Stability in competition. Econ. J. **39**, 41–57 (1929)
5. Drezner, Z., Suzuki, A., Drezner, T.: Locating multiple facilities in a planar competitive environment. J. Oper. Res. Soc. Jpn. **50**(3), 250–263 (2007)
6. Kress, D., Pesch, E.: Sequential competitive location on networks. Eur. J. Oper. Res. **217**(3), 483–499 (2012)
7. Drezner, T.: A review of competitive facility location in the plane. Logist. Res. **7**(1), 1–12 (2014). https://doi.org/10.1007/s12159-014-0114-z
8. Karakitsiou, A.: Modeling Discrete Competitive Facility Location. SO, Springer, Cham (2015). https://doi.org/10.1007/978-3-319-21341-5

9. Drezner, T., Drezner, Z., Kalczynski, P.: A cover-based competitive location model. J. of Heuristics **62**(1), 100–113 (2010)
10. Biesinger, B., Hu, B., Raidl, G.: A hybrid genetic algorithm with solution archive for the discrete (r|p)-centroid problem. J. of Heuristics **21**(3), 391–431 (2015)
11. Klamroth, K.: Single-Facility Location Problems with Barriers. SSOR, Springer, New York (2002). https://doi.org/10.1007/b98843
12. Katz, I.N., Cooper, L.: Facility location in the presence of forbidden regions, I: Formulation and the case of Euclidean distance with one forbidden circle. Eur. J. Oper. Res. **6**(2), 166–173 (1981)
13. Butt, S.E., Cavalier, T.M.: An efficient algorithm for facility location in the presence of forbidden regions. Eur. J. Oper. Res. **90**(1), 56–70 (1996)
14. McGarvey, R.G., Cavalier, T.M.: A global optimal approach to facility location in the presence of forbidden regions. Comput. Ind. Eng. **45**(1), 1–15 (2003)
15. Klamroth, K.: Planar weber location problems with line barriers. Optim. **49**(5–6), 517–27 (2001)
16. Klamroth, K., Wiecek, M.M.: A bi-objective median location problem with a line barrier. Oper. Res. **50**(4), 670–679 (2002)
17. Sarkar, A., Batta, R., Nagi, R.: Placing a finite size facility with a center objective on a rectangular plane with barriers. Eur. J. Oper. Res. **179**(3), 1160–1176 (2007)
18. Kelachankuttu, H., Batta, R., Nagi, R.: Contour line construction for a new rectangular facility in an existing layout with rectangular departments. Eur. J. Oper. Res. **180**(1), 149–162 (2007)
19. Canbolat, M.S., Wesolowsky, G.O.: The rectiline distance weber problem in the presence of a probabilistic line barrier. Eur. J. Oper. Res. **202**(1), 114–121 (2010)
20. Shiripour, S., Mahdavi, I., Amiri-Aref, M., Mohammadnia-Otaghsara, M., Mahdavi-Amiri, N.: Multi-facility location problems in the presence of a probabilistic line barrier: a mixed integer quadratic programming model. Int. J. Prod. Res. **50**(15), 3988–4008 (2012)
21. Akyüz, M.H.: The capacitated multi-facility weber problem with polyhedral barriers: efficient heuristic methods. Comput. Ind. Eng. **113**, 221–240 (2017)
22. Canbolat, M.S., Wesolowsky, G.O.: On the use of the varignon frame for single facility weber problems in the presence of convex barriers. Eur. J. Oper. Res. **217**(2), 241–247 (2012)
23. Mahmud, T.M.T.: Simulated Annealing Approach in Solving the Minimax Problem with Fixed Line Barrier. PhD thesis, Universiti Teknologi Malaysia (2013)
24. Gharravi, H.G.: Rectiline interdiction problem by locating a line barrier. Master's thesis, Middle East Technical University (2013)
25. Javadian, N., Tavakkoli-Moghaddam, R., Amiri-Aref, M., Shiripour, S.: Two metaheuristics for a multi-period minisum location-relocation problem with line restriction. Int. J. Adv. Manuf. Technol. **71**(5–8), 1033–1048 (2014)
26. Oguz, M., Bektas, T., Bennell, J.A.: Multicommodity flows and benders decomposition for restricted continuous location problems. Eur. J. Oper. Res. **266**(3), 851–863 (218)
27. Amiri-Aref, M., Shiripour, S., Ruiz-Hernández, D.: Exact and approximate heuristics for the rectiline Weber location problem with a line barrier. Comput. Oper. Res. **132**(1), 105293 (2021). https://doi.org/10.1016/j.cor.2021.105293
28. Bischoff, M., Fleischmann, T., Klamroth, K.: The multi-facility location-allocation problem with polyhedral barriers. Comput. Oper. Res. **36**(5), 1376–1392 (2009)
29. Bischoff, M., Klamroth, K.: An efficient solution method for weber problems with barriers based on genetic algorithms. Eur. J. Oper. Res. **177**(1), 22–41 (2007)

30. Drezner, Z.: Competitive location strategies for two facilities. Reg. Sci. Urban Econ. **12**, 485–493 (1982)
31. Davydov, I.A., Kochetov, Y.A., Carrizosa, E.: A local search heuristic for the (r|p)-centroid problem in the plane. Comput. Oper. Res. **52**, 334–340 (2014)
32. Berger, A., Grigoriev, A., Panin, A., Winokurow, A.: Location, pricing and the problem of Apollonius. Optim. Lett. **11**, 1797–1805 (2017)

Cascade Merge of Aircraft Flows as Mixed Integer Linear Programming Problem

Arseniy A. Spiridonov$^{(\boxtimes)}$ ⬥ and Sergey S. Kumkov ⬥

N.N. Krasovskii Institute of Mathematics and Mechanics of the Ural Branch
of the Russian Academy of Sciences (IMM UrB RAS),
16 S.Kovalevskaya Street, Yekaterinburg 620990, Russia
spiridonov.arseniy@gmail.com
https://www.imm.uran.ru/

Abstract. The problem of creating a non-conflict aircraft queue from several incoming flows is considered in the situation of a cascade merge. In this case, groups of flows merge at their joining points, and further the resultant flows merge at some other points. Such a consideration covers, for example, an entire scheme of incoming aircraft flows of an airport zone, as well as taking into account the arriving and departing aircraft. This problem is formalized in the framework of mixed integer linear programming and is solved using the optimization library Gurobi. An approach is proposed to control the permutations of aircraft of the same type within one flow. A series of numerical simulations is carried out. They include both some model examples illustrating the ideas of the algorithm and statistical modeling to estimate the performance of the proposed procedure.

Keywords: Aircraft flows merging · Minimal sate interval · Non-conflict queue · Variation interval · Mixed integer linear programming · Aircraft order · Minimal variation value

1 Introduction

Modern air traffic management involves movement of aircraft along airways, which can divide and join. Motion of an aircraft consists, as a rule, of passage of flight points, including merge points with other aircraft flows. At the join points of the routes, a problem of safe merging of incoming aircraft flows arises. The main criterion for the safety at a merge point is the presence of a sufficient gap between each pair of aircraft passing the merge point. Due to the nature of air traffic, this interval can be maintained by changing the trajectory and/or speed of the aircraft on the route section to the merge point.

The traditional formalization of the problem can be called *single staged*: only instants of arrival to the final merge point are considered (see, for example, [2,4,6–8,10] and references within). Often, such a point is the beginning of a

© The Author(s), under exclusive license to Springer Nature Switzerland AG 2022
Y. Kochetov et al. (Eds.): MOTOR 2022, CCIS 1661, pp. 132–146, 2022.
https://doi.org/10.1007/978-3-031-16224-4_9

glide path or a runway threshold. However, real air-route schemes often have a more complicated structure and not all flows are joined at a single merge point. Usually, the situation can be described as *cascade merge*, that is, several groups of 2–4 flows are joined at some points, and further the new resultant flows are joined at some other points. Also, in very overloaded airports, there can be also a cascade of three or even more consecutive merge points along some or all flows.

Despite of a very realistic formulation, there are a few papers studying it. We have found only work [9] where a very certain situation is considered. It includes two groups of aircraft flows, each joining at a point merge scheme. The two resultant flows join at some third point. With that, at the initial merge points, some strict optimizational methods are applied whereas the merge at the final point is provided by some heuristic methods.

In this paper, we suggest a strict formalization of the cascade merge problem in the framework of mixed integer linear programming (MILP). The suggested model is tested numerically by means of both some illustrative examples, which allow one to check the merging strategy, and statistical tests showing the performance of the solving procedure. The computational library `Gurobi` [3] is used during our computer simulations.

Simulations have been carried out on different computers. One of them is a desktop with a 6-cored Intel(R) Core(TM) i7-8700 CPU @ 3.20 GHz. Another one is the "Uran" supercomputer at the IMM UrB RAS. Each node of this computer is a 18-cored Intel(R) Xeon(R) Gold 6254 @ 3.10 GHz processor. The "Uran" supercomputer has been used only for collecting statistical data. Actually, during simulations, the solver from the `Gurobi` library was constrained by usage of 6 computational threads only, so, any computer with more or less powerful modern CPU can effectively perform computations of this kind.

The paper is organized as follows. Section 2 gives a general formalization of the cascade aircraft flow merge problem as a constrained finite-dimension optimization problem. In particular, Subsects. 2.4, 2.5 and 2.6 contain detailed description of the reduction of the initial problem to a MILP problem. Section 3 shows numerical results of computations of this problem. Another approach to process an aircraft collection is presented in Sect. 4, the results are compared with the results presented in Sect. 3. Finally, Sect. 5 concludes the paper as well as a reference list.

The numerical data used in the simulations (Sects. 3 and 4) are obtained from our colleagues from NITA, LLC (New Information Technologies in Aviation, Saint-Petersburg, Russia).

2   Problem Formalization

Usually, the problem of aircraft flows merging is studied as one-stage: airway routes of the flows merge at one merge point. However, in most cases, if we consider aircraft flows merging in the terminal area, the situation is different.

In Fig. 1, a scheme of the incoming flows of Koltsovo airport is presented. One can see that some primary flows are joined together at the points AKERA

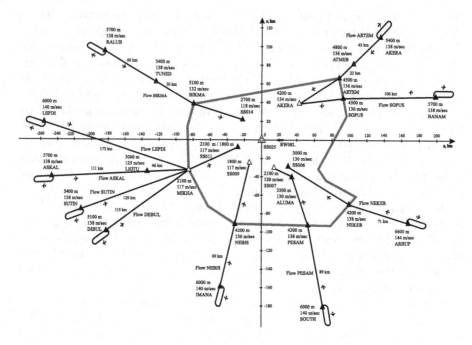

Fig. 1. A general scheme of the Koltsovo airport zone (Yekaterinburg, Russia)

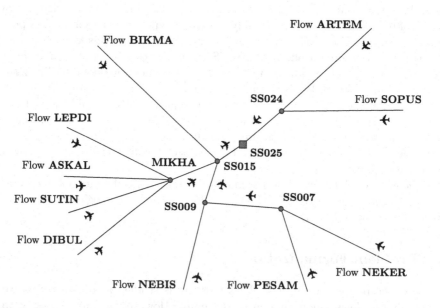

Fig. 2. The contracted scheme of Koltsovo airport. The merge point **SS025** (the grey-filled square) is the final merge point. Other points (the grey-filled circles) are intermediate merge points

and MIKHA. Also, two south-eastern flows join at the point SS007 and further with the southern one at the point SS009 (these two points are not shown in the scheme). Further, the north-western flow, south-western and south resultant ones join at a merge point scheme at some point SS016 (which is not shown in the general scheme). Finally, that resultant flow and the north-eastern resultant flow from the point AKERA join at the point SS025, which is the beginning of the glide path to the runway RW08L. Such a merge scheme reasonably can be named as *cascade*. In this scheme, not only the arrival instants at the final merge point SS025 should be planned, but also at other joining points AKERA, MIKHA, and SS016. With that, evidently, arrivals at the points of preliminary merge affect arrivals at further points along the flow.

For the problem of aircraft cascade merge, in contrast to the problem of one-stage (or single staged) merge, the topology of the airway scheme is important. However, one needs not the entire scheme, but only the merge points and legs between them. Such a scheme can be called *contracted*. For example, Fig. 2 shows the contracted scheme of Koltsovo airport. At the same time, information about the configuration of additional scheme elements, holding areas, path shortening and stretching elements, etc., is not essential in this problem. This information is taken into consideration by the minimal and maximal passage times of each leg.

The task of assigning new arrival instants is solved at each merge point. They are the solution of the problem and can be used by air-traffic managers (or by some automated systems) to control the motion of the aircraft along their routes.

In the framework of this paper, we make an important assumption that a contracted scheme is a tree (in the sense of the graph theory). If the contracted scheme is a directed acyclic graph or has loops then its MILP formalization seems to be much more complicated than for the case of a tree.

2.1   Input Data

Safety Information. We consider three types of aircraft: light, medium, and heavy denoted below by L, M, and H, respectively. The symbol σ_i means the type of the ith aircraft. Aircraft of different types have different characteristics.

The main demand to the resultant collection is presence of the minimal safe interval $\tau^{\text{safe}}_{\sigma_i,\sigma_j}$ between the ith and jth aircraft at each merge point P_k they pass both: if $t_i^{P_k} < t_j^{P_k}$, then $t_i^{P_k} + \tau^{\text{safe}}_{\sigma_i,\sigma_j} \leq t_j^{P_k}$. The values of the safe interval for every pair of the aircraft types are given in Table 1. The suggested formalization allows one to set own safety intervals for every merge point, however, we consider them the same for all points.

Topological Information. The contracted scheme is described by the following objects:

- the list $\{P_k\}$ of all points constituting the scheme; it includes both the initial points of all flows and the merge points;
- information about legs: what point follows after each point from the list, except the final point of the scheme;

Table 1. The minimal safe interval values $\tau^{\text{safe}}_{\sigma_i, \sigma_j}$ between aircraft of different types, seconds

The next aircraft type, σ_j	Light	Medium	Heavy
The previous aircraft type, σ_i			
Light	60	60	60
Medium	180	60	60
Heavy	180	120	120

– for each leg $P_{k'} P_{k''}$, the minimal, maximal, and nominal passage times $\underline{\theta}_{P_{k'} P_{k''}}$, $\overline{\theta}_{P_{k'} P_{k''}}$, and $\theta^{\text{nom}}_{P_{k'} P_{k''}}$. In the suggested formalization, it is assumed that these times are the same for aircraft of all types, however, it is not crucial for the formalization. Also, in the performed tests, it is assumed that for every leg $P_{k'} P_{k''}$,

$$\theta^{\text{nom}}_{P_{k'} P_{k''}} = 0.73 \cdot \underline{\theta}_{P_{k'} P_{k''}} + 0.27 \cdot \overline{\theta}_{P_{k'} P_{k''}}.$$

Aircraft Flows Description. Each aircraft flow is described by the initial point of the flow. All other points, which belong to it, and their order are taken from the topology of the contracted scheme.

Aircraft Description. For each aircraft, its type and its flow number are given. Also, the nominal instant $t^{\text{nom}}_{\text{enter},i}$ of its enter to the airport zone is defined. This instant together with the nominal passage times for all legs of the flow defines the nominal instant t^{nom}_i of arrival of the aircraft to the final point of the scheme. This instant is used in the optimality criterion (see below).

2.2   Optimality Criteria

The main feature of the resultant queue formed at the final point is its safety. However, there can be other demands to it. Usually, such demands are formalized as minimization of deviations of the assigned instants of aircraft arrivals to the final point from the nominal ones. This minimization implies minimization of aircraft fuel expenditures, total delay time, breaks with connected flights, collisions with airport services, etc.

A usual way to formalize these demands for one aircraft is minimization of the absolute value of the difference between the assigned and nominal arrival instants. The function can be asymmetric, that is, punishes differently accelerations and delays. The penalty for the entire set of aircraft, usually, is considered as a sum of individual penalties for all aircraft from the set.

In the performed tests, we consider two variants of the individual penalty functions:

$$f^{\text{s}}(t_i, t_i^{\text{nom}}, \sigma_i) = \beta_{\sigma_i} \cdot |t_i - t_i^{\text{nom}}|, \tag{1}$$

$$f^{\text{as}}(t_i, t_i^{\text{nom}}, \sigma_i) = \beta_{\sigma_i} \cdot \begin{cases} -K^- \cdot (t_i - t_i^{\text{nom}}), & \text{if } t_i < t_i^{\text{nom}}, \\ K^+ \cdot (t_i - t_i^{\text{nom}}), & \text{if } t_i \geq t_i^{\text{nom}}. \end{cases} \tag{2}$$

Here, t_i^{nom} is the nominal arrival instant of the ith aircraft to the final point of the scheme, t_i is the assigned arrival instant to the final point, β_{σ_i} is a weight coefficient of the variation cost depending on of the aircraft type. For function (2), K^- and K^+ are non-negative coefficients defining the penalty weight for early and late arrivals. Hereinafter, function (1) is referred as "symmetric", and function (2) is called "asymmetric".

In the numerical experiments, the coefficients are chosen as follows: $\beta_{\text{L}} = 1.0$, $\beta_{\text{M}} = 3.0$, $\beta_{\text{H}} = 5.0$, $K^- = 2$, $K^+ = 1$.

The general indicator of the optimality of the assigned arrival instants of the aircraft of the entire set is expressed as a sum of the penalties of the aircraft:

$$F(\{t_i\}, \{t_i^{\text{nom}}, \sigma_i\}) = \sum_{i=1}^{N} f(t_i, t_i^{\text{nom}}, \sigma_i) \to \min \tag{3}$$

Thus, the penalty functional depends on the instants of arrivals to the final point. Arrival instants at other merge points define only constraints of the optimization problem.

2.3   Aircraft Order Change Assumption

In this paper, an essential assumption is made about the aircraft order. Namely, that aircraft within a primary flow can change the mutual order only before the first merge point of this flow. It is also assumed that aircraft from two flows can change their order only at the first common merge point of these two flows, and at next common merge points, their order does not change.

This assumption is reasonable since the main tool for changing the aircraft order is holding areas when one aircraft leaves its route and lets another aircraft to go ahead. At the same time, holding areas, as a rule, are located far enough from the area where the routes join, before the first points of the flows merging.

The assumption allows us to limit the number of introduced integer variables, which determine the aircraft order. This is important for ensuring acceptable computation time since an increase in the number of integer variables significantly increases the computation time. Under this assumption, only two integer variables are introduced for each pair of aircraft, which determine their mutual order at all points of the contracted scheme common to the flows of these aircraft.

2.4 Arrival Instant Variables and Constraints

Let the kth flow pass through the points $P_{k,1}, P_{k,2}, \ldots, P_{k,N_k}$. Then, the variables $t_i^{P_{k,l}}$, $l = 1, \ldots, N_k$, are introduced for the ith aircraft belonging to the kth flow. The variable $t_i^{P_{k,l}}$ stands for the ith aircraft to the lth point of kth flow.

The variables $t_i^{P_{k,l-1}}$ and $t_i^{P_{k,l}}$ of passing two consecutive points $P_{k,l-1}$ and $P_{k,l}$ by the ith aircraft are subject to the constraints

$$t_i^{P_{k,l-1}} + \underline{\theta}_{P_{k,l-1},P_{k,l}} \leq t_i^{P_{k,l}} \leq t_i^{P_{k,l-1}} + \overline{\theta}_{P_{k,l-1},P_{k,l}}, \tag{4}$$

where $\underline{\theta}_{P_{j_i,k-1},P_{j_i,k}}$ and $\overline{\theta}_{P_{j_i,k-1},P_{j_i,k}}$ are the minimal and maximal times of passage of the leg between the points $P_{k,l-1}$ and $P_{k,l}$.

2.5 Taking into Account Aircraft Order and Safety Constraints

To take into account the order and safety intervals between aircraft at each merge point, an approach is used, which is quite commonly used now [1].

For each pair of ith and jth aircraft, two binary variables are introduced

$$\forall 1 \leqslant i < j \leqslant N \; : \; \delta_{i,j} = \begin{cases} 1, & \text{if } t_i^{P_k} < t_j^{P_k}, \\ 0, & \text{if } t_i^{P_k} > t_j^{P_k}, \end{cases} \quad \delta_{i,j} + \delta_{j,i} = 1.$$

These variables describe the order between the ith and jth aircraft at points P_k, which are passed by both these aircraft. A *binary variable* is an integer variable that can equal 0 or 1 only.

To ensure the aircraft order and safety intervals at each point P_k of the contracted scheme, for each pair of aircraft passing through this point, a pair of inequalities is defined

$$\begin{aligned} t_i^{P_k} - t_j^{P_k} + \tau_{i,j}^{\text{safe}} \cdot \delta_{i,j} - M \cdot \delta_{j,i} \leq 0, \\ t_j^{P_k} - t_i^{P_k} + \tau_{j,i}^{\text{safe}} \cdot \delta_{j,i} - M \cdot \delta_{i,j} \leq 0. \end{aligned} \tag{5}$$

Here, M is an "infinity", that is, some quite large positive value, $\tau_{i,j}^{\text{safe}}$ is the minimal safe time interval between the ith and jth aircraft according to their types.

The sense of these inequalities is the following. If $\delta_{i,j} = 1$ (and, consequently, $\delta_{j,i} = 0$), then the first inequality transforms into $t_i^{P_k} - t_j^{P_k} + \tau_{i,j}^{\text{safe}} \leq 0$, that is, $t_i^{P_k} + \tau_{i,j}^{\text{safe}} \leq t_j^{P_k}$, which just ensures that the ith aircraft passes the point P_k before the jth one and there is the safety interval $\tau_{i,j}^{\text{safe}}$ between these aircraft.

On the contrary, if $\delta_{i,j} = 0$ (and $\delta_{j,i} = 1$), then the first inequality turns into $t_i^{P_k} - t_j^{P_k} - M \leq 0$. This inequality is identically true for any reasonable values of $t_i^{P_k}$ and $t_j^{P_k}$ because of large positive value of M.

The second inequality in (5) behaves in the same way, except the change of the order of the indices i and j.

To reduce the possible number of binary variables $\delta_{i,j}$, several additional heuristics are used:

- restriction on the range of an aircraft repositioning relative to its nominal position in the initial queue; an aircraft place in the resultant queue cannot be further than 4 from its place in the initial queue of the nominal arrivals;
- prohibition of rearrangement of aircraft of the same type within each flow;
- prohibition of permutation attempts of aircraft whose variation intervals do not intersect, at least, at one their common point. The possible variation interval for the ith aircraft at some point is computed on the basis of the nominal entrance time $t_{\text{enter},i}^{\text{nom}}$ of the aircraft at the first point of its flow and possible times $\underline{\theta}_{P_{k'}P_{k''}}$, $\overline{\theta}_{P_{k'}P_{k''}}$ of passage of legs between the initial and current points of the flow.

These heuristics are implemented by imposing the constraint $\delta_{i,j} = 1$ for the pair of the ith and jth aircraft if the corresponding conditions are met and the ith aircraft should follow earlier than the jth one. Also, when the order of two aircraft is defined according to some heuristics, one can generate explicitly a "short" inequality of the type $t_i^{P_k} + \tau_{i,j}^{\text{safe}} \leq t_j^{P_k}$ instead of a pair of "long" inequalities (5) and the constraint $\delta_{i,j} = 1$.

2.6   Optimization Problem

Finally, we consider an optimization problem with the optimality criterion (3). The individual penalty function f is taken as symmetric (1) or asymmetric (2). The constraints for the problem are (4), (5).

Constraints (4) are generated for all aircraft and for all points, through which each certain aircraft passes.

Constraints (5) are generated for all pairs of aircraft (since any aircraft passes through the final point of the scheme) and for all points common for each certain pair of aircraft.

Since constraints (5) contain integer variables, the entire problem is solved by methods of mixed integer linear programming. Numerical solvers are taken from the library Gurobi [3].

Formally, presence of the absolute value function in the optimality criterion makes the problem non-linear. However, there are well-known approaches to reduce such a problem to a linear one (see, for example, [5]).

3   Numerical Results

In this section, some numerical results of solving problem (3), (4), (5) are presented. In Subsect. 3.1, a small contracted scheme is considered with some illustrative input data (including some small sets of aircraft). In Subsect. 3.2, results of statistical tests are given.

The computational program has been launched at one node of the supercomputer or a desktop computer with a multi-cored CPU, however, the internal structure of the Gurobi library assumes parallel concurrent run of several solving methods.

3.1   Model Examples

For two examples in this subsection, we consider a small contracted scheme shown in Fig. 3. There are 3 incoming aircraft flows. The first and second ones are joined at the point **MP**. The resultant flow and the third one are joined at the point **FP**.

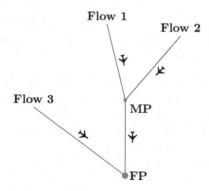

Fig. 3. The scheme example of the cascade merge problem. The grey-filled circles mean the merge points. **FP** is the final merge point

Example 1. Consider an example of 5 aircraft with the following input data:

$t^{\text{nom}}_{\text{enter},i}$	500	620	680	740	860
t^{nom}_i	2000	2120	2180	2240	2360
Type	H	H	L	H	H
Flow	1	2	3	1	2

The first and second lines of the table contain the nominal enter instants and nominal arrival instants at the final point **FP** of the aircraft (in seconds), correspondingly. The last two lines include the types and flow numbers of the aircraft.

There is a conflict situation between the second aircraft of the heavy type and the third aircraft of the light type. The minimal safe interval between them must be $\tau^{\text{safe}}_{2,3} = 180\,\text{s}$, however, the actual interval is $60\,\text{s}$ only.

The legs of the scheme have the following minimal and maximal passage times shown in seconds: [**Flow 1, MP**] : [500, 1000], [**Flow 2, MP**] : [500, 1000], [**MP, FP**] : [500, 1000], [**Flow 3, FP**] : [1000, 2000]. For better understanding, the passage times of the legs are chosen in such a way that the aircraft of all flows have equal possibilities for varying the arrival instants at the final point.

In the case of symmetric penalty function (1), the light aircraft is placed at the head of the queue, maintaining the minimum safety interval $\tau^{\text{safe}}_{\text{L,H}} = 60\,\text{s}$

before the first heavy vessel. The aircraft received the following arrival instants at the final merge point **FP**: $t_1 = 2000.00\,\mathrm{s}$, $t_2 = 2120.00\,\mathrm{s}$, $t_3 = 1940.00\,\mathrm{s}$, $t_4 = 2240.00\,\mathrm{s}$, $t_5 = 2360.00\,\mathrm{s}$.

In the case of asymmetric penalty function (2), the light aircraft is placed at the end of the queue with the minimum safe time interval $\tau_{\mathrm{H,L}}^{\mathrm{safe}} = 180\,\mathrm{s}$ before it. Since acceleration is penalized twice as much as delay, speeding up by 240 s and moving to the front of the queue is more expensive than delaying the light aircraft by 360 s and moving it to the back of the queue. The aircraft received the following arrival instants at the final merge point **FP**: $t_1 = 2000.00\,\mathrm{s}$, $t_2 = 2120.00\,\mathrm{s}$, $t_3 = 2540.00\,\mathrm{s}$, $t_4 = 2240.00\,\mathrm{s}$, $t_5 = 2360.00\,\mathrm{s}$.

Example 2. Now let us consider the same set of aircraft as in Example 1, but change the capability of time variation along the legs. New passage times are [**Flow 1, MP**] : [500, 1000], [**Flow 2, MP**] : [500, 1000]; [**MP, FP**] : [500, 1000], [**Flow 3, FP**] : [1500, 1500]. That is, the aircraft from the flow 3 (in particular, the light aircraft with the nominal position 3) have no the ability to vary their arrival instants at all. Thus, the conflict between the second and third aircraft is resolved using a variation of the heavy aircraft with the nominal positions 1 and 2.

For the cases of symmetric (1) and asymmetric (2) penalty functions, the solutions are the same. As a result of the algorithm, the aircraft received the following arrival instants: $t_1 = 1880.00\,\mathrm{s}$, $t_2 = 2000.00\,\mathrm{s}$, $t_3 = 2180.00\,\mathrm{s}$, $t_4 = 2240.00\,\mathrm{s}$, $t_5 = 2360.00\,\mathrm{s}$.

3.2   Statistical Tests

The developed algorithm is tested, in particular, for performance. Testing is carried out for the Koltsovo airport area (see Figs. 1 and 2). The contracted scheme has the following times in seconds of the fastest and slowest passages of the legs:

[Flow **BIKMA, SS015**] : [1483, 3534];   [Flow **PESAM, SS007**] : [990, 2950];
[Flow **ARTEM, SS024**] : [814, 2743];   [Flow **NEKER, SS007**] : [1140, 3122];
[Flow **SOPUS, SS024**] : [1069, 3039];   [**SS015, SS025**] : [96, 120];
[Flow **LEPDI, MIKHA**] : [1341, 3367];   [**SS024, SS025**] : [664, 1340];
[Flow **ASKAL, MIKHA**] : [1217, 3223];   [**MIKHA, SS015**] : [646, 1152];
[Flow **SUTIN, MIKHA**] : [1021, 2995];   [**SS009, SS015**] : [182, 782];
[Flow **DIBUL, MIKHA**] : [186, 3075];   [**SS007, SS009**] : [191, 227].
[Flow **NEBIS, SS009**] : [926, 2879];

For each aircraft flow of the contracted scheme, a value Δ is set. The value guarantees that the next aircraft in this flow will appear not earlier than after this time period. In other words, by setting the value Δ, one can regulate the intensity of the possible appearance of aircraft in each flow.

Traditionally, the western, northwestern, and southwestern directions at Koltsovo airport are the most intensive. The traffic from the south is slightly less. The smallest traffic is on the northern and eastern directions. The following values in seconds are used in the modelling: $\Delta_{\text{BIKMA}} = 210$, $\Delta_{\text{ARTEM}} = 990$, $\Delta_{\text{SOPUS}} = 900$, $\Delta_{\text{LEPDI}} = 210$, $\Delta_{\text{ASKAL}} = 330$, $\Delta_{\text{SUTIN}} = 300$, $\Delta_{\text{DIBUL}} = 450$, $\Delta_{\text{NEBIS}} = 600$, $\Delta_{\text{PESAM}} = 720$, $\Delta_{\text{NEKER}} = 600$.

Sets of 10, 15, 20, ..., 35 aircraft have been tested both with the uniform distribution of the aircraft types in the set and with some typical non-uniform distribution (3% light, 86% medium, 11% heavy aircraft).

Aircraft in a tested collection are randomly generated in the flows with the corresponding value Δ provided in such a way that the average density of the entire set varies from 10 aircraft per hour to 50 aircraft per hour in increments of 5 aircraft per hour.

The computations have been performed with a limitation on the processing time of the next set of aircraft of 30 s. If upon reaching the threshold in terms of computation time of the set, the optimal solution is not found, then this set is not considered as successfully processed.

Optimality of the assigned instants of aircraft arrival at the final merge point is estimated using symmetric (1) and asymmetric (2) criteria.

For each parameter combination, 1000 runs have been performed to collect statistical data (see Table 2).

Table 2. The computation results for combinations of the uniform and some non-uniform distributions of the types, the different sizes of aircraft ensembles, the symmetric and asymmetric penalty functions in the case of processing an entire ensemble

Distribution of types	Aircraft quantity	Penalty function	Successful launches	Average process time, s	Maximal process time, s
Uniform	15	symm	1000 (100%)	0.41	19.48
		asymm	1000 (100%)	0.4	7.76
	20	symm	896 (89.6%)	4.05	29.59
		asymm	901 (90.1%)	3.66	29.84
	25	symm	448 (44.8%)	8.6	29.59
		asymm	446 (44.6%)	7.94	29.89
Non-uniform	15	symm	1000 (100%)	0.18	18.41
		asymm	1000 (100%)	0.19	6.67
	20	symm	984 (98.4%)	1.04	28.39
		asymm	988 (98.8%)	1.01	26.64
	25	symm	890 (89.0%)	3.61	29.88
		asymm	847 (84.7%)	3.54	30.0

In general, the results of the algorithm can be assessed as good. The cases of 25 aircraft with the non-uniform type distribution (which are the closest to the real situation) are processed faster than 30 s in more than 80% of launches.

For parameter variants for which the number of successful launches is less than 100%, the value of the maximum process time is not too significant since there are algorithm launches with a process time greater than the set limit.

Unsuccessful launches are not taken into account in the calculation of the average and maximum process times.

In some cases, within the set time limit of 30 s, the algorithm has already found a solution close enough to the optimal one, but only refined it, exceeding the time limit. Probably, such candidate solutions can also be considered as "almost good". The received sets of arrival instants can be used by air-traffic controllers. It is only necessary to set in advance the value of the *gap* parameter calculated by typical MILP-solvers in the process of work (the relative difference between the values of the main criterion at the best found feasible solution and at the solution of the current relaxed linear programming problem), which would be an additional criterion for evaluating the solution when the time limit is reached.

The results of the algorithm for the sets of 25 aircraft with the uniform distribution of the aircraft types in the tested collections can be assessed as satisfactory only since the number of successful launches does not exceed 50%.

The shifts in the maximum process time for the symmetric penalty function on sets of 15 ships compared to the results for the asymmetric function can be explained as a result of the random number generator. Since the numbers of aircraft flows are assigned randomly, it is possible that in some runs a sufficient number of aircraft are generated (relative to the total number of aircraft in the set) in flows with a high input density (these are the flows **BIKMA, LEPDI, SUTIN, ASKAL**).

4   Sequential Solution of MILP Problem

The principle presented above for creating an optimal schedule of the aircraft arrival instants is the total processing of the entire set of aircraft. With that, the computation time naturally increases with an increase in the size of the processed ensemble. In this case, unacceptable 20–30 s of operation are achieved even with an ensemble size of 25 vessels in the case of the uniform distribution of the aircraft types (for example, see the results in Sect. 3, Table 2).

However, such a work mode is not typical for a real life when the procedure is started periodically. Between the runs, the situation cannot change significantly: 1–2 aircraft can land, 1–2 aircraft can enter the airport zone. Therefore, results from the previous run can be used as the initial approximation during the next one. It is reasonable to call this mode of work as *sequential*.

When one or more aircraft leave the ensemble in comparison with the previous run, then for all aircraft remaining in the ensemble, there are previously calculated optimal arrival instants. If some new aircraft appear, this information is available for all aircraft, except for the new ones.

To conduct such a testing, statistical experiments have been carried out. Each launch is performed according to the following algorithm:

1. **Generation stage**. At this stage, all aircraft are generated for further modelling. The whole set is sorted according to the aircraft nominal instants of entry into their flows. The nominal arrival instant is calculated from the generated nominal enter instant of an aircraft and the nominal travel time along the route.

2. **The stage of selection and calculation of the initial ensemble.** Some specified number (usually, 10) of aircraft from the generated set with the earliest entry times is added into the ensemble to be processed. This entire ensemble is calculated as a whole. To perform the calculation, a description of the corresponding MILP-problem is created.

3. **Adding remaining aircraft to the problem.** When the initial ensemble is computed, in a loop one by one, the rest aircraft are added to the group. With that, all variables and constraints connected with each next aircraft are added to the description of the problem. For those variables, which were in the model during the previous size of the ensemble, their found optimal values are set up as the initial ones for the next run.

   New variables associated with the added aircraft do not get their initial values. The library `Gurobi` allows this. In the process of finding a feasible solution, the values of these variables will be determined.

   Then the solving the new problem for the new aircraft ensemble is launched. In the case of an unsuccessful solution (out of the time limits), the processing of this generated set stops and it does not contribute any data to the statistical measurements. Otherwise, statistical data are accumulated, and the next aircraft is added to the ensemble.

4.1   Numerical Results

The results of the performed statistical tests are given in Table 3.

Table 3. The computation results for combinations of the uniform and some non-uniform distributions of the types, the different sizes of aircraft ensembles, the symmetric and asymmetric penalty functions in the case of sequential processing an ensemble

Distribution of types	Aircraft quantity	Penalty function	Successful launches	Average process time, s	Maximal process time, s
Uniform	15	symm	995 (99.5%)	0.31	4.36
		asymm	994 (99.4%)	0.36	3.48
	20	symm	864 (86.4%)	0.92	4.84
		asymm	855 (85.5%)	1.23	4.96
	25	symm	509 (50.9%)	1.80	4.88
		asymm	456 (45.6%)	1.87	4.99
Non-uniform	15	symm	1000 (100%)	0.16	1.97
		asymm	1000 (100%)	0.21	1.40
	20	symm	983 (98.3%)	0.43	4.68
		asymm	979 (97.9%)	0.66	4.37
	25	symm	928 (92.8%)	0.90	5.01
		asymm	877 (87.7%)	1.25	4.97

One can see that, on one hand, the average and the maximal computation times are significantly less than the ones for the case when the ensemble is processed entirely at once (see Table 2).

On the other hand, the percentage of successful launches is approximately the same. However, now this percent values concern not only the ensembles of some certain size, but all ensembles of this size and smaller ones. This is due to the applied procedure: if for some size of the ensemble, the computations are not finished during 30 s, then larger ensembles are not tested and this run is terminated.

Efficiency of the procedure is more or less justified by small values of the maximal computation time among all launches for every certain size of the ensemble.

5 Conclusion

The problem of cascade merging of aircraft flows is studied. The main peculiarity is that the uniqueness of the point where the safety passage should be provided is no longer assumed. Instead, there are two main assumptions of the formulation considered.

The first one is that the scheme of airways is a tree (in the sense of the graph theory). In other words, groups of flows merge at some points, after which the resulting merged flows, in their turn, merge at the next points of their paths. This setting allows to consider real airport schemes, as well as the situation with the flows of arriving and departing aircraft.

The second assumption is that inside one flow the aircraft can change their order only before the first merge point of this flow is passed. This allows one to reduce significantly the number of binary variables describing the order of the aircraft in the resultant queue. Decreasing the number of variables of this type allows the computation times to be acceptable.

The setting is formalized in terms of mixed integer linear programming. This approach makes possible to combine and simultaneously solve the discrete subproblem of finding optimal aircraft orders in merged flows and the continuous subproblem of assigning optimal arrival instants. The optimization library Gurobi is used to solve the obtained MILP problems.

The paper presents two methods for processing a set of aircraft. The first is the processing of the entire collection at once. The second method is sequential addition of a new aircraft to the processed set from the original one. The solution in the previous step is used as an initial approximation.

Numerical results are obtained for model examples. Statistical tests are also carried out for the Koltsovo airport scheme for various sets of parameters: the number of aircraft, the distribution of the aircraft types, various quality criteria. As expected, the sequential addition of aircraft and the use of the initial approximation give a significant increase in the speed of the algorithm. With that, the computational time of the program running in this or that regime is defined mostly by the number of aircraft to be processed, not by the airport itself. The situation with 25 aircraft studied among other variants corresponds to a significant load of an airport.

References

1. Beasley, J.E., Krishnamoorthy, M., Sharaiha, Y.M., Abramson, D.: Scheduling aircraft landings – the static case. Transp. Sci. **34**(2), 180–197 (2000). https://doi.org/10.1287/trsc.34.2.180.12302

2. Bennell, J.A., Mesgarpour, M., Potts, C.N.: Airport runway scheduling. 4OR – Q. J. Oper. Res. **9**(2), 115–138 (2011). https://doi.org/10.1007/s10288-011-0172-x

3. Gurobi Optimization, LLC: Gurobi Optimizer Reference Manual (2022). https://www.gurobi.com

4. Ikli, S., Mancel, C., Mongeau, M., Olive, X., Rachelson, E.: The aircraft runway scheduling problem: a survey. Comput. Oper. Res. **132**, 105336 (2021). https://doi.org/10.1016/j.cor.2021.105336

5. Soomer, M., Franx, G.: Scheduling aircraft landings using airlines' preferences. Eur. J. Oper. Res. **190**(1), 277–291 (2008). https://doi.org/10.1016/j.ejor.2007.06.017

6. Veresnikov, G.S., Egorov, N.A., Kulida, E.L., Lebedev, V.G.: Methods for solving of the aircraft landing problem. I. Exact solution methods. Autom. Remote. Control **80**(7), 1317–1334 (2019). https://doi.org/10.1134/S0005117919070099

7. Veresnikov, G.S., Egorov, N.A., Kulida, E.L., Lebedev, V.G.: Methods for solving of the aircraft landing problem. II. Approximate solution methods. Autom. Remote. Control **80**(8), 1502–1518 (2019). https://doi.org/10.1134/S0005117919080101

8. Vié, M.S., Zufferey, N., Leus, R.: Aircraft landing planning: past, present and future. In: Proceedings of the 19th Annual Congress of the French Operations Research Society (2018). https://archive-ouverte.unige.ch/unige:104854

9. You, L., Zhang, J., Yang, C., Hu, R.: Heuristic algorithm for aircraft arrival scheduling based on a point merge system. In: Proceedings of the 20th COTA International Conference of Transportation Professionals, Xi'an, China (2020). https://doi.org/10.1061/9780784483053.011

10. Zulkifli, A., Aziz, N.A.A., Aziz, N.H.A., Ibrahim, Z., Mokhtar, N.: Review on computational techniques in solving aircraft landing problem. In: Proceedings of the 2018 International Conference on Artificial Life and Robotics, Oita, Japan, pp. 128–131 (2018). https://doi.org/10.5954/ICAROB.2018.GS5-3

The Gap Between Cooperative and Non-cooperative Solutions for the Public-Private Partnership Model

Alexander Zyryanov, Yury Kochetov$^{(\boxtimes)}$ ⓘ, and Sergey Lavlinskii

Sobolev Institute of Mathematics, Novosibirsk, Russia
`jkochet@math.msc.ru`

Abstract. We study a new variant of the bi-level linear integer programming model for the strategic planning of the public-private partnership. In this model, we need to select and schedule some ecological, infrastructure, and production projects within a finite planning horizon. Two players, the leader (government) and the follower (private companies) make their own decisions sequentially under the weak budget constraints. We assume that an unused budget can be used later. The players try to maximize their own total profits. We show that this new bi-level problem is Σ_2^p-hard in cooperative and non-cooperative cases. Moreover, we claim that the absolute and relative gaps between the optimal cooperative and non-cooperative solutions can be arbitrarily large. Computational results for real-world instances for the Transbaikalian polymetallic fields and the gap between the cooperative and non-cooperative solutions are discussed.

Keywords: Bilevel programming · Metaheuristic · Computational complexity · Stackelberg game

1 Introduction

The bi-level programming problems already have a long history, hundreds of publications, several monographs, and specialized sessions at the international conferences. Nevertheless, a range of applications, especially for discrete settings, is constantly expanding, and interest in them is only growing. This is due to both the Stackelberg games for unequal right players that arise when modeling market competition, and algorithmic and computational complexity of resulting optimization problems. Some of them turn out to be Σ_2^p-hard, that is more difficult than well-known NP-complete problems unless P = NP.

In this paper, we study a new bi-level 0–1 linear programming problem that arises for modeling the public-private partnership (3P). The state (government) acts as the first player at the upper decision-making level (leader). Later on, the private companies (follower) make their own decisions at the lower level. Players try to maximize their profits within given budget constraints for industrial, ecological and infrastructure projects. In contrast to the previous research [9–11], we assume now that an unused budget in the

© The Author(s), under exclusive license to Springer Nature Switzerland AG 2022
Y. Kochetov et al. (Eds.): MOTOR 2022, CCIS 1661, pp. 147–160, 2022.
https://doi.org/10.1007/978-3-031-16224-4_10

current year can be used later. The first 3P models were introduced in [9–11], where the players decide to open or not the infrastructure, environmental and industrial projects, but do not choose when to do that. It is shown that the leader's problem is NPO-hard, and the follower's problem is NPO-complete. A local improvement method with a heuristic for starting solution is proposed. An additional instrument of state assistance to investor (tax benefits) is introduced in the model, which makes it a non-linear one. It is also shown that the leader's problem with tax benefits is Σ_2^p-hard. The paper [1, 13] generalized the model from [9–11] and allowed the players to choose the starting year for each project and designed a Tabu Search method for the bi-level problem without tax benefits.

In this paper, we consider the case without tax benefits but allow the players to choose the time for starting each project under the soft budget constraints. As in the previous cases, the resulting problem is ill-posed. In other words, for the given leader solution, the follower can have several optimal solutions. As a result, the leader can get different values of the objective function. Thus, we have to consider so-called cooperative and non-cooperative cases corresponding to "friendship" or "enmity" of the upper and lower levels. This approach makes it possible to obtain an interval for an optimal value of the leader's objective function under the follower's unpredictable behavior. We show that the new problem is Σ_2^p-hard in cooperative and non-cooperative cases [3, 4, 7, 10] and the absolute and relative gaps between the optimal cooperative and non-cooperative solutions can be arbitrarily large. Nevertheless, the computational experiments for semi-synthetic instances for the Transbaikalian polymetallic fields show that a positive gap is exotic but the maximal value is high enough.

The paper is organized as follows. Section 2 presents the problem formulation and exact bi-level 0–1 mathematical model. Section 3 describes the main properties of the problem and its computational complexity. In this section, we show that the absolute and relative gaps between the optimal cooperative and non-cooperative solutions can be arbitrarily large. Section 4 is devoted to the upper and lower bounds for the optimal solution of the problem. In Sect. 5, we present our computational results and Sect. 6 concludes the paper.

2   Problem Formulation

Let us consider a 3P model and its formulation as a bi-level 0–1 programming problem with flexible starting time for each project and participation of the private investor in the process of program developing of an industrial and raw material base utilization.

Model Input

- The set of industrial projects implemented by a private investor, starting moments and specific configuration of which the investor chooses to depend on what the state offers in the field of infrastructure projects;
- The set of infrastructure projects implemented by state, a specific list of which and starting moments the state chooses based on its assessments of efficiency in terms of prospects for long-term development of territory;

- The set of ecological projects required compensating for the ecological losses caused by the implementation of industrial projects; specific section of obligations for implementation of ecological projects between private investor and state is not defined in input and must be obtained at output of the model.

Model Output
- A schedule of projects starts of all types and the mechanism for sharing costs of implementing environmental projects between the state and the investor.

Let us introduce the following notation. Let NP, NE, NI be the number of industrial, ecological and infrastructure projects, T be the planning horizon, and indexes $i = 1, \ldots, NP, j = 1, \ldots, NI, k = 1, \ldots, NE, t = 1, \ldots, T, \tau = 1, \ldots, T$.

Boolean Variables

$z_{i\tau} = 1$, if investor starts industrial project i in year τ, and $z_{i\tau} = 0$ otherwise,

$x_{j\tau} = 1$, if state starts infrastructure project j in year τ, and $x_{j\tau} = 0$ otherwise,

$y_{k\tau} = 1$, if state starts ecological project k in year τ, and $y_{k\tau} = 0$ otherwise,

$u_{k\tau} = 1$, if investor starts ecological project k in year τ, and $u_{k\tau} = 0$ otherwise.

Industrial Project Parameters

$CP^t_{i\tau}$: The cash flow of industrial project i, which starts in year τ,

$EP^t_{i\tau}$: Ecological loss of industrial project i, which starts in year τ,

$DP^t_{i\tau}$: State income of industrial project i, which starts in year τ,

$ZP^t_{i\tau}$: Salary of industrial project i, which starts in year τ.

Infrastructure Project Parameters

$CI^t_{j\tau}$: The cost of infrastructure project j, which starts in year τ,

$EI^t_{j\tau}$: The ecological loss of infrastructure project j, which starts in year τ,

$DI^t_{j\tau}$: The other state income of infrastructure project j, which starts in year τ,

$ZI^t_{j\tau}$: Salary of infrastructure project j, which starts in year τ.

Ecological *Project Parameters*

$CE^t_{k\tau}$: The cost of ecological project k, which starts in year τ,

$DE^t_{k\tau}$: The ecological profit of ecological project k, which starts in year τ,

$ZE^t_{k\tau}$: Salary of ecological project k, which starts in year τ.

Projects Relationship Parameters

$\mu_{ij} = 1$, if industrial project i requires infrastructure project j, $\mu_{ij} = 0$ otherwise,

$\nu_{ik} = 1$, if industrial project i requires ecological project k, $\nu_{ik} = 0$ otherwise,

φ_{ij} is the minimum number of years before the start of industrial project i after the start of required infrastructure project j,

ψ_{ik} is the maximum number of years after the start of industrial project i before the start of required ecological project k,

γ_{ik} is the maximum number of years before the start of industrial project i after the start of required ecological project k,

State and Investor Discounts

θ^g, θ^{in} are the state and investor discounts, respectively.

State and Investor Budgets
b_t^g, b_t^{in} are the state and investor budgets in year t, respectively.

Bi-level 0–1 Linear Programming Model
The upper level problem is to maximize the total discounted profit of the pair "state-citizens":

$$P_l(x, y, z, u) = \sum_{t=1}^{T} \frac{\sum_{\tau=1}^{T} \left[\sum_{i=1}^{NP} (DP_{i\tau}^t + ZP_{i\tau}^t - EP_{i\tau}^t) z_{i\tau} + \sum_{k=1}^{NE} (DE_{k\tau}^t + ZE_{k\tau}^t) u_{k\tau} \right]}{(1 + \theta g)^t}$$

$$+ \sum_{t=1}^{T} \frac{\sum_{\tau=1}^{T} \left[\sum_{j=1}^{NI} \left(DI_{j\tau}^t + ZI_{j\tau}^t - EI_{j\tau}^t - CI_{j\tau}^t \right) x_{j\tau} + \sum_{k=1}^{NE} (DE_{k\tau}^t + ZE_{k\tau}^t - CE_{k\tau}^t) y_{k\tau} \right]}{(1 + \theta g)^t} \Rightarrow max$$

(1)

subject to:

$$\sum_{\tau=1}^{T} \left(\sum_{j=1}^{NI} CI_{j\tau}^t x_{j\tau} + \sum_{k=1}^{NE} CE_{k\tau}^t y_{k\tau} \right) \leq b_t^g, \quad t = 1, \ldots, T \tag{2}$$

$$\sum_{\tau=1}^{T} x_{j\tau} \leq 1, \quad j = 1, \ldots, NI \tag{3}$$

$$\sum_{\tau=1}^{T} y_{k\tau} \leq 1, \quad k = 1, \ldots, NE \tag{4}$$

$$x_{j\tau}, y_{k\tau} \in \{0, 1\} \tag{5}$$

where $(z, u) \in F^*(x, y)$ is the optimal solution to the lower-level problem.
The lower-level problem is to maximize discounted investor profit:

$$P_f(z, u) = \sum_{t=1}^{T} \frac{\sum_{\tau=1}^{T} \left(\sum_{i}^{NP} CP_{i\tau}^t z_{i\tau} - \sum_{k}^{NE} CE_{k\tau}^t u_{k\tau} \right)}{(1 + \theta^{in})^t} \Rightarrow max \tag{6}$$

subject to:

$$\sum_{\tau=1}^{T} \left(\sum_{k=1}^{NE} CE_{k\tau}^t u_{k\tau} - \sum_{i=1}^{NP} CP_{i\tau}^t z_{i\tau} \right) \leq b_t^{in}, \quad t = 1, \ldots, T \tag{7}$$

$$\sum_{\tau=1}^{T} z_{i\tau} \leq 1, \quad i = 1, \ldots, NP \tag{8}$$

$$\sum_{\tau=1}^{T} u_{k\tau} \leq 1, \quad k = 1, \ldots, NE \tag{9}$$

$$\sum_{\tau=1}^{T} (y_{k\tau} + u_{k\tau}) \leq 1, \quad k = 1, \ldots, NE \tag{10}$$

$$\sum_{\tau=1}^{T} x_{j\tau} \geq \mu_{ij} \sum_{\tau}^{T} z_{i\tau}, \quad j = 1, \ldots, NI \tag{11}$$

$$\sum_{\tau=1}^{T} (y_{k\tau} + u_{k\tau}) \geq \nu_{ik} \sum_{\tau=1}^{T} z_{i\tau}, \quad k = 1, \ldots, NE, \ i = 1, \ldots, NP \tag{12}$$

$$\sum_{\tau=1}^{T} (y_{k\tau} + u_{k\tau}) \leq \sum_{i=1}^{NP} \left(v_{ik} \sum_{\tau=1}^{T} z_{i\tau} \right), \quad k = 1, \ldots, NE, \ i = 1, \ldots, NP \quad (13)$$

$$x_{j\tau} + \mu_{ij} \sum_{\rho=1}^{\tau+\varphi_{ik}} z_{i\rho} \leq 1, \quad j = 1, \ldots, NI, \ i = 1, \ldots, NP, \ \tau = 1, \ldots, T \quad (14)$$

$$y_{k\tau} + u_{k\tau} + v_{ik} \sum_{1 \leq \rho < \tau - \psi_{ik}} z_{i\rho} \leq 1, \quad k = 1, \ldots, NE, \ i = 1, \ldots, NP, \ \tau = 1, \ldots, T \quad (15)$$

$$v_{ik} \sum_{\rho}^{\psi_{ik}} z_{i\rho} = 0, \quad k = 1, \ldots, NE, \ i = 1, \ldots, NP \quad (16)$$

$$y_{k\tau} + u_{k\tau} + v_{ik} \sum_{\tau + \gamma_{ik} < \rho \leq T} z_{i\rho} \leq 1, \quad k = 1, \ldots, NE, \ i = 1, \ldots, NP, \ \tau = 1, \ldots, T \quad (17)$$

$$u_{k\tau}, z_{i\tau} \in \{0, 1\} \quad (18)$$

In this model, the state maximizes an analogue of the net discounted income from the implementation of the entire development program. The objective function (1) includes the interests of the population and considers the economic gain from the creation of new workplaces and the environmental losses. Constraints (2) and (7) describe the budget constraints in the hard form [9–11] when the unused yearly budget burns out. Constraints (2a) and (7a) are soft form of the budget constraints (SBC). We carry the unused yearly budget over to the next years:

$$\sum_{t=1}^{\omega} \sum_{\tau=1}^{T} \left(\sum_{j=1}^{NI} CI_{j\tau}^t x_{j\tau} + \sum_{k=1}^{NE} CE_{k\tau}^t y_{k\tau} \right) \leq \sum_{t=1}^{\omega} b_t^g, \quad \omega = 1, \ldots, T \quad (2a)$$

$$\sum_{t=1}^{\omega} \sum_{\tau=1}^{T} \left(\sum_{k=1}^{NE} CE_{k\tau}^t u_{k\tau} - \sum_{i=1}^{NP} CP_{i\tau}^t z_{i\tau} \right) \leq \sum_{t=1}^{\omega} b_t^{in}, \quad \omega = 1, \ldots, T. \quad (7a)$$

Constraints (3), (4), (8), (9) and (10) guarantee that each ecological, infrastructure and industrial project can start at most once on the planning horizon. Constraints (11) and (12) guarantee the start of infrastructure and ecological projects required for industrial projects. Constraints (14) require the time lag before the start of industrial projects after the start of needed infrastructure projects. Constraints (15)–(17) guarantee the time window, when industrial projects can be started before or after required ecological projects.

Note that constraints (13) prohibit the starting of ecological projects for which corresponding industrial projects are not started. They include the state variables and can be moved to the upper level. These constraints are implemented by investor, and if the state has started an ecological project, then investor must support it by an appropriate industrial project. Such alternative formulation of the problem with constraint (13) at the upper level is more complicated and is not considered in this paper.

3 Main Properties and Computation Complexity

The 3P problem with hard or soft budget constraints is the ill-posed bi-level integer programming problem. The follower may have multiple solutions which lead to uncertainty in the definition of the leader's objective function. Thus, we consider the cooperative and non-cooperative cases [6, 7, 10].

Non-cooperative case is realized when the follower applies own optimal solution, which is the worst for the leader:

$$(x, y, z, u) = \underset{z,u \in F^*(x,y)}{argmin} \; P_l(x, y, z, u), \text{forgiven}(x, y).$$

Cooperative case is realized when the follower applies own optimal solution, which is the best for the leader:

$$(x, y, z, u) = \underset{z,u \in F^*(x,y)}{argmax} \; P_l(x, y, z, u)\text{forgiven}(x, y).$$

The following approach allows us to find the cooperative and non-cooperative solutions for given solution of the leader. Let $P_f^* := P_f(z, u), (z, u) \in F^*(x, y)$. For the cooperative case, we need to solve an auxiliary problem: $P_l(x, y, z, u) \Rightarrow max$ subject to (7)–(18) and an additional constraint $P_f(z, u) \geq P_f^*$. For the non-cooperative case, we reformulate the problem as follows: $P_l(x, y, z, u) \Rightarrow min$ subject to (7)–(18) and additional constraint $P_f(z, u) \geq P_f^*$.

Theorem 1. The 3P problem is Σ_2^P-hard in cooperative and non-cooperative cases for hard and soft budget constraints.

Proof. 1. Hard budget constraints.

Let us consider the Subset-Sum-Interval problem, which is Σ_2^P-hard [3]. There are positive integers $q_1, \ldots, q_l, R$ and r, where $r \leq l$. It is required to determine whether S such as $R \leq S < R + 2^r$ and for any $I \subseteq \{1, \ldots, l\} : \sum_{i \in I} q_i \neq S$.

Let us construct the following instance for 3P problem. We put $NP = l+2$ (industrial projects), $NE = r + 1$ (ecological projects), and $NI = 0$. For the first $l + 1$ industrial projects, no ecological projects are required. For the last $l + 2$ industrial project, all ecological projects are required. We put

$$T = 3; \tag{I}$$

$$CP_{i,1}^1 = -q_i, CP_{i,1}^3 = 2q_i, \quad i = 1, \ldots, l; \tag{II}$$

$$CP_{l+1,1}^1 = -\frac{1}{2}, CP_{l+1,1}^3 = 1; \tag{III}$$

$$\Delta = 2(R + 2^r) \text{ is a large number;} \tag{IV}$$

$$CP_{l+2,1}^3 = 2\Delta; \tag{V}$$

$$CP_{i,2}^t = CP_{i,3}^t = -\Delta, \quad i = 1, \ldots, NP, \ t = 1, \ldots, T; \tag{VI}$$

$$DP_{l+1,1}^3 = \Delta, DP_{l+2,1}^3 = 2\Delta; \tag{VIII}$$

$$CE_{1,1}^1 = CE_{1,1}^2 = R, CE_{k,1}^1 = CE_{k,1}^2 = 2^{k-2}, \quad k = 2, \ldots, NE; \tag{VIII}$$

$$CE_{k,2}^t = CE_{k,3}^t = \Delta, \quad k = 1 \ldots NE, \ t = 1, \ldots, T; \tag{IX}$$

$$b_t^g = R + 2^r - 1, \quad t = 1, \ldots, T; \tag{X}$$

$$b_1^{in} = R + 2^r - 1, \quad b_2^{in} = 2^r - 1. \tag{XI}$$

All other parameters of the model equal to 0.

According to the budget constraints, we have $z_{i\tau} = 0, y_{k\tau} = 0, u_{k\tau} = 0$, for all $= 1 \ldots NP, k = 1 \ldots NE, \tau = 2, 3$. It means that any industrial or ecological project can be opened in the first year only. Thus, the objective functions of the state (1) and the investor (6) take the form:

$$P_l(x, y, z, u) = \Delta z_{l+1,1} + 2\Delta z_{l+2,1} - \sum_{k=1}^{NE} \left(CE_{k,1}^1 + CE_{k,1}^2 \right) y_{k,1}; \tag{1'}$$

$$P_f(z, u) = \sum_{i=1}^{l} q_i z_{i,1} + \frac{1}{2} z_{l+1,1} + 2\wedge z_{l+2,1} - \sum_{k=1}^{NE} \left(CE_{k,1}^1 + CE_{k,1}^2 \right) u_{k,1}. \tag{6'}$$

Budget constraints (2) and (7) take the form:

$$\sum_{k=1}^{NE} CE_{k,1}^t y_{k,1} \leq R + 2^r - 1; \tag{2'}$$

$$\sum_{i=1}^{l} q_i z_{i,1} + \frac{1}{2} z_{l+1,1} + \sum_{k=1}^{NE} CE_{k,1}^1 u_{k,1} \leq R + 2^r - 1, \sum_{k=1}^{NE} CE_{k,1}^2 u_{k,1} \leq 2^r - 1. \tag{7'}$$

It can see from the structure of (1'), (6') and (2') (7'), $z_{l+2,1} = 1$ in the optimal solution of the state problem.

Investor budget constraint for the third year is not active. Constraints (3), (4), (8), (9), (11), (14)–(17) can be omitted too. Constraints (10), (12) and (13) take the form:

$$y_{k,1} + u_{k,1} \leq 1; y_{k,1} + u_{k,1} \geq z_{l+2,1}; y_{k,1} + u_{k,1} \leq z_{l+2,1}.$$

In other words, we have $y_{k,1} + u_{k,1} = 1$ because $z_{l+2,1} = 1$.

In this form, the problem is identical to the case from [10] where the polynomial reduction of the Subset-Sum-Interval problem to the similar 3P problem is shown. We present here the main idea of this reduction only.

To open the last industrial project $l + 2$, we need all ecological projects. The players must share the cost of the projects among themselves. The state can open some of them for a part S of own budget, $R \leq S < R + 2^r$. If $S < R$, then in the second year, the investor will not have enough budget to provide all the remaining projects. On the other hand, S can't be greater than the total cost of all such projects $R + 2^r - 1$.

In the first year after ecological projects opening, the investor will have budget S. He can spend it to open some of the first $l + 1$ industrial projects. Since the incomes of any subset of the first l industrial projects are greater than $1/2$, the investor will open it only if he cannot fully spend S on the first l projects. For the state, the $l + 1$ industrial project is quite profitable, and the state will try to choose such S that $\sum_{i \in I} q_i \neq S$ for any subset $I \subseteq \{1, \ldots, l\}$, if it is possible. Thus, we have answer "yes" in the Subset-Sum-Interval problem, if and only if the optimal solution to the 3P problem has value strictly greater than 2Δ. The cooperative and non-cooperative cases coincide here.

2. Soft budget constraints.

Note that the previous proof is valid for the state soft budget constraints. Thus, we need to check the case the investor soft budget constraints only. We modify our reduction by the following: $NP = l + 2 + r$, NI and NE remain the same and (I), (IV)–(X) are actual. Now we add new parameters:

$$CP_{i,1}^1 = -2^{i-l-3}, \ CP_{i,1}^3 = 2^{i-l-3} + \sum_{j=1}^{l} q_j, \quad i = l + 3, \ldots, l + 2 + r, \qquad \text{(XII)}$$

and replace (II), (III), (XI) by the following:

$$CP_{i,1}^2 = -q_i, \ CP_{i,1}^3 = 2q_i, \quad i = 1, \ldots, l; \qquad \text{(II}')$$

$$CP_{l+1,1}^2 = -\frac{1}{2}, \ CP_{l+1,1}^3 = 1; \qquad \text{(III}')$$

$$b_1^{in} = 2^r - 1, \ b_2^{in} = R + 2^r - 1. \qquad \text{(XI}')$$

Each of the last r industrial projects requires all ecological projects. All other model parameters are 0. Note that the cache flow in the first year for the last r projects allows us to get any value from $-2^r + 1$ to 0.

Given input data, the state objective function has the same form $(1')$ and the investor objective function has a new form:

$$P_f(z, u) = \sum_{i=1}^{l} q_i z_{i,1} + \frac{1}{2} z_{l+1,1} + 2\Delta z_{l+2,1} + \left(\sum_{i=1}^{l} q_i \right) \left(\sum_{i=l+3}^{l+2+r} z_{i,1} \right)$$
$$- \sum_{k=1}^{NE} \left(CE_{k,1}^1 + CE_{k,1}^2 \right) u_{k,1}. \qquad (6'')$$

The state budget constraints remain the same $(2')$. Investor budget constraints have new form:

$$\sum_{i=l+3}^{NP} 2^{(i-l-3)} z_{i,1} + \sum_{k=1}^{NE} CE_{k,1}^1 u_{k,1} \leq 2^r - 1, \ \sum_{i=l+3}^{NP} 2^{(i-l-3)} z_{i,1} + \sum_{i=1}^{l} q_i z_{i,1} + \frac{1}{2} z_{l+1,1}$$
$$+ \sum_{k=1}^{NE} \left(CE_{k,1}^1 + CE_{k,1}^2 \right) u_{k,1} \leq R + 2(2^r - 1). \qquad (7a')$$

Previous reasoning about the third-year investor budget constraints and the constraints (3), (4), (8)–(17) remains true in this case too.

As before, the players must share the cost of ecological projects among themselves. The state will open some projects for the part S of own budget, $R \leq S < R + 2^r$.

In the first year, after opening of all remaining ecological projects, the investor has a budget $S - R$. We can see from the structure of (6'') and (7a') that in the investor optimal solution, the investor completely spends $S - R$ on opening a suitable set of the last r industrial projects and transferring the budget to the second year doesn't occur. Thus, we return to the already considered case, but only in the second year. In the second year after maintaining all remaining environmental projects, the investor has a budget S which it spends on opening a suitable set of the first $l + 1$ industrial projects. If it fails to fully spend the budget for opening the first l industrial projects, it opens the $l + 1$ project. The $l + 1$ industrial project is quite profitable for the state, and the state will try to choose such S that $\sum_{i \in I} q_i \neq S$ for any $I \subseteq \{1, \ldots, l\}$, if it is possible. Thus, we have answer "yes" in the Subset-Sum-Interval problem if and only if the optimal solution to the 3P problem has value strictly greater than 2Δ. □

Theorem 2. The absolute and relative gaps between the cooperative and non-cooperative optimal solutions to the 3P problem can be arbitrarily large.

Proof. Let us consider the following instance for the 3P problem: $NP = 2, NE = NI = 0, T = 2; CP_{1,1}^1 = CP_{2,1}^1 = -1; CP_{1,1}^2 = CP_{2,1}^2 = 1; CP_{1,2}^2 = CP_{2,2}^2 = -1; b_1^{in} = 1, b_2^{in} = 0; EP_{1,1}^1 = l$, large number. All other parameters are equal to 0.

The objective functions of the state (1) and the investor (6) take the form:

$$P_l(x, y, z, u) = -lz_{1,1};$$

$$P_f(z, u) = \sum_{t=1}^{T} \sum_{\tau=1}^{T} \sum_{i}^{NP} CP_{i\tau}^t z_{i\tau}.$$

The investor budget constraints (7) take the form:

$$z_{1,1} + z_{2,1} \leq 1, -z_{1,1} - z_{2,1} + z_{1,2} + z_{2,2} \leq 0.$$

As we can see, the optimal value of the investor objective function $P_f(z, u)$ is 0. It is achieved if one of two industrial projects is opened in the first year. On the other hand, the state will get $-l$ if the first project is opened, which corresponds to the non-cooperative case, and 0 if the second project is opened (cooperative case). □

4 Upper and Lower Bounds

To obtain an upper bound (UB) to the optimal value of the 3P problem, we use the so-called high point relaxation [9, 10] and solve the single-level problem (1)–(18) without the investor objective function (6). Such a solution is called monopoly and assumes that the state controls both its own and investor actions. To obtain a lower bound, we need a feasible solution (x, y) to the state problem (1)–(5) where the optimal solution $(z, u) \in F^*(x, y)$ to the investor problem (6)–(18) is used.

To find a near optimal solution (x, y) to the bi-level 0-1optimization problem, the stochastic Tabu Search method by variables (x, y) has been developed [1, 2, 13]. Tabu Search belongs to the class of metaheuristics, in other words, to the class of general frameworks for designing heuristic algorithms that can be applied to almost each discrete optimization problem [11, 12]. All metaheuristics are iterative randomized procedures, and for many of them, asymptotic convergence of the best-found solution to the global optimum is established. Unlike the algorithms with performance guarantees, metaheuristics are not tied to the specifics of the problem. They are general iterative procedures that use randomization and self-learning elements, search intensification and diversification rules, adaptive control mechanisms, constructive heuristics, and local search methods. Metaheuristics include Simulated Annealing (SA), Tabu Search (TS), Genetic Algorithms (GA), Evolutionary Computation (EC), Variable Neighborhood Search (VNS), Ant Colony Optimization (ACO), Greedy Randomized Adaptive Search Procedure (GRASP) and others [11]. The idea of these methods assumes that the objective function has many local extrema, and it is impossible to check all feasible solutions. In such a situation, it is necessary to focus the search on the most promising parts of the feasible area. Each metaheuristic solves this problem in its own way.

Tabu search refers to trajectory methods. Trajectory methods leave a trajectory in the search space, a sequence of solutions, where each solution is neighbor to the previous one with respect to some neighborhood. TS scheme allows the algorithm to not stop at a local optimum, as prescribed in standard local improvement algorithm, but to travel from one local optimum to another in order to find the global optimum among them. The main mechanism that allows the algorithm to get out of the local optimum is a tabu list. It is modified according to the last few solutions, and prohibits a part of the neighborhood of current solution. The length of the tabu list determines its memory. The optimal solution to the lower level is found by the branch-and-bound method using GUROBI software. This approach has shown its effectiveness in solving many problems of bi-level programming [1, 2, 5, 11, 13].

Given the solution (x, y), we define the neighborhood $N(x, y)$ as all solutions obtained from the given one by applying one of the following operations: opening of a new project; closing of an opened project; shifting start of a project to new time.

To diversify search and reduce the running time of each iteration, we apply the randomized neighborhood: each neighboring solution is included in the randomized neighborhood $N_p(x, y)$ with a given probability p independently of each other [5, 7, 11]. Pseudocode of the TS algorithm to maximize the state objective function $P_l(x, y, z, u)$ can be described as follows.

TS Algorithm

1. Create a starting solution (x, y). Find an investor solution $(z, u) \in F^*(x, y)$.
   Define $P^* := P_l(x, y, z, u)$, $(x^*, y^*) := (x, y)$, $TabuList := \{(x^*, y^*)\}$.
2. Repeat until stopping condition is met:

a. Create randomized neighborhood $N_p(x, y)$;

b. Find the best feasible non-forbidden solution $\left(x', y', z', u'\right)$ in $N_p(x, y)$;

c. If $P^* < P_l\left(x', y', z', u'\right)$, then $P^* := P_l\left(x', y', z', u'\right)$, $(x^*, y^*) := \left(x', y'\right)$;

d. Put $(x, y) := \left(x', y'\right)$ and update *TabuList*.

3. Return the best found solution (x^*, y^*).

To get the cooperative or non-cooperative solutions, we need to use the investor cooperative or non-cooperative solutions respectively at Step 2.b. as we have described in Sect. 3. The starting solution is obtained from the high point relaxation. As the stopping condition, we use a number of iterations without improvement of the incumbent solution (x^*, y^*).

5 Computational Experiments

To compare the cooperative or non-cooperative cases, we use the test instances based on the 50 Transbaikalian polymetallic fields [8–10]. Each instance includes 50 industrial projects, 50 ecological projects, and 10 infrastructure projects. We vary the coefficients values for the state and the investor objective functions and the relationship of the industrial projects with the ecological and the infrastructure projects. Planning horizon, state and investor discounts are selected randomly and independently by uniform distribution, according to Table 1. The soft and hard budget constraints were chosen randomly and independently for each player.

Table 1. Min and max values for the uniform distribution of some model parameters

Parameter	Min	Max
T	1	20
θ^g	0	0.1
θ^{in}	0.05	0.5
b^g	1000	100000
b^{in}	1000	100000
p_f	0	100000

5.1 Cooperative or Non-cooperative Cases for the High Point Relaxation

We conduct 120 000 experiments for these semi-synthetic test instances. The first experiment deals with the high point relaxation with additional constraints for the investor objective function. We wish guarantee that this value is at least the given threshold p_f.

Then, for the obtained monopoly solution of the state, the investor problem is solved in the cooperative and non-cooperative manner.

Table 2 presents the computational results for the high point relaxation. It is interesting to note that minimum ratio of the optimal state values in non-cooperative case to the cooperative case achieved 0.849. But these values coincide in 96% for these instances. The ration is at most 0.997 for 2% of instances and at most 0.994 for 1%.

Table 2. The ratio of the non-cooperative to cooperative optimal values

Percentile	Non-cooperative/cooperative
0.001	0.885
0.01	0.924
0.1	0.964
1	0.994
2	0.997
4	0.999

5.2 Tabu Search

The second computational experiment deals with the TS algorithm for the 3P problem. We use GUROBI solver to compute cooperative and non-cooperative investor solutions for the best found solution for the state. We use the same test instances based on the 50 Transbaikalian polymetallic fields. Monopoly solution from the previous subsection is applied as the starting solution for the TS algorithm. The maximal number of iteration is 500, maximum iterations without improvement of the incumbent solution is 100, the length of the Tabu list is 10. We ignore 90% of neighboring solutions at Step 2.b. to accelerate the search, the parameter $p = 0.1$ for the randomized neighborhood $N_p(x, y)$. Table 3 presents computational results for the different combinations of the hard and soft budget constraints (*SBC*) and cooperative (*Coop*) and non-cooperative (*Nonc*) cases. The column *State SBC* shows the variant of the budget constraints for the state, 0 corresponds to the hard variant, 1 corresponds to the soft variant. The column *Inv. SBC* shows the same information for the investor. The columns *LB* and *UB* present the lower and upper bounds obtained by the high point relaxation. The column *S* indicates the state value (1) for the best found solution by the TS algorithm. Finally, the last column δ shows the deviation of the best found solution from the upper bound, $\delta = (UB-S)/UB$. As we can see, in most cases, the TS algorithm can find a solution with at most 5% from the upper bound. Each line here corresponds to one run for an instance.

Table 3. Computational results for the Tabu Search

θ^g	θ^{in}	b^g	b^{in}	State SBC	Inv. SBC	Case	LB	UB	S	δ
0.01	0.1	50	50	0	0	Coop	2161.61	2681.49	2616.03	0.02
0.01	0.1	50	50	0	0	Nonc	2161.61	2681.49	2572.57	0.04
0.01	0.1	50	200	0	1	Coop	3174.47	3624.46	3515.76	0.03
0.01	0.1	50	200	0	1	Nonc	3174.47	3624.46	3514.70	0.03
0.01	0.2	50	200	0	1	Coop	3184.56	3519.32	3422.00	0.03
0.01	0.2	50	200	0	1	Nonc	3184.56	3519.32	3414.83	0.03
0.05	0.1	50	50	0	0	Coop	1258.78	1539.18	1474.63	0.04
0.05	0.1	50	50	0	0	Nonc	1258.78	1539.18	1499.87	0.03
0.05	0.1	50	200	0	1	Coop	1935.22	2193.09	2136.39	0.03
0.05	0.1	50	200	0	1	Nonc	1935.22	2193.09	2157.40	0.02
0.05	0.1	100	100	1	0	Coop	2287.56	2728.59	2602.77	0.05
0.05	0.1	100	100	1	0	Nonc	2287.56	2728.59	2641.57	0.03
0.05	0.2	50	50	0	0	Coop	1076.57	1536.74	1500.37	0.02
0.05	0.2	50	50	0	0	Nonc	1076.57	1536.74	1473.54	0.04
0.05	0.2	200	50	1	1	Coop	1171.67	2074.38	1962.24	0.05
0.05	0.2	200	50	1	1	Nonc	1171.67	2074.38	1951.27	0.06

6   Conclusions

We have considered a new variant of the bi-level 0-1 programming model for the strategic planning of the public-private partnership. It is shown that the problem is Σ_2^P-hard in cooperative and non-cooperative cases for the hard and soft budget constraints. Moreover, we show that the absolute and relative gaps between the cooperative and non-cooperative optimal solutions can be arbitrarily large. We conduct computational experiments on the semi-synthetic test instances for the Transbaikalian polymetallic fields. In most part of the experiments, the cooperative and non-cooperative cases coincide. But in 1% of experiments, the difference is at least 0.5%. In extremal cases, this difference can achieve 10% and even 15%. In the most cases, the TS algorithm can find near optimal solutions with relative deviation from the upper bound of the high point relaxation at most 5%.

Acknowledgement. The research is carried out within the framework of the state contract of the Sobolev Institute of Mathematics (FWNF-2022-0019).

References

1. Alekseeva, E., Kochetov, Yu., Plyasunov, A.: An exact method for the discrete (r|p)centroid problem. J. Glob. Optim. **63**(3), 445–460 (2015)

2. Alekseeva, E., Kochetov, Yu.: Matheuristics and exact methods for the discrete (r|p)centroid problem. In: Talbi, E.-G. (ed.) Metaheuristics for Bi-level Optimization. SCI, vol. 482, pp. 189–219. Springer, Heidelberg (2013). https://doi.org/10.1007/978-3-642-37838-6_7
3. Caprara, A., Carvalho, M., Lodi, A., Woeginger, G.J.: A study on the computational complexity of the bilevel knapsack problem. SIAM J. Optim. **24**(2), 823–838 (2014)
4. Davydov, I., Kochetov, Yu., Plyasunov, A.: On the complexity of the (r|p)-centroid problem in the plane. TOP **22**(2), 614–623 (2014)
5. Davydov, I.A., Kochetov, Y.A., Carrizosa, E.: A local search heuristic for the (r|p)centroid problem in the plane. Comput. Oper. Res. **52**, 334–340 (2014)
6. Dempe, S., Zemkoho, A.: Bilevel Optimization. Advances and Next Challenges. SOIA, vol. 161. Springer, Cham (2020). https://doi.org/10.1007/978-3-030-52119-6
7. Iellamo, S., Alekseeva, E., Chen, L., Coupechoux, M., Kochetov, Yu.: Competitive location in cognitive radio networks. 4OR **13**(1), 81–110 (2014). https://doi.org/10.1007/s10288-014-0268-1
8. Lavlinskii, S.M., Panin, A.A., Plyasunov, A.V.: Comparison of models of planning public-private partnership. J. Appl. Ind. Math. **10**(3), 356–369 (2016). https://doi.org/10.1134/S1990478916030066
9. Lavlinskii, S.M., Panin, A.A., Plyasunov, A.V.: A bilevel planning model for public–private partnership. Autom. Remote Control **76**(11), 1976–1987 (2015). https://doi.org/10.1134/S0005117915110077
10. Lavlinskii, S.M., Panin, A.A., Plyasunov, A.V.: The Stackelberg model in territorial planning. Autom. Remote Control **80**, 286–296 (2019)
11. Talbi, E.-G. (ed.): Metaheuristics. From Design to Implementation. Computer Science, Wiley, New York (2009)
12. Talbi, E.-G. (ed.): Metaheuristics for Bi-level Optimization. SCI, vol. 482. Springer, Heidelberg (2013). https://doi.org/10.1007/978-3-642-37838-6_7
13. Zyryanov, A.A., Kochetov, Y.A., Lavlinskii, S.M.: Stochastic local search for the strategic planning public-private partnership. CEUR Workshop Proc. **2098**, 446–463 (2018)

Mathematical Programming

Distributed Methods with Absolute Compression and Error Compensation

Marina Danilova[1,2] and Eduard Gorbunov[2(✉)]

[1] Institute of Control Sciences of RAS, Moscow, Russia
[2] Moscow Institute of Physics and Technology, Moscow, Russia
eduard.gorbunov@phystech.edu

Abstract. Distributed optimization methods are often applied to solving huge-scale problems like training neural networks with millions and even billions of parameters. In such applications, communicating full vectors, e.g., (stochastic) gradients, iterates, is prohibitively expensive, especially when the number of workers/nodes is large. Communication compression is a powerful approach to alleviating this issue, and, in particular, methods with biased compression and *error compensation* are extremely popular due to their practical efficiency. Sahu et al. (2021) [30] propose a new analysis of Error Compensated SGD (EC-SGD) for the class of absolute compression operators showing that in a certain sense, this class contains optimal compressors for EC-SGD. However, the analysis was conducted only under the so-called (M, σ^2)-bounded noise assumption. In this paper, we generalize the analysis of EC-SGD with absolute compression to the arbitrary sampling strategy and propose the first analysis of Error Compensated Loopless Stochastic Variance Reduced Gradient method (EC-LSVRG) [10] with absolute compression for (strongly) convex problems. Our rates improve upon the previously known ones in this setting. Numerical experiments corroborate our theoretical findings.

Keywords: Distributed optimization · Compressed communication · Error compensation

1 Introduction

In the recent few years, distributed optimization methods has been receiving a lot of attention from various research communities and, especially, from the machine learning one. This can be explained by the need of training deep learning models with billions of parameters on the hundreds of gigabytes of data [4] (and sometimes even this is not enough [16]). Clearly, such problems cannot be solved in a reasonable time on a single yet powerful machine [23]. Next, distributed methods are literally the only possible choices in such applications like Federated Learning (FL) [15,19,20], where the data is privately stored on multiple devices.

The research was supported by Russian Science Foundation grant (project No. 21-71-30005).

Due to the huge dimensions of corresponding problems and large number of workers in the networks naïve methods like centralized Parallel SGD [37] suffer from the so-called *communication bottleneck*. This phenomenon means that a method spends much more time on communication rounds than on computations. A natural and popular way of addressing this issue is *communication compression* [32] – a technique that uses special compression operators called *compressors* applied to the information that devices send through the network.

The works on distributed methods with compression usually focus either on unbiased compressors [1,14,25] like RandK or ℓ_2-quantization (e.g., see [3]) or on biased compressors [3,10,32,33] like TopK. Although the world of unbiased compressors has richer theory, the methods with biased compressors are very popular due to their efficiency in practice. However, to make them convergent one has to apply additional tricks on top of SGD, e.g., *error-compensation* [3,32–34].

Error Compensated SGD (EC-SGD) was proposed in [32] where the authors demonstrated its efficiency in practical tasks, but the first theoretical analysis of EC-SGD was given in [33] and tightened in [34]. This analysis was extended in various directions including (but not limited to) decentralized communications [18], arbitrary sampling and variance reduction (with the first linearly convergent variants) [10], acceleration [28], and also some prominent alternatives were proposed [14,29]. However, the compression operators in these papers are usually assumed to be δ-*contractive*.[1]

Recently, the authors of [30] developed a new analysis of EC-SGD with *absolute compressors* [30,36], i.e., such (stochastic) operators $\mathcal{C}$ that for some $\Delta \geq 0$ the inequality $\mathbb{E}[\|\mathcal{C}(x) - x\|^2] \leq \Delta^2$ holds for all $x \in \mathbb{R}^d$, where $\mathbb{E}[\cdot]$ denotes an expectation. In particular, they proved that this class contains special operators called hard-threshold sparsifiers that are optimal in view of total error minimization (a special quantity arising in the analysis of EC-SGD) for any *fixed* sequence of errors. Moreover, the authors of [30] derived convergence rates for EC-SGD with absolute compressors under (M, σ^2)-bounded noise assumption and illustrate the theoretical and practical benefits of absolute compressors compared to δ-contractive ones. However, several fruitful directions were unexplored for EC-SGD with absolute compressor including more general analysis of the standard version of the method and variants with variance reduction.

1.1   Main Contributions

◇ **Unified Analysis of EC-SGD with Absolute Compressors.** We propose a generalized analysis of EC-SGD with absolute compression covering different stochastic estimators under various assumptions. In particular, we consider the simplified version of the parametric assumption from [10] (see Assumption 1 and the discussion after) and derive a general result on the convergence of EC-SGD with absolute compressors (Theorem 1). The considered assumption covers various setups including the one from [30], the derived result gives sharp rates.

---

[1] The mapping (possibly stochastic) $\mathcal{C} : \mathbb{R}^d \to \mathbb{R}^d$ is called δ-*contractive* compressor if there exists $\delta \in (0, 1]$ such that $\mathbb{E}[\|\mathcal{C}(x) - x\|^2] \leq (1 - \delta)\|x\|^2$ for all $x \in \mathbb{R}^d$.

Algorithm 1. Error-Compensated Stochastic Gradient Descent (EC-SGD)

Input: starting point x^0, stepsize $\gamma > 0$, number of iterations $K \geq 0$

1: Set $e_i^0 = 0$ for all $i = 1, \ldots, n$
2: **for** $k = 0, \ldots, K - 1$ **do**
3:     Server broadcasts x^k to all workers
4:     **for** $i = 1, \ldots, n$ in parallel **do do**
5:         Compute stochastic gradient g_i^k and send $v_i^k = \gamma \mathcal{C}\left(\frac{e_i^k + \gamma g_i^k}{\gamma}\right)$ to the server
6:         Update error-vector: $e_i^{k+1} = e_i^k + \gamma g_i^k - v_i^k$
7:     **end for**
8:     Server gathers $v_1^k, v_2^k, \ldots, v_n^k$ from all workers and computes $v^k = \frac{1}{n}\sum_{i=1}^n v_i^k$
9:     Set $x^{k+1} = x^k - v^k$
10: **end for**

$\diamond$ **EC-SGD with Absolute Compression and Arbitrary Sampling.** To illustrate the flexibility of our approach, we propose the first analysis of EC-SGD with absolute compression and arbitrary sampling. The derived bounds are superior to the ones from [30] under certain assumptions.

$\diamond$ **EC-LSVRG with Absolute Compression and Arbitrary Sampling.** As a special case of our theoretical framework, we obtain the analysis of EC-LSVRG [10] with absolute compression. In contrast to [10], we handle non-uniform sampling in EC-LSVRG. The derived rate for EC-LSVRG with absolute compression has the leading term proportional to $1/K^2$, while the leading term for EC-SGD is proportional to $1/K$, where K is the total number of iterations.

$\diamond$ **Numerical Experiments.** We conduct several numerical experiments to support our theory and compare the performance of EC-LSVRG with hard-threshold and TopK sparsifiers. The numerical results corroborate our theoretical findings and highlight the benefits of using absolute compressors.

1.2   Preliminaries

Problem. We consider a classical centralized optimization problem

$$\min_{x \in \mathbb{R}^d} \left\{ f(x) = \frac{1}{n} \sum_{i=1}^n f_i(x) \right\}, \tag{1}$$

where the information defining differentiable functions $f_1, \ldots, f_n : \mathbb{R}^d \to \mathbb{R}$ is distributed among n workers/clients/devices connected with parameter-server in a centralized way. In particular, (stochastic) gradients of function f_i are available to client i only. Throughout the work, we assume that the solution x^* of problem (1) is unique and that the function f is convex and μ-quasi strongly convex [26], where the later is a relaxation of strong convexity meaning that

$$\forall x \in \mathbb{R}^d \quad f(x^*) \geq f(x) + \langle \nabla f(x), x^* - x \rangle + \frac{\mu}{2}\|x - x^*\|^2, \quad \mu \geq 0. \tag{2}$$

We assume that $f(x)$ is L-smooth: $\|\nabla f(x) - \nabla f(y)\| \leq L\|x - y\|$ for all $x, y \in \mathbb{R}^d$.

Compression. In this work, we focus on *absolute compression operators* [30,36].

Definition 1. *The mapping (possibly stochastic) $C : \mathbb{R}^d \rightarrow \mathbb{R}^d$ is called absolute compression operator/absolute compressor if there exists $\Delta \geq 0$ such that*

$$\mathbb{E}\left[\|C(x) - x\|^2\right] \leq \Delta^2, \quad \forall x \in \mathbb{R}^d. \tag{3}$$

An example of absolute compressor is hard-threshold sparsifier $C_{\mathrm{HT}}(x)$ [6,30,35] defined as $[C_{\mathrm{HT}}(x)]_i = [x]_i$ if $|[x]_i| \geq \lambda$ and $[C_{\mathrm{HT}}(x)]_i = 0$ otherwise for some $\lambda \geq 0$, where $[\cdot]_i$ denotes the i-th component of the vector. One can show that $C_{\mathrm{HT}}(x)$ satisfies (3) with $\Delta = \lambda\sqrt{d}$. Other examples include (stochatsic) rounding schemes with bounded error [12] and scaled integer rounding [31].

Paper Organization. In Sect. 2, we formulate our general result on the convergence of EC-SGD with absolute compression. Next, we provide particular examples of the variants of EC-SGD fitting our framework – EC-SGD with arbitrary sampling (Sect. 3) and EC-LSVRG (Sect. 4) – and discuss the convergence guarantees obtained for them. Finally, in Sect. 5, we discuss the results of numerical experiments supporting our theoretical findings. The proofs are delegated to the Appendix.

2   Unified Analysis

In our analysis, we rely on a simplified version of Assumption 3.3 from[2] [10].

Assumption 1 (Key Parametric Assumption). *For all $k \geq 0$ the average of stochastic gradients used in EC-SGD (Algorithm 1) is an unbiased estimate of $\nabla f(x^k)$, i.e., for $g^k = \frac{1}{n}\sum_{i=1}^n g_i^k$ we have $\mathbb{E}_k[g^k] = \nabla f(x^k)$ for all $k \geq 0$, where $\mathbb{E}_k[\cdot]$ denotes an expectation w.r.t. the randomness coming from iteration k. Moreover, there exist non-negative parameters $A, B, C, D_1, D_2 \geq 0$, $\rho \in (0,1]$, and sequence of (possibly random) variables $\{\sigma_k^2\}_{k\geq 0}$ such that for all $k \geq 0$ the iterates produced by EC-SGD and the objective function f satisfy*

$$\mathbb{E}_k\left[\|g^k\|^2\right] \leq 2A\left(f(x^k) - f(x^*)\right) + B\sigma_k^2 + D_1, \tag{4}$$

$$\mathbb{E}_k\left[\sigma_{k+1}^2\right] \leq (1-\rho)\sigma_k^2 + 2C\left(f(x^k) - f(x^*)\right) + D_2. \tag{5}$$

As it is shown in [10], the above assumption is very general and covers various algorithms in different settings. In Sects. 3 and 4, we consider two particular examples when Assumption 1 is satisfied. In all known special cases, parameters A and C are typically related to the smoothness properties of the problem, σ_k^2 describes the variance reduction process (with "rate" ρ), D_1 and D_2 are remaining noises not handled by variance reduction, and B is some constant.

Under Assumption 1 we derive the following result in Appendix A.

---

[2] The unified analysis of stochastic first-order methods was proposed in [8] for quasi strongly convex problems. After that, this idea was extended to the case of convex functions [17], methods with error feedback [10] and local updates [9], and to the methods for solving variational inequalities and min-max problems [2,7].

Theorem 1. *Let function f be convex, μ-quasi strongly convex (with unique solution x^*), L-smooth, and Assumption 1 hold. Assume that $0 < \gamma \leq 1/4(A+CF)$, where $F = 4B/3\rho$. Then, for all $K \geq 0$ the iterates produced by* EC-SGD *(Algorithm 1) with absolute compression operator $\mathcal{C}$ (see Definition 1) satisfy*

$$\mathbb{E}\left[f(\overline{x}^K) - f(x^*)\right] \leq \frac{(1-\eta)^{K+1}2\mathbb{E}[T_0]}{\gamma} + 2\gamma\left(D_1 + FD_2 + 3L\gamma\Delta^2\right), \text{ if } \mu > 0, \quad (6)$$

$$\mathbb{E}\left[f(\overline{x}^K) - f(x^*)\right] \leq \frac{2\mathbb{E}[T_0]}{\gamma(K+1)} + 2\gamma\left(D_1 + FD_2 + 3L\gamma\Delta^2\right), \text{ if } \mu = 0, \quad (7)$$

where $\overline{x}^K = \frac{1}{W_K}\sum_{k=0}^{K} w_k x^k$, $w_k = (1-\eta)^{-(k+1)}$, $\eta = \min\{\gamma\mu/2, \rho/4\}$, $W_K = \sum_{k=0}^{K} w_k$, and $T_0 = \|x^0 - x^*\|^2 + F\gamma^2\sigma_0^2$

Upper bounds (6) and (7) establish convergence to some neighborhood of the solution (in terms of the functional suboptimality). Applying Lemmas I.2 and I.3 from [9], we derive the convergence rates to the exact optimum.

Corollary 1. *Let the assumptions of Theorem 1 hold. Then, there exist choices of the stepsize γ such that* EC-SGD *(Algorithm 1) with absolute compressor guarantees $\mathbb{E}[f(\overline{x}^K) - f(x^*)]$ of the order*

$$\widetilde{\mathcal{O}}\left((A+CF)\mathbb{E}[\hat{T}_0]\exp\left(-\min\left\{\frac{\mu}{A+CF}, \rho\right\}K\right) + \frac{D_1+FD_2}{\mu K} + \frac{L\Delta^2}{\mu^2 K^2}\right), \quad (\mu > 0) \quad (8)$$

$$\mathcal{O}\left(\frac{(A+CF)R_0^2}{K} + \frac{R_0^2\sqrt{B\mathbb{E}[\sigma_0^2]}}{K\sqrt{\rho}} + \sqrt{\frac{R_0^2(D_1+FD_2)}{K}} + \frac{L^{1/3}R_0^{4/3}\Delta^{2/3}}{K^{2/3}}\right), \quad (\mu = 0) \quad (9)$$

when $\mu = 0$, where $R_0 = \|x^0 - x^*\|$, $\hat{T}_0 = R_0^2 + \frac{F}{16(A+CF)^2}\sigma_0^2$.

This general result allows to obtain the convergence rates for all methods satisfying Assumption 1 via simple plugging of parameters A, B, C, D_1, D_2 and ρ in the upper bounds (8) and (9). For example, due to such a flexibility we recover the results from [30], where the authors assume that each f_i is convex and L_i-smooth, f is μ-quasi strongly convex[3], and stochastic gradients have (M, σ^2)-bounded noise, i.e., $\mathbb{E}_k[\|g_i^k - \nabla f_i(x^k)\|^2] \leq M\|\nabla f_i(x^k)\|^2 + \sigma^2$ for all $i \in [n]$. In the proof of their main results (see inequality (24) in [30]), they derive an upper-bound for $\mathbb{E}_k[\|g^k\|^2]$ implying that Assumption 1 is satisfied in this case with $A = L + \frac{M\max_{i\in[n]} L_i}{n}$, $B = 0$, $\sigma_k^2 \equiv 0$, $D_1 = \frac{2M\zeta_*^2+\sigma^2}{n}$, $C = 0$, $D_2 = 0$, where $\zeta_*^2 = \frac{1}{n}\sum_{i=1}^{n} \|\nabla f_i(x^*)\|^2$ measures the heterogeneity of local loss functions at the solution. Plugging these parameters in Corollary 1 we recover[4] the rates from [30]. In particular, when $\mu > 0$ the rate is

$$\widetilde{\mathcal{O}}\left((L + \frac{M\max_{i\in[n]} L_i}{n})R_0^2 \exp\left(-\frac{\mu}{L+\frac{M\max_{i\in[n]} L_i}{n}}K\right) + \frac{2M\zeta_*^2+\sigma^2}{\mu n K} + \frac{L\Delta^2}{\mu^2 K^2}\right). \quad (10)$$

Below we consider two other examples when Assumption 1 is satisfied.

---

[3] Although the authors of [30] write in Sect. 3.1 that all f_i are μ-strongly convex, in the proofs, they use convexity of f_i and quasi-strong monotonicity of f.

[4] When $\mu = 0$, our result is tighter than the corresponding one from [30].

3   Absolute Compression and Arbitrary Sampling

Consider the case when each function f_i has a finite-sum form, i.e., $f_i(x) = \frac{1}{m}\sum_{j=1}^{m} f_{ij}(x)$, which is a classical situation in distributed machine learning. Typically, in this case, workers sample (e.g., uniformly with replacement) some batch of functions from their local finite-sums to compute the stochastic gradient. To handle a wide range of sampling strategies, we follow [11] and consider a stochastic reformulation of the problem:

$$f(x) = \mathbb{E}_{\xi \sim \mathcal{D}}\left[f_\xi(x)\right], \quad f_\xi(x) = \frac{1}{n}\sum_{i=1}^{n} f_{\xi_i}(x), \quad f_{\xi_i}(x) = \frac{1}{m}\sum_{j=1}^{m}\xi_{ij}f_{ij}(x), \quad (11)$$

where $\xi = (\xi_1^\top, \ldots, \xi_n^\top)$ and random vector $\xi_i = (\xi_{i1}, \ldots, \xi_{im})^\top$ defines the sampling strategy with distribution $\mathcal{D}_i$ such that $\mathbb{E}[\xi_{ij}] = 1$ for all $i \in [n]$, $j \in [m]$. We assume that functions $f_{\xi_i}(x)$ satisfy *expected smoothness* property [10,11].

Assumption 2 (Expected Smoothness). *Functions $f_1, \ldots, f_n$ are $\mathcal{L}$-smooth in expectation w.r.t. distributions $\mathcal{D}_1, \ldots, \mathcal{D}_n$. That is, there exists constant $\mathcal{L} > 0$ such that for all $x \in \mathbb{R}^d$ and for all $i = 1, \ldots, n$*

$$\mathbb{E}_{\xi_i \sim \mathcal{D}_i}\left[\|\nabla f_{\xi_i}(x) - \nabla f_{\xi_i}(x^*)\|^2\right] \leq 2\mathcal{L}\left(f_i(x) - f_i(x) - \langle \nabla f_i(x^*), x - x^* \rangle\right).$$

One can show that this assumption (and reformulation itself) covers for a wide range of situations [11]. For example, when all functions f_{ij} are convex and L_{ij}-smooth, then for the classical uniform sampling we have $\mathbb{P}\{\xi_i = me_j\} = \frac{1}{m}$ and $\mathcal{L} = \mathcal{L}_{US} = \max_{i\in[n],j\in[m]} L_{ij}$, where $e_j \in \mathbb{R}^m$ denotes the j-th vector in the standard basis in $\mathbb{R}^m$. Moreover, importance sampling $\mathbb{P}\{\xi_i = \frac{m\overline{L}_i}{L_{ij}}e_j\} = \frac{L_{ij}}{m\overline{L}_i}$, where $\overline{L}_i = \frac{1}{m}\sum_{j=1}^{m} L_{ij}$, also fits Assumption 2 with $\mathcal{L} = \mathcal{L}_{IS} = \max_{i\in[n]}\overline{L}_i$, which can be significantly smaller than $\mathcal{L}_{US}$.

Next, it is worth mentioning that Assumption 2 and (M, σ^2)-bounded noise assumption used in [30] cannot be compared directly, i.e., in general, none of them is stronger than another. However, in contrast to (M, σ^2)-bounded noise assumption, Assumption 2 is satisfied whenever $f_{ij}(x)$ are convex and smooth.

Consider a special case of EC-SGD (Algorithm 1) applied to the stochastic reformulation (11), i.e., let $g_i^k = \nabla f_{\xi_i^k}(x^k)$, where ξ_i^k is sampled from $\mathcal{D}_i$ independently from previous steps and other workers. Since this version of EC-SGD supports arbitrary sampling we will call it EC-SGD-AS. In this setup, we show[5] that EC-SGD-AS fits Assumption 1 (the proof is deferred to Appendix B).

Proposition 1. *Let f be L-smooth, f_i have finite-sum form, and Assumption 2 hold. Then the iterates produced by EC-SGD-AS satisfy Assumption 1 with $A = L + 2\mathcal{L}/n$, $B = 0$, $D_1 = \frac{2\sigma_*^2}{n} = \frac{2}{n^2}\sum_{i=1}^{n}\mathbb{E}[\|\nabla f_{\xi_i}(x^*) - \nabla f_i(x^*)\|^2]$, $\sigma_k^2 \equiv 0$, $\rho = 1$, $C = 0$, $D_2 = 0$.*

---

[5] Proposition 1 is a refined version of Lemma J.1 from [10].

Plugging the parameters from the above proposition in Theorem 1 and Corollary 1, one can derive convergence guarantees for EC-SGD-AS with absolute compression operator. In particular, our general analysis implies the following result.

Theorem 2. *Let the assumptions of Proposition 1 hold. Then, there exist choices of stepsize $0 < \gamma \le (4L + 8\mathcal{L}/n)^{-1}$ such that* EC-SGD-AS *with absolute compressor guarantees* $\mathbb{E}[f(\overline{x}^K) - f(x^*)]$ *of the order*

$$\widetilde{\mathcal{O}}\left(\left(L + \tfrac{\mathcal{L}}{n}\right)R_0^2 \exp\left(-\tfrac{\mu}{L+\mathcal{L}/n}K\right) + \tfrac{\sigma_*^2}{\mu nK} + \tfrac{L\Delta^2}{\mu^2 K^2}\right), \quad \text{when } \mu > 0, \qquad (12)$$

$$\mathcal{O}\left(\tfrac{(L+\mathcal{L}/n)R_0^2}{K} + \sqrt{\tfrac{\sigma_*^2 R_0^2}{nK}} + \tfrac{L^{1/3} R_0^{4/3} \Delta^{2/3}}{K^{2/3}}\right), \quad \text{when } \mu = 0. \qquad (13)$$

Consider the case when $\mu > 0$ (similar observations are valid when $\mu = 0$). In these settings, Assumption 2 is satisfied whenever f_{ij} are convex and smooth without assuming (M, σ^2)-bounded noise assumption used in [30]. Moreover, our bound (12) has better $\mathcal{O}(1/K)$ decaying term than bound (10) from [30]. In particular, when $\sigma_*^2 = 0$, i.e., workers compute full gradients our bound has $\mathcal{O}(1/K^2)$ decaying leading term while for (10) the leading term decreases as $\mathcal{O}(1/K)$, when $\zeta_*^2 = 0$ (local functions has different optima). Next, when $\mu > 0$, the best-known bound for EC-SGD-AS for δ-contractive compressors is (see [10])

$$\widetilde{\mathcal{O}}\left(AR_0^2 \exp\left(-\tfrac{\mu}{A}K\right) + \tfrac{\sigma_*^2}{\mu nK} + \tfrac{L(\sigma_*^2 + \zeta_*^2/\delta)}{\delta\mu^2 K^2}\right), \qquad (14)$$

where $A = \mathcal{L} + \tfrac{\max_{i \in [n]} L_i + \sqrt{\delta \max_{i \in [n]} L_i \mathcal{L}}}{\delta}$ (this parameter can be tightened using the independence of the samples on different workers, which we use in our proofs). The second terms from (12) and (14) are the same while the third terms are different. Although, these results are derived for different classes of compressors, one can compare them for particular choices of compressions. In particular, for hard-threshold and Top1 compressors the third term in (12) is proportional to $d\lambda/K^2$, while the corresponding term from (14) is proportional to $(d\sigma_*^2 + d^2\zeta_*^2)/K^2$. When $\lambda = \mathcal{O}(1)$ (e.g., see Fig. 1) and ζ_*^2 is large enough the bound (12) is more than d times better than (14).

4   Absolute Compression and Variance Reduction

In the same setup as in the previous section, we consider a variance-reduced version of EC-SGD called Error Compensated Loopless Stochastic Variance-Reduced Gradient (EC-LSVRG) from [10]. This method is a combination of LSVRG [13,21] and EC-SGD and can be viewed as Algorithm 1 with

$$g_i^k = \nabla f_{\xi_i^k}(x^k) - \nabla f_{\xi_i^k}(w^k) + \nabla f_i(w^k) \qquad (15)$$

$$w^{k+1} = \begin{cases} x^k, & \text{with probability } p, \\ w^k, & \text{with probability } 1-p, \end{cases} \qquad w^0 = x^0, \qquad (16)$$

where ξ_i^k is sampled from $\mathcal{D}_i$ independently from previous steps and other workers, and probability p of updating w^k is usually taken as $p \sim 1/m$. Such choice of p ensures that full gradients $\nabla f_i(w^k)$ are computed rarely meaning that the expected number of $\nabla f_{\xi_i}(x)$ computations per iteration is the same as for EC-SGD-AS (up to the constant factor). We point out that EC-LSVRG was studied for the contractive compressors and uniform sampling [10] (although it is possible to generalize the proofs from [10] to cover EC-LSVRG with arbitrary sampling as well). As we show next, EC-LSVRG with arbitrary sampling satisfies Assumption 1 (the proof is deferred to Appendix B).

Proposition 2. *Let f be L-smooth, f_i have finite-sum form, and Assumption 2 hold. Then the iterates produced by EC-LSVRG satisfy Assumption 1 with $A = L + \frac{2\mathcal{L}}{n}$, $B = \frac{2}{n}$, $D_1 = 0$, $\sigma_k^2 = 2\mathcal{L}(f(w^k) - f(x^*))$, $\rho = p$, $C = p\mathcal{L}$, $D_2 = 0$.*

Due to the variance reduction, noise terms D_1 and D_2 equal zero for EC-LSVRG allowing the method to achieve better accuracy with constant stepsize than EC-SGD-AS. Plugging the parameters from the above proposition in Theorem 1 and Corollary 1, one can derive convergence guarantees for EC-LSVRG with absolute compression operator. In particular, our general analysis implies the following result.

Theorem 3. *Let the assumptions of Proposition 2 hold. Then, there exist choices of stepsize $0 < \gamma \le \gamma_0 = (4L + 152\mathcal{L}/3n)^{-1}$ such that EC-LSVRG with absolute compressor guarantees $\mathbb{E}[f(\overline{x}^K) - f(x^*)]$ of the order[6]*

$$\widetilde{\mathcal{O}}\left(\left(L + \frac{\mathcal{L}}{n}\right)\widetilde{T}_0 \exp\left(-\min\left\{\frac{\mu}{L+\mathcal{L}/n}, \frac{1}{m}\right\}K\right) + \frac{L\Delta^2}{\mu^2 K^2}\right), \quad \text{when } \mu > 0, \quad (17)$$

$$\mathcal{O}\left(\frac{(L+\mathcal{L}/n)R_0^2}{K} + \frac{\sqrt{m\mathcal{L}L}R_0^2}{\sqrt{n}K} + \frac{L^{1/3}R_0^{4/3}\Delta^{2/3}}{K^{2/3}}\right), \quad \text{when } \mu = 0, \quad (18)$$

where $\widetilde{T}_0 = \|x^0 - x^*\|^2 + \frac{64m}{3n}\gamma_0^2\mathcal{L}(f(x^0) - f(x^*))$.

Consider the case when $\mu > 0$ (similar observations are valid when $\mu = 0$). As expected, the bound (17) does not have terms proportional to any kind of variance of the stochastic estimator. Therefore, the leading term in the complexity bound for EC-LSVRG decreases as $\mathcal{O}(1/K^2)$, while EC-SGD-AS has $\mathcal{O}(1/K)$ leading term. Next, in case of δ-contractive compression operator the only known convergence rate for EC-LSVRG [10] has the leading term $\mathcal{O}(\frac{L\zeta_*^2}{\delta^2 \mu^2 K^2})$. Using the same arguments as in the discussion after Theorem 2, one can that the leading term in the case of hard-threshold sparsifier can be more than d times better than the leading term in the case of Top1 compressor.

---

[6] We take into account that due to L-smoothness of f we have $T_0 = \|x^0 - x^*\|^2 + \frac{64m}{3n}\gamma^2\mathcal{L}(f(x^0) - f(x^*)) \le (1 + \frac{32m\mathcal{L}L\gamma^2}{3}n)\|x^0 - x^*\|^2$.

5   Numerical Experiments

We conduct several numerical experiments to support our theory, i.e., we tested the methods on the distributed logistic regression problem with ℓ_2-regularization:

$$\min_{x\in\mathbb{R}^d}\left\{f(x)=\frac{1}{nm}\sum_{i=1}^{n}\sum_{j=1}^{m}\underbrace{\ln\left(1+\exp\left(-y_i\langle a_{ij},x\rangle\right)\right)+\frac{l_2}{2}\|x\|^2}_{f_{ij}(x)}\right\},\qquad(19)$$

where vectors $\{a_{ij}\}_{i\in[n],j\in[m]}\in\mathbb{R}^d$ are the columns of the matrix of features $\mathbf{A}^\top$, $\{y_i\}_{i=1}^{n}\in\{-1,1\}$ are labels, and $l_2\geq 0$ is a regularization parameter. One can show that f_{ij} is l_2-strongly convex and L_{ij}-smooth, and f is L-smooth with $L_{ij}=l_2+\|a_{ij}\|^2/4$ and $L=l_2+\lambda_{\max}(\mathbf{A}^\top\mathbf{A})/4nm$, where $\lambda_{\max}(\mathbf{A}^\top\mathbf{A})$ is the largest eigenvalue of $\mathbf{A}^\top\mathbf{A}$. In particular, we took 3 datasets – a9a ($nm=32000$, $d=123$)[7], gisette ($nm=6000$, $d=5000$), and w8a ($nm=49700$, $d=300$) – from LIBSVM library [5]. Out code is based on the one from [10]. Each dataset was shuffled and equally split between $n=20$ workers. In the first two experiments, we use only hard-threshold sparsifier and in the third experiment, we also consider TopK compressor. The results are presented in Fig. 1. The methods were run for S epochs, where the values of S are given in the titles of corresponding plots. We compare the methods in terms of total number of bits that each worker sends to the server (on average).

Experiment 1: EC-SGD with and without importance sampling. In this experiment, we tested EC-SGD with hard-threshold sparsifier and two different sampling strategies: uniform sampling (US) and importance sampling (IS), described in Sect. 3. We chose $l_2=10^{-4}\cdot\max_{i\in[n]}\overline{L}_i$. Stepsize was chosen as $\gamma_{\mathrm{US}}=(L+n^{-1}\max_{i\in[n],j\in[m]}L_{ij})^{-1}$ and $\gamma_{\mathrm{IS}}=(L+n^{-1}\max_{i\in[n]}\overline{L}_i)^{-1}$ for the case of US and IS respectively, which are the multiples of the maximal stepsizes that our theory allows for both cases. Following [30], we took parameter λ as $\lambda=5000\sqrt{\varepsilon/d^2\gamma_{\mathrm{US}}}$ for $\varepsilon=10^{-3}$. We observe that EC-SGD behaves similarly in both cases for a9a ($L\approx1.57$, $\max_{i\in[n]}\overline{L}_i\approx3.47$, $\max_{i\in[n],j\in[m]}L_{ij}\approx3.5$) and gisette ($L\approx842.87$, $\max_{i\in[n]}\overline{L}_i\approx1164.89$, $\max_{i\in[n],j\in[m]}L_{ij}\approx1201.51$), while for w8a ($L\approx0.66$, $\max_{i\in[n]}\overline{L}_i\approx3.05$, $\max_{i\in[n],j\in[m]}L_{ij}\approx28.5$) EC-SGD with IS achieves good enough accuracy much faster than with US. This is expected since for w8a $\max_{i\in[n]}\overline{L}_i$ is almost 10 times smaller than $\max_{i\in[n],j\in[m]}L_{ij}$. That is, as our theory implies, importance sampling is preferable when $\max_{i\in[n]}\overline{L}_i\ll\max_{i\in[n],j\in[m]}L_{ij}$.

Experiment 2: EC-SGD vs EC-LSVRG. Next, we compare EC-SGD and EC-LSVRG to illustrate the benefits of variance reduction for error-compensated methods with absolute compression. Both methods were run with stepsize $\gamma=1/\max_{i\in[n],j\in[m]}L_{ij}$ and $\lambda=5000\sqrt{\varepsilon/d^2\gamma}$ for $\varepsilon=10^{-3}$. In all cases, EC-LSVRG

---

[7] We take the first 32000 and 49700 samples from a9a and w8a to get a multiple of $n=20$.

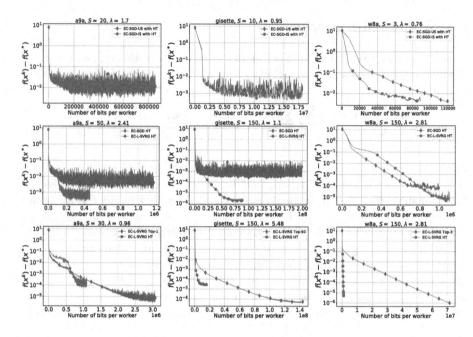

Fig. 1. Trajectories of EC-SGD with uniform and importance samplings sampling and hard-threshold sparsifier, EC-LSVRG with hard-threshold and TopK sparsifiers. The row i of plots corresponds to Experiment i described in Sect. 5.

achieves better accuracy of the solution than EC-SGD that perfectly corroborates our theoretical results.

Experiment 3: EC-LSVRG with hard-threshold and TopK sparsifiers. Finally, to highlight the benefits of hard-threshold (HT) sparsifier compared to TopK sparsifier we tested EC-LSVRG with both compressors. We chose stepsize as $\gamma = 1/\max_{i\in[n],j\in[m]} L_{ij}$, $K \approx d/100$ (see the legends of plots in row 3 from Fig. 1), and $\lambda = \alpha\sqrt{\varepsilon/d^2\gamma}$, where $\varepsilon = 10^{-3}$ and $\alpha = 2000$ for a9a, $\alpha = 25000$ for gisette, $\alpha = 5000$ for w8a. We observe that EC-LSVRG with HT achieves a reasonable accuracy (10^{-3} for a9a, 10^{-4} for gisette, and 10^{-5} for w8a) faster than EC-LSVRG with TopK for all datasets. In particular, EC-LSVRG with HT significantly outperforms EC-LSVRG with TopK on w8a dataset.

A   Missing Proofs from Sect. 2

In the analysis, we use auxiliary iterates that are never computed explicitly during the work of the method: $\widetilde{x}^k = x^k - e^k$, where $e^k = \frac{1}{n}\sum_{i=1}^{n} e_i^k$. These iterates are usually called *perturbed* or *virtual* iterates [22,24]. They are used in many previous works on error feedback [10,30,33,34]. One of the key properties these iterates satisfy is the following recursion:

$$\widetilde{x}^{k+1} = x^{k+1} - e^{k+1} = x^k - v^k - e^k - \gamma g^k + v^k = \widetilde{x}^k - \gamma g^k, \qquad (20)$$

where we use $e^{k+1} = e^k + \gamma g^k - v^k$, which follows from $e_i^{k+1} = e_i^k + \gamma g_i^k - v_i^k$ and definitions of e^k, g^k, and v^k.

A.1 Proof of Theorem 1

Our proof is close to the ones from [10,30]. Recursion (20) implies

$$\|\widetilde{x}^{k+1} - x^*\|^2 = \|\widetilde{x}^k - x^*\|^2 - 2\gamma\langle\widetilde{x}^k - x^*, g^k\rangle + \gamma^2\|g^k\|^2.$$

Next, we take conditional expectation $\mathbb{E}_k[\cdot]$ from the above inequality and apply unbiasedness of g^k, inequality (4) from Assumption 1, and μ-quasi strong convexity of f, i.e., $-\langle x^k - x^*, \nabla f(x^k)\rangle \overset{(2)}{\leq} -(f(x^k) - f(x^*)) - \frac{\mu}{2}\|x^k - x^*\|^2$:

$$\mathbb{E}_k\left[\|\widetilde{x}^{k+1} - x^*\|^2\right] \leq \|\widetilde{x}^k - x^*\|^2 - \gamma\mu\|x^k - x^*\|^2 + 2\gamma\langle x^k - \widetilde{x}^k, \nabla f(x^k)\rangle$$
$$-2\gamma(1 - A\gamma)\left(f(x^k) - f(x^*)\right) + B\gamma^2\sigma_k^2 + \gamma^2 D_1 \quad (21)$$

Applying $\|a - b\|^2 \geq \frac{1}{2}\|a\|^2 - \|b\|^2$, which holds for any $a, b \in \mathbb{R}^d$, we get

$$-\gamma\mu\|x^k - x^*\|^2 \leq -\tfrac{\gamma\mu}{2}\|\widetilde{x}^k - x^*\|^2 + \gamma\mu\|\widetilde{x}^k - x^k\|^2. \quad (22)$$

To estimate the inner product, we use Fenchel-Young inequality $\langle a, b\rangle \leq \frac{\alpha}{2}\|a\|^2 + \frac{1}{2\alpha}\|b\|^2$ holding for any $a, b \in \mathbb{R}^d$, $\alpha > 0$ together with standard inequality $\|\nabla f(x^k)\|^2 \leq 2L(f(x^k) - f(x^*))$, which holds for any L-smooth function f [27]:

$$2\gamma\langle x^k - \widetilde{x}^k, \nabla f(x^k)\rangle \leq 2\gamma L\|x^k - \widetilde{x}^k\|^2 + \tfrac{\gamma}{2L}\|\nabla f(x^k)\|^2$$
$$\leq 2\gamma L\|\widetilde{x}^k - x^k\|^2 + \gamma\left(f(x^k) - f(x^*)\right). \quad (23)$$

Plugging upper bounds (22) and (23) in (21), we derive

$$\mathbb{E}_k\left[\|\widetilde{x}^{k+1} - x^*\|^2\right] \overset{\mu \leq L}{\leq} \left(1 - \tfrac{\gamma\mu}{2}\right)\|\widetilde{x}^k - x^*\|^2 - \gamma(1 - 2A\gamma)\left(f(x^k) - f(x^*)\right)$$
$$+B\gamma^2\sigma_k^2 + \gamma^2 D_1 + 3L\gamma\|e^k\|^2.$$

Since the definition of e^{k+1} and Jensen's inequality imply that for all $k \geq 0$

$$\|e^{k+1}\|^2 \leq \frac{1}{n}\sum_{i=1}^{n}\|e_i^k + \gamma g_i^k - v_i^k\|^2 = \frac{\gamma^2}{n}\sum_{i=1}^{n}\left\|\frac{e_i^k + \gamma g_i^k}{\gamma} - \mathcal{C}\left(\frac{e_i^k + \gamma g_i^k}{\gamma}\right)\right\|^2 \leq \gamma^2\Delta^2,$$

we have (taking into account that $\|e^0\|^2 = 0 \leq \gamma^2\Delta^2$)

$$\mathbb{E}_k\left[\|\widetilde{x}^{k+1} - x^*\|^2\right] \leq \left(1 - \tfrac{\gamma\mu}{2}\right)\|\widetilde{x}^k - x^*\|^2 - \gamma(1 - 2A\gamma)\left(f(x^k) - f(x^*)\right)$$
$$+B\gamma^2\sigma_k^2 + \gamma^2 D_1 + 3L\gamma^3\Delta^2.$$

Summing up the above inequality with $F\gamma^2$-multiple of (5) and introducing new notation $T_k = \|\widetilde{x}^k - x^*\|^2 + F\gamma^2\sigma_k^2$, we obtain

$$\mathbb{E}_k[T_{k+1}] \leq \left(1 - \tfrac{\gamma\mu}{2}\right)\|\widetilde{x}^k - x^*\|^2 + F\gamma^2\left(1 - \rho + \tfrac{B}{F}\right)\sigma_k^2$$
$$-\gamma(1 - 2(A + CF)\gamma)\left(f(x^k) - f(x^*)\right)$$
$$+\gamma^2\left(D_1 + FD_2 + 3L\gamma\Delta^2\right)$$
$$\overset{F = 4B/3\rho}{\leq} (1 - \eta)T_k + \gamma^2\left(D_1 + FD_2 + 3L\gamma\Delta^2\right) - \tfrac{\gamma}{2}\left(f(x^k) - f(x^*)\right),$$

where in the last step we use $\left(1 - \frac{\gamma\mu}{2}\right)\|\widetilde{x}^k - x^*\|^2 + \left(1 - \frac{\rho}{4}\right)F\gamma^2\sigma_k^2 \le (1 - \eta)\left(\|\widetilde{x}^k - x^*\|^2 + F\gamma^2\sigma_k^2\right) = (1 - \eta)T_k$, where $\eta = \min\{\gamma\mu/2, \rho/4\}$, and $0 < \gamma \le 1/4(A+CF)$. Rearranging the terms and taking full expectation, we derive

$$\frac{\gamma}{2}\mathbb{E}\left[f(x^k) - f(x^*)\right] \le (1 - \eta)\mathbb{E}[T_k] - \mathbb{E}[T_{k+1}] + \gamma^2\left(D_1 + FD_2 + 3L\gamma\Delta^2\right).$$

Summing up the above inequality for $k = 0, \ldots, K$ with weights $w_k = (1 - \eta)^{-(k+1)}$ and using Jensen's inequality $f(\overline{x}^K) \le \frac{1}{W_K}\sum_{k=0}^{K}w_k f(x^k)$, where $\overline{x}^K = \frac{1}{W_K}\sum_{k=0}^{K}w_k x^k$, $W_K = \sum_{k=0}^{K}w_k$, and $w_k = (1 - \eta)w_{k-1}$, we get

$$\mathbb{E}\left[f(\overline{x}^K) - f(x^*)\right] \le \frac{2\mathbb{E}[T_0]}{\gamma W_k} + 2\gamma\left(D_1 + FD_2 + 3L\gamma\Delta^2\right), \tag{24}$$

where we use $\sum_{k=0}^{K}\left(w_{k-1}\mathbb{E}[T_k] - w_k\mathbb{E}[T_{k+1}]\right) = w_{-1}\mathbb{E}[T_0] - w_{k+1}\mathbb{E}[T_{K+1}] \le w_{-1}\mathbb{E}[T_0] = \mathbb{E}[T_0]$. Finally, it remains to notice that (24) implies (6) and (7). Indeed, when $\mu > 0$, we have $W_K \ge w_K = (1 - \eta)^{-(K+1)}$, and when $\mu = 0$, we have $W_K = K + 1$, since $\eta = 0$.

B   Missing Proofs from Sects. 3 and 4

B.1   Proof of Proposition 1

Independence of $\xi_1^k, \ldots, \xi_n^k$ for fixed history, variance decomposition, and standard inequality $\|\nabla f(x^k)\|^2 \le 2L(f(x) - f(x^*))$ [27] imply

$$\mathbb{E}_k\left[\|g^k\|^2\right] = \mathbb{E}_k\left[\|g^k - \nabla f(x^k)\|^2\right] + \|\nabla f(x^k)\|^2$$

$$\le \mathbb{E}_k\left[\left\|\frac{1}{n}\sum_{i=1}^{n}\nabla f_{\xi_i^k}(x^k) - \nabla f_i(x^k)\right\|^2\right] + 2L\left(f(x^k) - f(x^*)\right)$$

$$= \frac{1}{n^2}\sum_{i=1}^{n}\mathbb{E}_k\left[\|\nabla f_{\xi_i^k}(x^k) - \nabla f_i(x^k)\|^2\right] + 2L\left(f(x^k) - f(x^*)\right)$$

$$\le \frac{2}{n^2}\sum_{i=1}^{n}\mathbb{E}_k\left[\|\nabla f_{\xi_i^k}(x^k) - \nabla f_{\xi_i^k}(x^*) - (\nabla f_i(x^k) - \nabla f_i(x^*))\|^2\right]$$

$$+ \frac{2}{n^2}\sum_{i=1}^{n}\mathbb{E}_k\left[\|\nabla f_{\xi_i^k}(x^*) - \nabla f_i(x^*)\|^2\right] + 2L\left(f(x^k) - f(x^*)\right),$$

where in the last step we use that $\|a + b\|^2 \le 2\|a\|^2 + 2\|b\|^2$ for all $a, b \in \mathbb{R}^d$. Since the variance is upper-bounded by the second moment, we have

$$\mathbb{E}_k\left[\|g^k\|^2\right] \le \frac{2}{n^2}\sum_{i=1}^{n}\mathbb{E}_k\left[\|\nabla f_{\xi_i^k}(x^k) - \nabla f_{\xi_i^k}(x^*)\|^2\right] + 2L\left(f(x^k) - f(x^*)\right) + \frac{2\sigma_*^2}{n}$$

$$\le 2\left(L + \frac{2\mathcal{L}}{n}\right)\left(f(x^k) - f(x^*)\right) + \frac{2\sigma_*^2}{n},$$

where the second inequality follows from Assumption 2 and $f(x) = \frac{1}{n}\sum_{i=1}^{n}f_i(x)$. The derived inequality implies that Assumption 1 holds with $A = L + \frac{2\mathcal{L}}{n}$, $B = 0$, $D_1 = \frac{2\sigma_*^2}{n} = \frac{2}{n^2}\sum_{i=1}^{n}\mathbb{E}[\|\nabla f_{\xi_i}(x^*) - \nabla f_i(x^*)\|^2]$, $\sigma_k^2 \equiv 0$, $\rho = 1$, $C = 0$, $D_2 = 0$.

B.2 Proof of Proposition 2

We start with deriving an upper-bound for $\mathbb{E}_k[\|g^k\|^2]$. Similarly to the proof of Proposition 1, we use independence of $\xi_1^k, \ldots, \xi_n^k$ for fixed history, variance decomposition, and standard inequality $\|\nabla f(x^k)\|^2 \leq 2L(f(x) - f(x^*))$ [27]:

$$
\mathbb{E}_k\left[\|g^k\|^2\right] = \mathbb{E}_k\left[\|g^k - \nabla f(x^k)\|^2\right] + \|\nabla f(x^k)\|^2
$$

$$
\leq \mathbb{E}_k\left[\left\|\frac{1}{n}\sum_{i=1}^n \nabla f_{\xi_i^k}(x^k) - \nabla f_{\xi_i^k}(w^k) + \nabla f_i(w^k) - \nabla f_i(x^k)\right\|^2\right]
$$

$$
+ 2L\left(f(x^k) - f(x^*)\right)
$$

$$
= \frac{1}{n^2}\sum_{i=1}^n \mathbb{E}_k\left[\left\|\nabla f_{\xi_i^k}(x^k) - \nabla f_{\xi_i^k}(w^k) - (\nabla f_i(x^k) - \nabla f_i(w^k))\right\|^2\right]
$$

$$
+ 2L\left(f(x^k) - f(x^*)\right).
$$

Since the variance is upper-bounded by the second moment and $\|a + b\|^2 \leq 2\|a\|^2 + 2\|b\|^2$ for all $a, b \in \mathbb{R}^d$, we have

$$
\mathbb{E}_k\left[\|g^k\|^2\right] \leq \frac{2}{n^2}\sum_{i=1}^n \mathbb{E}_k\left[\left\|\nabla f_{\xi_i^k}(x^k) - \nabla f_{\xi_i^k}(x^*)\right\|^2\right]
$$

$$
+ \frac{2}{n^2}\sum_{i=1}^n \mathbb{E}_k\left[\left\|\nabla f_{\xi_i^k}(w^k) - \nabla f_{\xi_i^k}(x^*)\right\|^2\right] + 2L\left(f(x^k) - f(x^*)\right)
$$

$$
\leq 2\left(L + \frac{2\mathcal{L}}{n}\right)\left(f(x^k) - f(x^*)\right) + 4\mathcal{L}\left(f(w^k) - f(x^*)\right), \tag{25}
$$

where the third inequality follows from Assumption 2 and $f(x) = \frac{1}{n}\sum_{i=1}^n f_i(x)$. Using the definitions of σ_{k+1}^2 and w^{k+1}, we derive an upper bound for $\mathbb{E}_k[\sigma_{k+1}^2]$:

$$
\mathbb{E}_k[\sigma_{k+1}^2] = 2\mathcal{L}\mathbb{E}_k\left[f(w^{k+1}) - f(x^*)\right]
$$

$$
= (1 - p)2\mathcal{L}\left(f(w^k) - f(x^*)\right) + 2p\mathcal{L}\left(f(x^k) - f(x^*)\right). \tag{26}
$$

Inequalities (25) and (26) imply that Assumption 1 holds with $A = L + \frac{2\mathcal{L}}{n}$, $B = \frac{2}{n}$, $D_1 = 0$, $\sigma_k^2 = 2\mathcal{L}(f(w^k) - f(x^*))$, $\rho = p$, $C = p\mathcal{L}$, $D_2 = 0$.

References

1. Alistarh, D., Grubic, D., Li, J., Tomioka, R., Vojnovic, M.: QSGD: communication-efficient SGD via gradient quantization and encoding. In: Advances in Neural Information Processing Systems, pp. 1709–1720 (2017)
2. Beznosikov, A., Gorbunov, E., Berard, H., Loizou, N.: Stochastic gradient descent-ascent: unified theory and new efficient methods. arXiv preprint arXiv:2202.07262 (2022)
3. Beznosikov, A., Horváth, S., Richtárik, P., Safaryan, M.: On biased compression for distributed learning. arXiv preprint arXiv:2002.12410 (2020)

4. Brown, T., et al.: Language models are few-shot learners. Adv. Neural Inf. Process. Syst. **33**, 1877–1901 (2020)
5. Chang, C.C., Lin, C.J.: LIBSVM: a library for support vector machines. ACM Trans. Intell. Syst. Technol. **2**(3), 1–27 (2011)
6. Dutta, A., et al.: On the discrepancy between the theoretical analysis and practical implementations of compressed communication for distributed deep learning. In: Proceedings of the AAAI Conference on Artificial Intelligence, vol. 34, pp. 3817–3824 (2020)
7. Gorbunov, E., Berard, H., Gidel, G., Loizou, N.: Stochastic extragradient: general analysis and improved rates. arXiv preprint arXiv:2111.08611 (2021)
8. Gorbunov, E., Hanzely, F., Richtárik, P.: A unified theory of SGD: variance reduction, sampling, quantization and coordinate descent. In: The 23rd International Conference on Artificial Intelligence and Statistics (AISTATS 2020) (2020)
9. Gorbunov, E., Hanzely, F., Richtárik, P.: Local SGD: unified theory and new efficient methods. In: International Conference on Artificial Intelligence and Statistics, pp. 3556–3564. PMLR (2021)
10. Gorbunov, E., Kovalev, D., Makarenko, D., Richtárik, P.: Linearly converging error compensated SGD. Adv. Neural Inf. Process. Syst. **33** (2020)
11. Gower, R.M., Loizou, N., Qian, X., Sailanbayev, A., Shulgin, E., Richtárik, P.: SGD: general analysis and improved rates. In: International Conference on Machine Learning, pp. 5200–5209 (2019)
12. Gupta, S., Agrawal, A., Gopalakrishnan, K., Narayanan, P.: Deep learning with limited numerical precision. In: International Conference on Machine Learning, pp. 1737–1746. PMLR (2015)
13. Hofmann, T., Lucchi, A., Lacoste-Julien, S., McWilliams, B.: Variance reduced stochastic gradient descent with neighbors. In: Advances in Neural Information Processing Systems, pp. 2305–2313 (2015)
14. Horváth, S., Kovalev, D., Mishchenko, K., Stich, S., Richtárik, P.: Stochastic distributed learning with gradient quantization and variance reduction. arXiv preprint arXiv:1904.05115 (2019)
15. Kairouz, P., et al.: Advances and open problems in federated learning. arXiv preprint arXiv:1912.04977 (2019)
16. Kaplan, J., McCandlish, S., et al.: Scaling laws for neural language models. arXiv preprint arXiv:2001.08361 (2020)
17. Khaled, A., Sebbouh, O., Loizou, N., Gower, R.M., Richtárik, P.: Unified analysis of stochastic gradient methods for composite convex and smooth optimization. arXiv preprint arXiv:2006.11573 (2020)
18. Koloskova, A., Stich, S., Jaggi, M.: Decentralized stochastic optimization and gossip algorithms with compressed communication. In: International Conference on Machine Learning, pp. 3478–3487 (2019)
19. Konečný, J., McMahan, H.B., Yu, F.X., Richtárik, P., Suresh, A.T., Bacon, D.: Federated learning: strategies for improving communication efficiency. arXiv preprint arXiv:1610.05492 (2016)
20. Konečný, J., McMahan, H.B., Yu, F., Richtárik, P., Suresh, A.T., Bacon, D.: Federated learning: strategies for improving communication efficiency. In: NIPS Private Multi-Party Machine Learning Workshop (2016)
21. Kovalev, D., Horváth, S., Richtárik, P.: Don't jump through hoops and remove those loops: SVRG and Katyusha are better without the outer loop. In: Proceedings of the 31st International Conference on Algorithmic Learning Theory (2020)

22. Leblond, R., Pedregosa, F., Lacoste-Julien, S.: Improved asynchronous parallel optimization analysis for stochastic incremental methods. J. Mach. Learn. Res. **19**(1), 3140–3207 (2018)
23. Li, C.: Openai's gpt-3 language model: a technical overview. Blog Post (2020)
24. Mania, H., Pan, X., Papailiopoulos, D., Recht, B., Ramchandran, K., Jordan, M.I.: Perturbed iterate analysis for asynchronous stochastic optimization. SIAM J. Optimiz. **27**(4), 2202–2229 (2017)
25. Mishchenko, K., Gorbunov, E., Takáč, M., Richtárik, P.: Distributed learning with compressed gradient differences. arXiv preprint arXiv:1901.09269 (2019)
26. Necoara, I., Nesterov, Y., Glineur, F.: Linear convergence of first order methods for non-strongly convex optimization. Math. Program. **175**(1), 69–107 (2019)
27. Nesterov, Yurii: Lectures on Convex Optimization. SOIA, vol. 137. Springer, Cham (2018). https://doi.org/10.1007/978-3-319-91578-4
28. Qian, X., Richtárik, P., Zhang, T.: Error compensated distributed SGD can be accelerated. arXiv preprint arXiv:2010.00091 (2020)
29. Richtárik, P., Sokolov, I., Fatkhullin, I.: Ef21: a new, simpler, theoretically better, and practically faster error feedback. Adv. Neural Inf. Process. Syst. **34** (2021)
30. Sahu, A., Dutta, A., M Abdelmoniem, A., Banerjee, T., Canini, M., Kalnis, P.: Rethinking gradient sparsification as total error minimization. Adv. Neural Inf. Process. Syst. **34** (2021)
31. Sapio, A., et al.: Scaling distributed machine learning with in-network aggregation. In: 18th USENIX Symposium on Networked Systems Design and Implementation (NSDI 21), pp. 785–808 (2021)
32. Seide, F., Fu, H., Droppo, J., Li, G., Yu, D.: 1-bit stochastic gradient descent and its application to data-parallel distributed training of speech DNNs. In: Fifteenth Annual Conference of the International Speech Communication Association (2014)
33. Stich, S.U., Cordonnier, J.B., Jaggi, M.: Sparsified SGD with memory. In: Advances in Neural Information Processing Systems, pp. 4447–4458 (2018)
34. Stich, S.U., Karimireddy, S.P.: The error-feedback framework: better rates for SGD with delayed gradients and compressed updates. J. Mach. Learn. Res. **21**, 1–36 (2020)
35. Strom, N.: Scalable distributed DNN training using commodity GPU cloud computing. In: Sixteenth Annual Conference of the International Speech Communication Association (2015)
36. Tang, H., Yu, C., Lian, X., Zhang, T., Liu, J.: DoubleSqueeze: parallel stochastic gradient descent with double-pass error-compensated compression. In: International Conference on Machine Learning, pp. 6155–6165 (2019)
37. Zinkevich, M., Weimer, M., Li, L., Smola, A.: Parallelized stochastic gradient descent. Adv. Neural Inf. Process. Syst. **23** (2010)

An Experimental Analysis of Dynamic Double Description Method Variations

Sergey O. Semenov[ID] and Nikolai Yu. Zolotykh[(✉)][ID]

Mathematics of Future Technologies Center, Lobachevsky State University of Nizhny Novgorod, Gagarin ave. 23, Nizhny Novgorod 603950, Russia
{sergey.semenov,nikolai.zolotykh}@itmm.unn.ru

Abstract. In this study, we perform a comparative experimental analysis of several modifications of the dynamic double description method for generating the extreme rays of a polyhedral cone. The modifications under consideration include using graph adjacency test, bit pattern trees and maintaining the set of all adjacent extreme rays. Results of computational experiments on several classes of problems are presented.

Keywords: System of linear inequalities · Convex hull · Cone · Polyhedron · Double description method

1 Introduction

It is well-known that any convex polyhedron $P \subseteq F^d$, where F is an ordered field, can be represented in any of the following two ways [13,17]:

(1) as the set $P = \{x \in F^d : Ax \leq b\}$ of solutions to a system of linear inequalities, where $A \in F^{m \times d}$, $b \in F^m$, $m \in \mathbb{N}$ (*facet description*);
(2) as the sum of the conical hull of a set of vectors $v_1, \ldots, v_s$ in F^d and the convex hull of a set of points $w_1, \ldots, w_n$ in F^d, where $s, n \in \mathbb{N}$ (*vertex description*).

The problem of finding the representation (1) given the representation (2) is called *the convex hull problem*. According to the classical theorem of Weyl, this problem is equivalent to the problem of constructing the representation (2) given the representation (1). These two problems are referred to as *finding the dual representation of a polyhedron.*

The problem of constructing the dual representation of a convex polyhedron plays a central role in the theory of linear inequality systems and computational geometry [5,17]. The importance of studying this problem is also emphasized by the fact that it has a variety of applications, the most common of which are linear and integer programming [13], combinatorial optimization [17] and global optimization [10].

Likewise, there are two ways to represent a polyhedral cone:

The article was prepared under financial support of Russian Science Foundation grant No. 21-11-00194.

(1) as the solution set of a homogeneous system of linear inequalities $C = \{x \in F^d : Ax \geq 0\}$, where $A \in F^{m \times d}, m \in \mathbb{N}$. The system of linear inequalities $Ax \geq 0$ is then said to *define* the cone C.

(2) as the conical hull of a set of vectors $v_1, \ldots, v_s$:

$$C = \{x = \alpha_1 v_1 + \alpha_2 v_2 + \cdots + \alpha_s v_s : \alpha_i \geq 0 \ (i = 1, \ldots, s)\}$$

or, in matrix form:

$$C = \{x = \alpha V, \ \alpha \in F^s, \ \alpha \geq 0\},$$

where $v_1, v_2, \ldots, v_s$ are rows of matrix $V \in F^{s \times d}, s \in \mathbb{N}$ and α is a row vector. The set of vectors $v_1, \ldots, v_s$ are said to *generate* the cone C.

From an algorithmic or theoretical standpoint it is easier to consider these problems only for polyhedral cones. There is a standard method for reducing the problem of finding the dual representation for convex polyhedra to the corresponding one for polyhedral cones. For example, in order to find the representation (2) for a polyhedron $P = \{x \in F^d : Ax \leq b\}$ it is sufficient to solve the corresponding problem for the polyhedral cone $\{x = (x_0, x_1, \ldots, x_d) \in F^{d+1} : bx_0 - Ax \geq 0, x_0 \geq 0\}$, and then set $x_0 = 1$ [17].

There are several known algorithms for solving these problems. One of the most popular ones is *the double description method* (DDM) [12], also known as Chernikova's algorithm [6]. The double description method generally outperforms the other algorithms when applied to degenerate inputs and/or outputs [1].

In this paper we explore the impact of several known double description method variations in the context of the *dynamic* version of the problem [14]. In this version of the problem the full primal representation is not known in advance, instead, it is fed into the algorithm row-by-row, adding a single inequality on each iteration, which renders many of the heuristics proposed by various authors unusable. This work builds on the initial dynamic double description method proposed and implemented as part of [14].

A non-zero vector $u \in C$ is referred to as a *ray* of the cone C. Two rays u and v are *equal* (written as $u \simeq v$) if for some $\alpha > 0$ it is true that $u = \alpha v$. A ray $u \in C$ is said to be *extreme* if the condition $u = \alpha v + \beta w$, where $\alpha \geq 0, \beta \geq 0$ and $v, w \in C$ implies that $u \simeq v \simeq w$. The set of extreme rays of an acute cone is also called the *skeleton* of the cone. The skeleton is the minimal generating system of an acute cone. Suppose that P is a convex subset of F^d, and for some $a \in F^d, \alpha \in F$, it holds that $P \subseteq \{x : ax \leq \alpha\}$. Then $P \cap \{x : ax = \alpha\}$ is called a *face* of the set P. Two different extreme rays u and v of an acute cone C are said to be *adjacent*, if no minimal face containing both rays contains any other extreme rays of the cone C. The skeleton of C is denoted by $U(C)$, while the set of all edges or pairs of adjacent extreme rays is denoted by $E(C)$. Let $Z(u) = \{i : A_i u = 0\}$, i.e. the set of indices of active inequalities for ray u.

2   Double Description Method

The double description method [12] is an incremental algorithm that is based on maintaining both primal and dual representations while adding a new row of the former and updating the latter on each iteration. Its dynamic variation [14] adds the additional restriction that the full primal representation is not known in advance and the input is fed into the algorithm row-by-row on each iteration, which limits the heuristics that can be applied.

Generally, it is enough to consider what happens to the generating system on one iteration of the algorithm. With the addition of a new inequality $au \geq 0$ to the system each extreme ray in its prior dual representation $U(C)$ falls into one of these three subsets:

$$U_+ = \{u \in U(C) : au > 0\}$$
$$U_- = \{u \in U(C) : au < 0\}$$
$$U_0 = \{u \in U(C) : au = 0\}$$

Then the minimal generating system of the new cone $C' = C \cap \{x : ax \geq 0\}$ is $U_+ \cup U_- \cup U_\pm$, where:

$$U_\pm = \{w = (au)v - (av)u : u \in U_+, v \in U_-, \{u, v\} \in E(C)\}$$

Finding the set $E(C)$, i.e. testing extreme rays for adjacency, appears to be the most computation-intensive part of the algorithm. Various approaches have been proposed in order to alleviate this overhead [8,9,16,18]. It should be noted that since we consider the dynamic version of the problem, variations of the algorithm that either modify the order primal representation rows are considered in (e.g. [7,18]) or otherwise rely on the fact that the primal representation is known in advance (such as reducing the number of adjacency tests proposed in [8]) are not available. Here we consider modifications that affect the test of two extreme rays for adjacency and the moment when the extreme rays are tested.

3   Dynamic DDM Modifications

The combinatorial adjacency test was originally proposed in [12]. It is formulated by the following proposition:

Proposition 1. Let $(u, v) \in U(C)$. Then $\{u, v\} \in E(C)$ if and only if $\nexists\ w \in U(C) \setminus \{u, v\} : Z(u) \cap Z(v) \subseteq Z(w)$.

Another necessary, but not sufficient, condition for adjacency of two extreme rays is the following:

Proposition 2. If $\{u, v\} \in E(C)$, then $|Z(u) \cap Z(v)| \geq r - 2$, where $r = rank(A)$.

This condition has been considered by many authors (e.g. [6,8,11]). Numerous experiments suggest that this condition has to be checked first in order to significantly reduce the running time of the algorithm by filtering out some of the adjacency candidate pairs earlier. We consider this two-step combinatorial adjacency test as baseline: the first step, which checks the condition from the Proposition 2, will be referred to as the *narrowing* step and the second one, which checks the condition from the Proposition 1, will be called the *verification* step. All of the adjacency test variations considered in this paper follow this two-step approach. The baseline combinatorial adjacency test is presented as the function FINDADJACENTCOMBINATORIAL.

function FINDADJACENTCOMBINATORIAL(I_+, I_-, I_0)
   $E \leftarrow \emptyset$
   for $i_+ \in I_+$
      for $i_- \in I_-$
         // Narrowing step
         if $|Z(i_+) \cap Z(i_-)| \geq r - 2$
            // Verification step
            if $\nexists\, i \in I_+ \cup I_i \cup I_0, i \neq i_+ \wedge i \neq i_- : Z(i_+) \cap Z(i_-) \subseteq Z(i)$
               $E \leftarrow E \cup \{(i_+, i_-)\}$
   return E

The adjacency test modifications considered in this paper involve using data structures to speed up one or both of the two steps. One such modification called the graph test was originally proposed in [18]:

Proposition 3. *Let G be a simple graph, where the vertices are elements of $U(C)$ and edges $E(G) = \{\{u, v\} : |Z(u) \cap Z(v)| \geq r - 2|\}$. Then $\{u, v\} \in E(C)$ if and only if $\nexists\, w \in U(C) \setminus \{u, v\} : \{u, w\} \in E(G) \wedge \{v, w\} \in E(G) \wedge Z(u) \cap Z(v) \subseteq Z(w)$.*

It should be noted that the graph used in this adjacency test does not necessarily involve using a data structure since it can be constructed implicitly, like how it was originally proposed in [18]. However, since such an implementation leads to rebuilding edges of the graph, it does not utilize the full power of this modification [19]. For this reason we employ explicit graph construction. This approach is outlined in FINDADJACENTGRAPH.

function FINDADJACENTGRAPH(I_+, I_-, I_0)
   Construct graph $G = \{(i_1, i_2) : |Z(i_1) \cap Z(i_2)| \geq r - 2; i_1, i_2 \in I_+ \cup I_- \cup I_0\}$
   for $i_+ \in I_+$
      // Narrowing step
      for $i_- \in I_-$ such that $(i_+, i_-) \in G$
         // Verification step
         if $\nexists\, i, (i_+, i) \in G \wedge (i_-, i) \in G : Z(i_+) \cap Z(i_-) \subseteq Z(i)$
            $E \leftarrow E \cup \{(i_+, i_-)\}$
   return E

Another modification of the combinatorial adjacency test involves usage of bit pattern trees [9,15,16]. This data structure is a generalization of k-d trees where every inner node has a disjunction bit pattern associated with it that serves as a descriptor for all data elements stored in that subtree. In terms of narrowing step, if the bit pattern fails the check, then that means that all of the elements in the subtree fail it as well and therefore can be ignored. In terms of verification step, if the bit pattern does not represent a counterexample, then none of the elements in the subtree do either. This allows the algorithm to skip over whole branches of the tree and can potentially reduce the number of checks the algorithm has to make at each step.

Bit pattern tree construction is outlined in CONSTRUCTBPT. TRAVERSE-NARROWINGBPT and TRAVERSEVERIFICATIONBPT demonstrate checking the narrowing step and the verification step conditions respectively.

function CONSTRUCTBPT(U')
    c - BPT leaf capacity
    if $|U'| <= c$
        return leaf node storing U
    Choose branching bit ℓ
    $U_1' \leftarrow \{u : u \in U, \ell \in Z(u)\}$
    $U_2' \leftarrow \{u : u \in U, \ell \notin Z(u)\}$
    return node with bit pattern $B = \bigcup_{u' \in U'} Z(u')$, pointing to
        CONSTRUCTBPT(U_1') and CONSTRUCTBPT(U_2')

function TRAVERSENARROWINGBPT(u_+, n)
    // n is current node
    if n is a leaf that stores some U':
        return $\{u_- \in U' : Z(u_+) \cap Z(u_-) \geq r - 2\}$
    $B \leftarrow$ bit pattern of n
    if $Z(u_+) \cap B \geq r - 2$
        $n_1, n_2 \leftarrow$ children of n
        return TRAVERSENARROWINGBPT(u_+, n_1) $\cap$
            TRAVERSENARROWINGBPT(u_+, n_2)
    return {}

function TRAVERSEVERIFICATIONBPT(u_+, u_-, n)
    // n is current node
    if n is a leaf that stores some U':
        return $\nexists i \in U', u \neq u_+, u \neq u_- : Z(u_+) \cap Z(u_-) \subseteq Z(u)$
    $B \leftarrow$ bit pattern of n
    if $Z(u_+) \cap Z(u_-) \subseteq B$
        $n_1, n_2 \leftarrow$ children of n
        return TRAVERSEVERIFICATIONBPT(u_+, u_-, n_1) $\wedge$
            TRAVERSEVERIFICATIONBPT(u_+, u_-, n_2)
    return true

FINDADJACENTBPT represents the full bit pattern tree based adjacency test, however, it should be noted that the use of bit pattern trees on each step is not dependant on the other, so they can be applied together or separately by reverting to the baseline approach on either step.

function FINDADJACENTBPT(U_+, U_-, U_0)
  NBPT $\leftarrow$ CONSTRUCTBPT(U_-)
  VBPT $\leftarrow$ CONSTRUCTBPT$(U_+ \cup U_- \cup U_0)$
  $E \leftarrow \emptyset$
  for $u_+ \in U_+$
    // Narrowing step
    for $u_- \in$ TRAVERSENARROWINGBPT$(u_+, NBPT)$
      // Verification step
      if TRAVERSEVERIFICATIONBPT$(u_+, u_-, VBPT)$
        $E \leftarrow E \cup \{(u_+, u_-)\}$
  return E

In addition, we also consider a variation of the double description method algorithm where the set of adjacent extreme rays is maintained and rebuilt immediately after the list of extreme rays is updated [8]. Its dynamic variation is described in [14] and will be referred to as M1 version of the algorithm. All of the adjacency tests variations described earlier can be easily adapted to this modification by iterating over $i_1, i_2 \in \{I_\pm, I_\pm\}$ instead of $(i_+, i_-) \in \{I_+, I_-\}$.

4   Computational Results

A C++ implementation of the modifications described above has been developed. The computational experiments were performed on a computer with Intel(R) Core(TM)i7-8700K CPU at 3.70 GHz, Microsoft Windows 10 operating system, using the Microsoft Visual Studio 2017 compiler with default optimization options. Bit pattern tree implementation used leaf capacity value of 4 and branching strategy for a balanced tree.

The experiments were run on several classes of problems: cyclic polytopes with fixed dimension, dwarfed cubes, products of simplices and random datasets with fixed dimensions generated with a uniform distribution. The figures in this section present results of the baseline combinatorial adjacenecy test, combinatorial test with bit-pattern tree based narrowing, combinatorial test with bit pattern tree based verification, fully bit pattern tree based combinatorial test, graph test and the M1 version of the graph test. M1 versions of other adjacency tests are omitted, since on almost all problem instances the M1 modification was detrimental to the overall running time of the algorithm with all adjacency tests but the graph one, the only exception being mit729-9 dataset where it was beneficial across the board.

All of the algorithms show similar performance numbers on the cyclic polytope problems (Fig. 1) with the non-M1 version of the algorithm with the graph test being slightly ahead at the largest values of n.

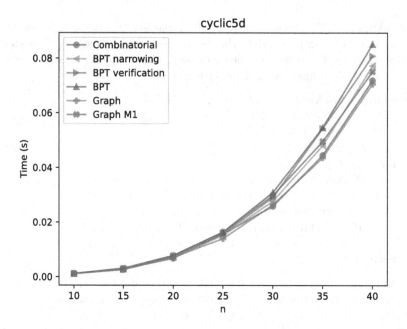

Fig. 1. Running time of dynamic DDM modifications on cyclic polytopes with $d = 5$.

On the dwarfed cube datasets (Fig. 2) bit pattern tree based verification shows the best results at larger dimensions, while using bit pattern trees on the narrowing step introduces additional overhead without any noticeable benefit, since most pairs of extreme rays pass the first check. This is also where the impact that using M1 modification has on the graph test is quite significant.

On the simplex product datasets (Fig. 3) the best results are shown by the baseline combinatorial, graph and bit pattern tree based narrowing adjacency tests. Notably, this class of problems and cyclic polytopes are the only cases considered where the M1 modification harms the performance of the algorithm on all adjacency test variations.

On the randomly generated datasets (Fig. 4) the use of bit pattern trees brings significant performance improvements on both steps of the adjacency test, making it the best performing variation of the algorithm on all cases in this class.

Finally, we ran experiments on several standard well-known problems (Table 1 and Table 2). On both complete cut cone and complete cut polytope problem instance the use of bit pattern trees showed excellent performance gains on both steps of the adjacency test, placing it far ahead of the other variations. mit729-9 dataset is unique in that M1 modification improves performance with all adjacency tests, while prodst62 is the only problem where baseline combinatorial test outperforms any of the modifications.

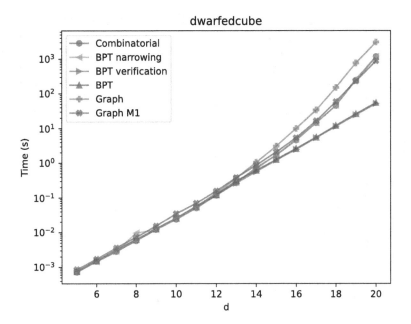

Fig. 2. Running time of dynamic DDM modifications on dwarfed cubes.

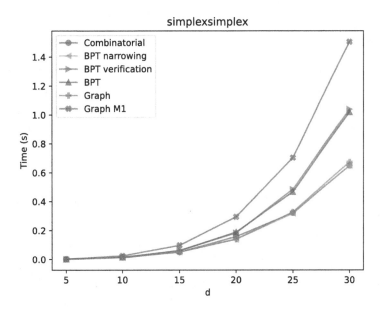

Fig. 3. Running time of dynamic DDM modifications on products of simplices.

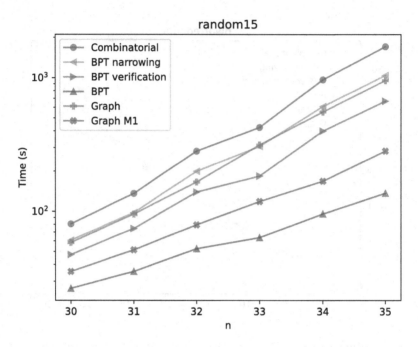

Fig. 4. Running time of dynamic DDM modifications on random datasets with $d = 15$ generated using a uniform distribution.

Table 1. Performance of the adjacency test modifications with the non-M1 version of the algorithm (s)

Dataset	Combinatorial	Graph	BPTNarrow	BPTVerify	BPT
ccc7	3903.72	2809.95	3111.3	1245.23	**390.938**
ccp7	4863.41	3465.9	3653.55	1615.45	**483.554**
mit729-9	151.444	982.239	**111.19**	204.187	166.909
prodst62	**15.3299**	26.4044	16.7268	40.3941	42.4681

Table 2. Performance of the adjacency test modifications with the M1 version of the algorithm (s)

Dataset	Combinatorial	Graph	BPTNarrow	BPTVerify	BPT
ccc7	Timeout (>7200)	2657.8	Timeout (>7200)	2585.91	2058.64
ccp7	Timeout (>7200)	2067.23	Timeout (>7200)	3020.08	2426.25
mit729-9	137.996	130.861	134.526	133.797	140.933
prodst62	88.9306	84.1389	102.097	82.8531	99.0562

5   Conclusion

This paper compares three modifications of the double description method and their possible combinations when applied to the dynamic version of the problem: the graph adjacency test, the bit pattern tree based combinatorial adjacency test and maintaining the edge set to generate new extreme rays. The results of computational experiments on several classes of problems show varying effectiveness of these modifications, with the bit pattern tree based approaches being ahead of the rest in most instances considered.

References

1. Avis, D., Bremner, D., Seidel, R.: How good are convex hull algorithms? Comput. Geomet. **7**(5–6), 265–301 (1997)
2. Avis, D., Fukuda, K.: A pivoting algorithm for convex hulls and vertex enumeration of arrangements and polyhedra. Discr. Comput. Geomet. **8**(3), 295–313 (1992). https://doi.org/10.1007/BF02293050
3. Bastrakov, S.I., Churkin, A.V., Zolotykh, N.Yu.: Accelerating Fourier-Motzkin elimination using bit pattern trees. In: Optimization Methods and Software, pp. 1–14 (2020)
4. Bastrakov, S.I., Zolotykh, N.Yu.: Fast method for verifying Chernikov rules in Fourier-Motzkin elimination. Comput. Math. Math. Phys. **55**(1), 160–167 (2015). https://doi.org/10.1134/S0965542515010042
5. Chernikov, S.: Linear Inequalities. Nauka, Moscow (1968)
6. Chernikova, N.V.: Algorithm for finding a general formula for the non-negative solutions of a system of linear equations. USSR Comput. Math. Math. Phys. **4**(4), 151–158 (1964)
7. Fernandez, F., Quinton, P.: Extension of Chernikova's algorithm for solving general mixed linear programming problems. Doctoral dissertation, INRIA (1988)
8. Fukuda, K., Prodon, A.: Double description method revisited. In: Deza, M., Euler, R., Manoussakis, I. (eds.) CCS 1995. LNCS, vol. 1120, pp. 91–111. Springer, Heidelberg (1996). https://doi.org/10.1007/3-540-61576-8_77
9. Genov, B.: The convex hull problem in practice: improving the running time of the double description method. Universitat Bremen, PhD dissertation (2015)
10. Horst, R., Pardalos, P.M., Van Thoai, N.: Introduction to global optimization. Springer, New York (2000)
11. Le Verge, H.: A note on Chernikova's algorithm. Doctoral dissertation, INRIA (1992)
12. Motzkin, T., Raiffa, H., Thompson, G., Thrall, R.: The double description method. In: Kuhn, H., Tucker, A.W. (eds.) Contributions to Theory of Games, vol. 2. Princeton University Press, Princeton (1953)
13. Schrijver, A.: Theory of Linear and Integer Programming. Wiley (1998)
14. Semenov, S.O., Zolotykh, N.Yu.: A dynamic algorithm for constructing the dual representation of a polyhedral cone. In: Khachay, M., Kochetov, Y., Pardalos, P. (eds.) MOTOR 2019. LNCS, vol. 11548, pp. 59–69. Springer, Cham (2019). https://doi.org/10.1007/978-3-030-22629-9_5
15. Terzer, M., Stelling, J.: Accelerating the computation of elementary modes using pattern trees. In: Bücher, P., Moret, B.M.E. (eds.) WABI 2006. LNCS, vol. 4175, pp. 333–343. Springer, Heidelberg (2006). https://doi.org/10.1007/11851561_31

16. Terzer, M., Stelling, J.: Large-scale computation of elementary flux modes with bit pattern trees. Bioinformatics **24**, 2229–2235 (2008)
17. Ziegler, G.M.: Lectures on Polytopes, vol. 152. Springer, New York (2012). https://doi.org/10.1007/978-1-4613-8431-1
18. Zolotykh, N.Yu.: New modification of the double description method for constructing the skeleton of a polyhedral cone. Comput. Math. Math. Phys. **52**(1), 146–156 (2012)
19. Zolotykh, N.Yu., Bastrakov, S.I.: Two variations of graph test in double description method. Comput. Appl. Math. **38**(3), 1–9 (2019). https://doi.org/10.1007/s40314-019-0862-0
20. Zolotykh, N.Yu., Kubarev, V.K., Lyalin, S.S.: Double description method over the field of algebraic numbers. Vestn. Udmurtsk. Univ. Mat. Mekh. Komp. Nauki **28**(2), 161–175 (2018) (in Russian)

Conditions for the Fixed Points Existence of Locally Indecomposable Subhomogeneous Maps

Alexander I. Smirnov$^{(\boxtimes)}$![ORCID] and Vladimir D. Mazurov

Krasovskii Institute of Mathematics and Mechanics UB RAS, Ekaterinburg, Russia
`asmi@imm.uran.ru`

Abstract. Various concepts of map indecomposability used in mathematical economics are analyzed along with a classical concept of Morishima-Nikaido. We also introduce a weakening of the classical notion of map indecomposability—the notion of local indecomposability at point and on set of points. The sufficiency of this weakened property for some spectral properties of positively homogeneous maps obtained earlier under the assumption of their classical indecomposability, is shown. Maps classes are given for which the classical notion of indecomposability and its proposed weakening coincide. Necessary and sufficient conditions for the existence of nonzero fixed points for indecomposable subhomogeneous and concave maps are obtained.

Keywords: Indecomposable map · Local indecomposable map · Positively homogeneous map · Subhomogeneous map · Standard interference map · Fixed points

1 Introduction and Basic Definitions

Used notation:

$\mathbb{R}^q_+$—the nonnegative orthant of $\mathbb{R}^q$; $\mathbb{R}_+ = \mathbb{R}^1_+$; $\mathbb{Z}$—the set of integers; $\mathbb{Z}_+$—the set of nonnegative integers; $\overline{m,n} = \{i \in \mathbb{Z}_+ \mid m \le i \le n\}$;
$x \lneqq y$ means $x \le y$, $x \ne y$; $\mathrm{co}\,(M)$—the convex hull of M;
R_x—a positive ray from origin containing $x \gneqq 0$: $R_x = \{\alpha x \mid 0 < \alpha < +\infty\}$.

We briefly write $x = (x_i)$ instead of $x = (x_1, x_2, \ldots, x_q)$. Iterations of F are denoted by $F^t(x)$ $(t = 1, 2, \ldots)$, $F^0(x) \equiv x$.

We consider a discrete dynamical system of the form

$$x_{t+1} = F(x_t), \qquad t = 0, 1, 2, \ldots, \tag{1}$$

on the nonnegative cone $\mathbb{R}^q_+$ of $\mathbb{R}^q$. This system is assumed to have a trivial equilibrium state. For the map F that generates it, this means the presence of a zero fixed point: $F(0) = 0$. For some classes of maps, the asymptotic properties of the process (1) are well studied; often they are related to spectral properties of the map F. Thus, in mathematical economics, there is a well-developed

Y. Kochetov et al. (Eds.): MOTOR 2022, CCIS 1661, pp. 189–203, 2022.
https://doi.org/10.1007/978-3-031-16224-4_13

spectral theory for monotone positively homogeneous maps of the first degree, which generalizes the main statements of spectral theory for linear maps [5, 6]. Another generalization direction of linear operators spectral theory is the nonlinear Perron-Frobenius theory. The properties of a map indecomposability, which are generalizations of that for a nonnegative matrix, turn out to be essential for proofs some key statements in nonlinear Perron-Frobenius theory.

The determining factor for characterizing the asymptotic properties of system (1) is the presence or absence of its nontrivial equilibrium states. Therefore, it is so important to study the conditions for the existence of fixed points of certain maps classes.

The main class of maps under consideration is the class of *subhomogeneous maps* (definitions will be given below). We follow here the terminology of monographs [1, 2] containing classical results on nonlinear Perron-Frobenius theory. Maps of this kind (called differently in different areas of research—sublinear maps, standard interference maps, concave operators etc.) are widely used both in theoretical research and in applications.

One of the new areas of application for subhomogeneous maps are research on resource allocation in wireless communication networks; in this area of research, the term *standard interference map* is used for subhomogeneous maps [10]. From a mathematical point of view, one of the main objectives of these studies is to answer the question of whether the standard interference map has a fixed point. In particular, positive concave maps are standard interference maps, and they need to have relatively simple criteria for the fixed points existence [9].

Another important application of our results belongs to the urgent problem of optimal nondestructive exploitation of ecological populations. In a series of papers on this topic (see references in [11]), conditions for the existence of nondestructive controls in the problem of maximal effect exploitation an ecological population are obtained. In particular, this applies to a populations modelled by a nonlinear generalization of so called Leslie's model, widely used by ecologists. The conditions for the existence of fixed points of subhomogeneous and concave maps, which will be obtained below, for this model are the conditions for the existence of a nontrivial equilibrium of the exploited population. In this series of papers, the properties of nondestructive feasible and optimal controls in cases of an indecomposable and decomposable of the model step operator. These results are important in constructing specific models of ecological populations, since they make it possible to adequately choose the class of models used.

The goal of this paper is to consider various variants of the indecomposability concepts for maps defined on $\mathbb{R}_+^q$ and to carry out their comparative analysis. In addition, we intend to show the usefulness of our generalization of a map indecomposability classical concept—of local indecomposability, for the analysis of properties certain maps classes, first of all, classes of positively homogeneous and subhomogeneous maps. Using our generalizations of a map indecomposability concept, we also intend to obtain criteria for nonzero fixed points existence of subhomogeneous and concave maps.

Let us give the necessary definitions. We assume everywhere below that $F \in \{\mathbb{R}_+^q \mapsto \mathbb{R}_+^q\}$ $(q > 1)$, $F(0) = 0$.

Definition 1. *A map F is called monotone increasing if*

$$\forall x \in \mathbb{R}_+^q : x \leq y \Rightarrow F(x) \leq F(y). \tag{2}$$

Accordingly, in the case of the opposite inequality in conclusion of implication (2), the map F is called monotone decreasing.

A map F is called strongly monotone if

$$\forall x \in \mathbb{R}_+^q : x \lneqq y \Rightarrow F(x) < F(y). \tag{3}$$

For brevity, a monotone increasing map will also be called simply *monotone*; accordingly, a monotone decreasing map will usually be called simply *decreasing*.

Definition 2. *A map F is called positively homogeneous if*

$$F(\alpha x) = \alpha F(x) \quad (\forall \alpha \geq 0, \ x \in \mathbb{R}_+^q). \tag{4}$$

The main requirement for the map that generates the dynamical system (1) in this paper is its subhomogeneity.

Definition 3. *A map F is called subhomogeneous if*

$$F(\alpha x) \geq \alpha F(x) \quad (\forall \alpha \in (0, 1), \ x \in \mathbb{R}_+^q). \tag{5}$$

The class of subhomogeneous maps contains, in particular, the classes of positively homogeneous and concave maps.

Monotone subhomogeneous map is continuous on $\operatorname{int} \mathbb{R}_+^q$, and has a continuous extension to the entire cone $\mathbb{R}_+^q$ [2, Theorem 5.1.5]; therefore we assume in what follows that the subhomogeneous maps under consideration are *continuous on the whole of* $\mathbb{R}_+^q$.

The property (5), which determines subhomogeneity, is equivalent to each of the following conditions:

$$F(\beta x) \leq \beta F(x) \quad (\forall \beta \in (1, +\infty), \ x \in \mathbb{R}_+^q), \tag{6}$$

$$0 < \alpha_1 \leq \alpha_2 \Rightarrow (\alpha_1)^{-1} F(\alpha_1 x) \geq (\alpha_2)^{-1} F(\alpha_2 x). \tag{7}$$

The property (7) means that for every fixed $x \in \mathbb{R}_+^q$ the map $F_x(\alpha) = (\alpha)^{-1} F(\alpha x)$ is decreasing in $\alpha > 0$ and, therefore, positively homogeneous maps

$$F_0(x) = \lim_{\alpha \to +0} \alpha^{-1} F(\alpha x), \quad F_\infty(x) = \lim_{\alpha \to +\infty} \alpha^{-1} F(\alpha x), \tag{8}$$

can be defined, which are majorant and minorant of $F(x)$, respectively, on $\mathbb{R}_+^q$:

$$F_\infty(x) \leq F(x) \leq F_0(x) \quad (\forall x \in \mathbb{R}_+^q). \tag{9}$$

To analyze the stability of nontrivial equilibrium states (if they exist) for system (1) with a subhomogeneous monotone map F as the step operator, it is essential that such map always has an invariant segment. Indeed, for any fixed point $\bar{x} > 0$ by (5), (6) we have: $\alpha \bar{x} \leq F(\alpha \bar{x}) \leq F(\beta \bar{x}) \leq \beta \bar{x}$ $(\forall \alpha \in (0, 1), \ \beta > 1)$, i.e. segment $[\bar{x}, \bar{y}] = [\alpha \bar{x}, \beta \bar{x}]$ is invariant for F.

Remark 1. Note that $F_0(x)$ may have improper coordinates. The concept of a dominant eigenvalue for positively homogeneous maps needs to be generalized to the case of partially improper maps. It is natural to use the approximation approach for this, using a convergent sequence of proper maps.

Suppose that a map F, which is positively homogeneous, has an improper coordinate: $f_i(x) = +\infty$ ($\forall x \in \text{int } R_+^q$) and is the pointwise limit of some increasing sequence of positively homogeneous maps: $\lim\limits_{n \to +\infty} F_n(x) = F(x)$, where $F_{n+1}(x) \geq F_n(x)$ ($\forall x \in R_+^q$, $n = 1, 2, \ldots$). If $\lim\limits_{n \to +\infty} \lambda(F_n) = +\infty$, then we presume, by definition, that $\lambda(F) = +\infty$.

Such an understanding of the infinite value for dominant eigenvalue justified, since under these assumptions $F(x) \geq F_n(x)$ and, according to [6, Theorem 10.3], $\lambda(F) \geq \lambda(F_n) \to +\infty$.

2 Indecomposability of a Matrix and Its Generalizations

Depending on the specifics of the relevant sections in Perron-Frobenius nonlinear theory, various definitions of the maps indecomposability are used. The natural common starting point of these definitions is the fact that they are different generalizations of indecomposability concept for linear map—in the linear case, all types of a map indecomposability must mean the indecomposability of a map matrix.

Definition 4. *A nonnegative matrix $A = [a_{i,j}]$ of order $q > 1$ is called decomposable if*

$$\exists I \subseteq \overline{1,q}, \ \varnothing \neq I \neq \overline{1,q}: \quad a_{i,j} = 0 \quad (\forall i \notin I, j \in I).$$

Otherwise, the matrix A is called indecomposable.

One of the generalization directions for the indecomposability concept is the transition to infinite-dimensional spaces. The use of the indecomposability property for operators in Banach spaces goes back to the studies of V. Ya. Stetsenko and V. A. Semiletov. In their scheme for nonlinear operators [12], indecomposability is equivalent to the at zero indecomposability (the corresponding definitions will be given below).

For operators in general topological spaces, another approach is used to generalize the concept of indecomposability, which uses the fact that a necessary and sufficient property of the matrix decomposability is the presence of proper invariant subspaces for the corresponding linear map. Here we will be more interested in generalizations of the indecomposability concept to nonlinear maps in $\mathbb{R}^q$. In mathematical economics, the definition of a map indecomposability introduced by M. Morishima [5] and H. Nikaido [6] is generally accepted, which was used by them to characterize the spectral properties of monotone positively homogeneous maps. Let's give this definition in a slightly different form, using the following partition of coordinates:

$$I^+(x,y) = \{j \in \overline{1,q} \mid x_j > y_j\}, \quad I^0(x,y) = \{j \in \overline{1,q} \mid x_j = y_j\},$$

$$I^+(x) = \{i \in \overline{1,q} \mid x_i > 0\}, \qquad I^0(x) = \{i \in \overline{1,q} \mid x_i = 0\}. \tag{10}$$

Definition 5. *A map F is called decomposable if*

$$\exists x,y \in \mathbb{R}_+^q: \quad x \gneqq y, \quad I^0(x,y) \neq \varnothing, \quad I^0(x,y) \subseteq I^0(F(x),F(y)). \tag{11}$$

Accordingly, a map F is called indecomposable if

$$\forall x,y \in \mathbb{R}_+^q: \quad x \gneqq y, \ I^0(x,y) \neq \varnothing \Rightarrow I^0(x,y) \backslash I^0(F(x),F(y)) \neq \varnothing.$$

For monotone maps this indecomposability definition can be refined:

$$\forall x,y \in \mathbb{R}_+^q: \quad x \gneqq y, \ I^0(x,y) \neq \varnothing \Rightarrow I^0(x,y) \cap I^+(F(x),F(y)) \neq \varnothing. \tag{12}$$

For the sake of convenience and brevity, we call below the map F, according to Definition 5, decomposable or indecomposable in the sense of Morishima-Nikaido, *MN-decomposable* or *MN-indecomposable*, respectively, depending on whether (11) or (12) is fulfilled.

Below we also give a definition of at point indecomposability, so a map that is indecomposable in the sense of (12) we (for obvious reasons) also call *globally indecomposable* (on $\mathbb{R}_+^q$).

Note that a strongly monotone map (see Definition 1) is globally indecomposable. Obviously, a linear map defined by an indecomposable matrix (see Definition 4) is globally indecomposable in the sense of this definition.

As we will see below, to prove some assertions for positively homogeneous and subhomogeneous maps, a local version of indecomposability property suffices. Therefore, it is justified to use the following definition [3].

Definition 6. *A map F is called decomposable at the point $y \in \mathbb{R}_+^q$, if*

$$\exists x \in \mathbb{R}_+^q: \quad x \gneqq y, \quad I^0(x,y) \neq \varnothing, \quad I^0(x,y) \subseteq I^0(F(x),F(y)). \tag{13}$$

The map, decomposable at every point of the set M is called decomposable on M. Accordingly monotone map F is called indecomposable at point of y, if

$$\forall x \in \mathbb{R}_+^q: \quad x \geq y, \ I^0(x,y) \neq \varnothing \Rightarrow I^0(x,y) \cap I^+(F(x),F(y)) \neq \varnothing. \tag{14}$$

The map, indecomposable at every point of M is called indecomposable on M.

We consider separately the case of the indecomposability at point $y = 0$.

Definition 7. *A map F is called decomposable at zero, if*

$$\exists x \in \mathbb{R}_+^q: \quad x \gneqq 0, \quad I^0(x) \neq \varnothing, \quad I^0(x) \subseteq I^0(F(x)). \tag{15}$$

Accordingly monotone map F is called indecomposable at zero, if

$$\forall x \in \mathbb{R}_+^q: \quad x \gneqq 0, \ I^0(x) \neq \varnothing \Rightarrow I^0(x) \cap I^+(F(x)) \neq \varnothing. \tag{16}$$

The global indecomposability, by definition, entails indecomposability at any point of $\mathbb{R}_+^q$ and, in particular, indecomposability at zero. For linear maps, the converse is also true, i.e. the concept of at zero indecomposability coincides with the global indecomposability concept (which, in turn, coincides with the indecomposability concept of a linear map matrix).

Remark 2. Another class of maps for which the concepts of indecomposability at zero and global indecomposability coincide is the class of concave positively homogeneous maps (a detailed presentation of such maps theory is given in [1]).

Let us show that the *MN-decomposability* of a concave positively homogeneous map implies its decomposability at zero. Indeed, for a concave positively homogeneous function f, from the definition of concavity we obtain for the middle of a segment with ends y and $x - y$:

$$x \geq y \Rightarrow f(x - y) \leq f(x) - f(y). \tag{17}$$

If the map F is *MN-decomposable*, then $f_i(\bar{x}) = f_i(\bar{y})$ ($\forall i \in I_0(\bar{x}, \bar{y})$)) for some $\bar{x}, \bar{y}$, where $x \gneqq y$. Denote $\bar{z} = \bar{x} - \bar{y}$, then $\bar{z} \gneqq 0$ and $I_0(\bar{z}) \neq \varnothing$. Next, it follows from (17) that $f_i(\bar{z}) \leq f_i(\bar{x}) - f_i(\bar{y}) = 0$, i.e. $f_i(\bar{z}) = 0$ for $i \in I_0(\bar{z})$. But then the map F is decomposable at zero.

Thus, the indecomposability at zero of a concave positively homogeneous map guarantees global indecomposability. In the same way, it can be shown that the decomposability at zero of a convex positive homogeneous map actually guarantees a stronger property—decomposability at any point of $\mathbb{R}_+^q$.

For nonlinear maps, the concepts of at zero indecomposability and global indecomposability, generally speaking, do not coincide. These concepts are also different for positively homogeneous maps. Let us give the corresponding examples.

Example 1. Consider the map $F(x) = (f_1(x), f_2(x))$, where

$$f_1(x) = f_2(x) = \max\{x_1, x_2\}.$$

For any $\bar{x}$ with coordinates $\bar{x}_1 = 0$, $\bar{x}_2 > 0$ we obtain $F(\bar{x}) = (\bar{x}_2, \bar{x}_2)$, and the condition $f_1(\bar{x}) = 0$ results in equality $\bar{x}_2 = 0$, i.e. $\bar{x}_1 = \bar{x}_2 = 0$, which contradicts the assumption $\bar{x}_2 > 0$. The same is true for situation $\bar{x}_1 > 0$, $\bar{x}_2 = 0$. So, the assumption that this map is decomposable at zero yields the contradiction; hence, this map is indecomposable at zero.

This map is also indecomposable on the ray $R = \{y = (y_1, y_2) \mid y_1 = y_2 > 0\}$. Indeed, for any $\bar{y} \in R$ and $\bar{x}$ with coordinates $\bar{x}_1 = \bar{y}_1$, $\bar{x}_2 > \bar{y}_2$, we obtain $f_1(\bar{x}) = f_2(\bar{x}) = \bar{x}_2 > \bar{y}_1 = f_1(\bar{y})$ by virtue of $\bar{x}_2 > \bar{y}_2 = \bar{y}_1 = \bar{x}_1$, i.e. $f_1(\bar{x}) > f_1(\bar{y})$. Similarly, for $\bar{x}_1 > \bar{y}_1$, $\bar{x}_2 = \bar{y}_2$, we have $f_2(\bar{x}) > f_2(\bar{y})$. By (14), this map is indecomposable at indicated points.

Now let's find the points in which this map is decomposable. Take any $\bar{y} = (\bar{y}_1, \bar{y}_2)$ such that $\bar{y}_1 < \bar{y}_2$. Then for $\bar{x} = (\bar{y}_2, \bar{y}_2)$ we have $f_1(\bar{x}) = f_2(\bar{x}) = \bar{y}_2$, $f_1(\bar{y}) = f_2(\bar{y}) = \bar{y}_2$, so $\bar{x}_2 = \bar{y}_2$ and $f_2(\bar{x}) = f_2(\bar{y}_2)$. By (13), this map is decomposable on the set $\{y = (y_1, y_2) \mid 0 \leq y_1 < y_2\}$. The case $y_1 > y_2$ is considered similarly. Thus, by Definition 6, this map is decomposable on the set $\{y = (y_1, y_2) \geq 0 \mid y_1 \neq y_2\}$.

Example 2. Consider the map $F(x) = (f_1(x), f_2(x))$, where

$$f_1(x) = f_2(x) = \sqrt{x_1, x_2}.$$

This positively homogeneous map is strongly monotone on $\text{int}\,\mathbb{R}_+^q$ (see (3)), and therefore cannot be decomposable on $\text{int}\,\mathbb{R}_+^q$. On the other hand, since $f_1(x) = f_2(x) = 0$ for all x that have at least one zero coordinate, this map is decomposable on $\mathbb{R}_+^q \setminus \text{int}\,\mathbb{R}_+^q$ and, in particular, is decomposable at zero.

If monotone positively homogeneous map $F(x) = (f_i(x))$ is decomposable at the point $\bar{x} \gneqq 0$, then it is also decomposable at any point $\bar{y} = \alpha\bar{x}$ for $\alpha > 0$. Indeed, the existence of x satisfying the conditions $x \geq \bar{x}$, $x_i = \bar{x}_i$, $f_i(x) = f_i(\bar{x})$ for some $i \in \overline{1,q}$ results in the existence of $y = \alpha x$ for which $y \geq \bar{y}$, $y_i = \bar{y}_i$, $f_i(y) = f_i(\bar{y})$.

Thus, the set of points, at which the positively homogeneous map is decomposable, together with each nonzero point x contains the positive ray R_x defined by it, starting from the origin.

A similar situation is valid for the set of points at which a positively homogeneous map is indecomposable. Examples 1, 2 demonstrate these features.

Thus, we can deem that the concept of local indecomposability for a monotone positively homogeneous map refers to positive rays starting from the origin.

To characterize the property of indecomposability at zero, we need the following simple but useful assertions. Recall that the set $I^0(x)$ is defined in (10).

Lemma 1. *A subhomogeneous monotone map F is decomposable at zero if and only if*

$$\exists a \in \mathbb{R}_+, \bar{x} \in \mathbb{R}_+^q: \varnothing \neq I^0(\bar{x}) \neq \overline{1,q}, \quad F(\bar{x}) \leq a\bar{x}. \tag{18}$$

Since in linear case condition (18) becomes the necessary and sufficient condition for the map matrix indecomposability [6, Lemma 7.4], Lemma 1 implies for linear map the equivalence of properties at zero indecomposability and map matrix indecomposability. Thus, the notion of at zero map indecomposability is indeed a generalization of the matrix indecomposability notion.

Lemma 2. *Let the function $f(x)$ be subhomogeneous and monotone increasing on $\mathbb{R}_+^q$. If $f(\bar{x}) = 0$ for some $\bar{x} \gneqq 0$, then $f(x) = 0$ on $\{x \in \mathbb{R}_+^q \mid I^0(x) \supseteq I^0(\bar{x})\}$.*
In particular, if $f(\bar{x}) = 0$ for some $\bar{x} > 0$, then $f(x) \equiv 0$ on $\mathbb{R}_+^q$.

Corollary 1. *A subhomogeneous monotone map F is decomposable at zero if and only if there exists $\bar{x} \gneqq 0$, for which*

$$\varnothing \neq I^0(\bar{x}) \subseteq I^0(F(\bar{x})) \subseteq I^0(F^2(\bar{x})) \subseteq \dots . \tag{19}$$

Proof. The sufficiency of condition (19) for decomposability F at zero is obvious.

Let F be decomposable at zero. The first inclusion in chain (19) is the definition of at zero decomposability; we prove the rest inclusions. We have $f_i(\bar{x}) = 0$ for $i \in I^0(\bar{x}) \subseteq I^0(F(\bar{x}))$. By Lemma 2, then $f_i(f(\bar{x})) = 0$ for $i \in I^0(F(\bar{x}))$, i.e. $I^0(F(\bar{x})) \subseteq I^0(F^2(\bar{x}))$. Applying mathematical induction, we obtain (19).

A indecomposability concept similar to given by Definition 7 was introduced in a slightly different form by Yu. Oshime [8]. Let us first introduce some terms.

The partition $I \cup J = \overline{1,q}$ of suffixes is called *proper*, if $I \neq \varnothing$, $J \neq \varnothing$, $I \cap J = \varnothing$. Denote $x_I = (x_{i_1}, \ldots, x_{i_m})$, $x_J = (x_{j_1}, \ldots, x_{j_n})$, where $I = \{i_1, \ldots, i_m\}$, $J = \{j_1, \ldots, j_n\}$. Then conditionally we can write x in the form $x = (x_I, x_J)$; this also applies to $F(x)$.

Definition 8. *A map F is called indecomposable if for any proper partition $I \cup J = \overline{1,q}$ of suffixes $F_J(x_I, 0_J) \neq 0_J$ for all $x_I > 0$.*

Comparing Definitions 7 and 8, we see that Oshima's map indecomposability is nothing but at zero indecomposability. Using this definition of indecomposability, in [8] some generalizations of the classical statements about the spectral properties of positively homogeneous and subhomogeneous maps are proved, sometimes with additional assumptions about subadditivity and real analyticity of the considered maps.

Another kind of a map indecomposability was used by U. Krause [1].

Definition 9. *A map $F(x) = (f_i(x))$ is called indecomposable if the condition*

$$\forall I \subseteq \overline{1,q} \quad (\varnothing \neq I \neq \overline{1,q}) \quad \exists i \in I, j \notin I \colon\ f_i(e_j) > 0 \tag{20}$$

is satisfied, where e_j are unit vectors (i.e., vectors that have exactly one nonzero coordinate equal to one).

Accordingly, the map F is decomposable if

$$\exists I \subseteq \overline{1,q} \quad (\varnothing \neq I \neq \overline{1,q}) \colon\quad f_i(e_j) = 0 \quad (\forall i \in I, j \notin I). \tag{21}$$

To distinguish the concepts introduced by the Definition 9 from those previously defined, we call the map F, which is Krause's decomposable or indecomposable, *K-decomposable* or *K-indecomposable*, respectively, depending on whether the property (20) or property (21) holds.

In the class of monotone subhomogeneous maps on $\mathbb{R}_+^q$ for $q > 2$, the indecomposability at zero is weaker than K-indecomposability. Indeed, if F is decomposable at zero, then there exists a nonzero vector $\bar{x} = (\bar{x}_i)$ with $\bar{x}_i = 0$ and $f_i(\bar{x}) = 0$ for $i \in I = I^0(\bar{x})$, $\varnothing \neq I \neq \overline{1,q}$. By Lemma 2, then $f_i(\alpha \bar{x}) = 0$ for all $\alpha \geq 0$, so we can assume that $\bar{x}_j \leq 1$ for $j \notin I$. But then due to the monotonicity and subhomogeneity of $f_i(x)$ by inequality (7), we obtain:

$$0 = f_i(\bar{x}) = (\bar{x}_j)^{-1} f_i(\bar{x}) \geq f_i((\bar{x}_j)^{-1}\bar{x}) \geq f_i(e_j) \quad (\forall i \in I, j \notin I),$$

i.e. $f_i(e_j) = 0$ for all $i \in I, j \notin I$, so, the map F is K-decomposable.

At the same time, there are examples of indecomposable at zero maps that are K-decomposable.

In contrast to the relation between the properties of indecomposability at zero and K-indecomposability, the relation between the properties of global indecomposability and K-indecomposability in the class of positively homogeneous maps is not so definite and depends on additional properties of the maps under consideration. Thus, in the subclass of convex positively homogeneous maps, the global indecomposability property is weaker than the K-indecomposability property; in the subclass of concave positively homogeneous maps, this relation is reversed [1].

Remark 3. The property of at zero indecomposability for monotone subhomogeneous maps guarantees the positivity of any nonzero fixed point. Indeed, the assumption $I^0(\bar{x}) \neq \varnothing$ for some $\bar{x} \gneq 0$, by $F(\bar{x}) = \bar{x}$ gives the equality $I^0(\bar{x}) = I^0(F(\bar{x}))$, which contradicts the indecomposability of F at zero (see (15) and (16)).

This also applies to nonnegative eigenvectors of monotone positively homogeneous maps. The corresponding eigenvalues are positive in this case.

In [6, Theorem 10.4 of Chap. 3] it is proved that the eigenvectors and eigenvalues of positively homogeneous map are positive under the assumption of global indecomposability. The proof actually uses the at zero indecomposability only. The assumptions of the second part of this theorem about the uniqueness of the eigenray (containing eigenvectors) can also be weakened—instead of global indecomposability, it suffices to require indecomposability on the set of eigenvectors.

The map given in Example 1 illustrates this situation. Its dominant eigenvalue is equal to one, his eigenray $\{y = (y_1, y_2) \mid y_1 = y_2 > 0\}$ is unique, and this map, as shown above, is indecomposable only at zero and on this ray.

Despite the fact that this map does not satisfy the assumptions of the aforementioned theorem, its conclusion about the eigenray uniqueness, as we see, is valid, and this is due precisely to the indecomposability at zero and on the set of eigenvectors of this map.

Thus, the introduction of the concept of local indecomposability contributes to a more subtle analysis of the reasons for the fulfillment of some properties of the maps under consideration.

The importance of indecomposability at zero for the eigenray uniqueness is illustrated by Example 2: here, in addition to the dominant eigenvalue, also equal to one, there is one more eigenvalue equal to zero, whose eigenvectors are all vectors that have exactly one zero coordinate.

In [5, Appendix, Theorem 3], the positiveness of nonnegative eigenvectors and eigenvalues, as well as the uniqueness of the positive eigenvalue and eigenray for monotone positively homogeneous map under the assumption of its global indecomposability was established. It turns out that for the uniqueness of the positive eigenvalue, it suffices that the map is indecomposable only at zero.

Theorem 1. *Let F be a monotone positively homogeneous map indecomposable at zero, and $F(x) = \lambda x$ for some $x \gneq 0$, $\lambda \geq 0$. Then λ is the dominant eigenvalue of the map F: $\lambda = \lambda(F)$, and $\lambda > 0$.*

Proof. Under the Theorem assumptions the corresponding to the dominant eigenvalue $\lambda(F)$, is positive, along with eigenvector x (see Remark 3). It follows that $\lambda(F) > 0$, since otherwise $F(y) = 0$ for $y > 0$, which contradicts the indecomposability F at zero. Similarly, we get $\lambda > 0$.

Suppose now that $\lambda \neq \lambda(F)$; then, by definition of the dominant eigenvalue, $\lambda < \lambda(F)$. For the eigenvector y, the equality $F(y) = \lambda(F)y$ also holds. This equality implies $x \neq y$. Let $\alpha = \max\{y_i/x_i \mid i \in \overline{1,q}\}$, then $y \leq \alpha x$, at that $I^0(\alpha x, y) \neq \varnothing$. From inequality $y \leq \alpha x$, due to the monotonicity of F, we obtain: $\lambda(F)y = F(y) \leq F(\alpha x) = \lambda \alpha x$, or $y \leq \lambda'\alpha x$, where $\lambda' = \lambda/\lambda(F)$. Since $\lambda' < 1$, this means that $y < \alpha x$, i.e. $I^0(\alpha x, y) = \varnothing$. The obtained contradiction shows that $\lambda = \lambda(F)$. The Theorem has been proven.

The essentiality of at zero indecomposability assumption in Theorem 1 is demonstrated by Example 1.

3 Criteria for the Existence of Fixed Points for Subhomogeneous and Concave Maps

Here we obtain conditions for the existence of monotone subhomogeneous maps and refine these conditions for an important particular case of subhomogeneous maps—for concave maps.

We assume that there are no completely zero components of $F(x) = (f_i(x))$:

$$\forall i \in \overline{1,q} \;\; \exists x \in \mathbb{R}_+^q: \quad f_i(x) > 0.$$

In this case, the image of a positive vector for a subhomogeneous monotone map is also positive:

$$F(\mathrm{int}\,\mathbb{R}_+^q) \subseteq \mathrm{int}\,\mathbb{R}_+^q.$$

Indeed, assuming the opposite, we obtain the existence of $i_0 \in \overline{1,q}$ and $\bar{x} > 0$ satisfying the condition $f_{i_0}(\bar{x}) = 0$, which, by Lemma 2, has as a consequence the identity $f_{i_0}(x) \equiv 0$.

If the set of positive fixed points for a subhomogeneous monotone map is nonempty, then the question arises about the structure of this set. We see from (4) that the set of fixed points of the positively homogeneous map, along with each positive fixed point $\bar{x}$, contains the entire positive ray $R_{\bar{x}}$ outgoing from origin and passing through this point. A subhomogeneous map is a generalization of a positively homogeneous map (compare Definitions 2, 3) and, naturally, may inherit some of its properties. Indeed, it was shown that if such ray contains more than one positive fixed point of subhomogeneous map, then the set of its fixed points also contains some continuous part of this ray [2, Lemma 6.5.5].

Just as for positively homogeneous maps, for subhomogeneous maps the question of the uniqueness of the positive ray with fixed points outgoing from origin is important. It can be shown that the global map indecomposability guarantees the uniqueness of such ray containing the positive fixed points of this map.

In [4], weaker sufficient conditions have studied that guarantee the uniqueness of the ray with positive fixed points, using the local indecomposability assumption. It turned out that such sufficient conditions are at zero indecomposability for concave on $\mathbb{R}_+^q$ map [4, Corollary 3], and indecomposability on the positive fixed points set, for a monotone subhomogeneous map [4, Corollary 4].

Note that this is in complete agreement with the fact that for positively homogeneous concave maps, the indecomposability only at zero implies its global indecomposability (see Remark 2), which has, as a consequence, the uniqueness of the ray with positive fixed points outgoing from origin.

These results make it possible to study the structure of fixed points set for concave maps in more detail. Let us first give some auxiliary assertions. We present the following assertion without proof.

Lemma 3. *Let the function $f(x)$ be subhomogeneous on $\mathbb{R}_+^q$, $f(0) = 0$ and $\bar{x} \gneqq 0$. Then the following properties are fulfilled:*

$$f(\alpha_0 \bar{x}) = \alpha_0 f(\bar{x}), \ \ 0 < \alpha_0 < 1 \Rightarrow f(\alpha \bar{x}) = \alpha f(\bar{x}) \ \ (\forall \alpha \in [\alpha_0, 1]);$$

$$f(\beta_0 \bar{x}) = \beta_0 f(\bar{x}), \ \ \beta_0 > 1 \Rightarrow f(\beta \bar{x}) = \beta f(\bar{x}) \ \ (\forall \beta \in [1, \beta_0]).$$

If $f(x)$ is concave on $\mathbb{R}_+^q$, $f(0) = 0$ and $\bar{x} \gneqq 0$, then the following property holds:

$$f(\alpha_0 \bar{x}) = \alpha_0 f(\bar{x}) \ (\alpha_0 \neq 0, \neq 1) \Rightarrow f(\alpha \bar{x}) = \alpha f(\bar{x}) \ \ (\forall \alpha \in [0, \alpha_0]).$$

We now turn to sufficient conditions for the existence of a fixed point for a monotone subhomogeneous map. We consider only the case of a proper map F_0 when $\lambda(F_0) < +\infty$; the case $\lambda(F_0) = +\infty$ is considered similarly based on the approximation of F_0 by proper maps, as it is done, for example, in [8, Propostion 2.17] (see Remark 1). Let us introduce the following notation:

$$M_F(x) = \{\alpha > 0 \mid F(\alpha x) = \alpha F(x)\}, \tag{22}$$

$$\alpha_F(x) = \inf M_F(x), \quad \beta_F(x) = \sup M_F(x). \tag{23}$$

By Lemma 3, $M_F(x)$ either contains only unity, or is a segment of positive length, which is the section of positive homogeneity of F.

Denote by K_0 and K_∞ the cones of eigenvectors corresponding to the dominant eigenvalues $\lambda(F_0)$ and $\lambda(F_\infty)$ of maps F_0, F_∞, respectively:

$$K_0 = \{x \gneqq 0 \mid F_0(x) = \lambda(F_0) x \}, \quad K_\infty = \{x \gneqq 0 \mid F_\infty(x) = \lambda(F_\infty) x \}.$$

Lemma 4. *If the map $F(x) = (f_i(x))$ is subhomogeneous, then the following properties are valid:*

$$\exists x \gneqq 0 \colon \alpha_F(x) = 0 \Leftrightarrow F_0(x) = F(x) \Leftrightarrow F(\alpha x) = \alpha F(x) \ \ (\forall \alpha \in [0, 1]), \tag{24}$$

$$\exists x \gneqq 0 \colon \beta_F(x) = +\infty \Leftrightarrow F_\infty(x) = F(x) \Leftrightarrow F(\beta x) = \beta F(x) \ (\forall \beta \in [1, +\infty)). \tag{25}$$

Proof. Let us prove (24). The equality $\alpha_F(x) = \inf M_F(x) = 0$, by Lemma 3 results in the inclusion $M_F(x) \supset (0, 1)$; so, $F(\alpha x) = \alpha F(x) \ (\forall \alpha \in (0, 1)$. But then, by definition (8), $F_0(x) = \lim\limits_{\alpha \to +0} \alpha^{-1} F(\alpha x) = F(x)$.

The converse is also true. Indeed, it follows from (5), (7), (8) that $F(x) \le \alpha^{-1}F(\alpha x) \le F_0(x)$ $(\forall \alpha \in (0,1))$. Hence, the equality $F_0(x) = F(x)$ results in $F(\alpha x) = \alpha F(x)$ $(\forall \alpha \in (0,1))$. But then by (22) $M_F(x) \supset (0,1)$ and $\alpha_F(x) = \inf M_F(x) = 0$.

Property (25) is proved in exactly the same way using the inequalities $F_\infty(x) \le \beta^{-1}F(\beta x) \le F(x)$ $(\forall \beta \in (1, +\infty))$.

Note that in the following assertion, indecomposability at zero of F is not assumed, so the nonzero fixed point of F need not be positive.

Theorem 2. *Let the map F $(F(0) = 0)$ be subhomogeneous and monotone on $\mathbb{R}^q_+$. If one of the conditions*

$$\lambda(F_0) = 1, \; \exists x \in K_0 : \; \alpha_F(x) = 0, \tag{26}$$

$$\lambda(F_\infty) < 1 < \lambda(F_0), \tag{27}$$

$$\lambda(F_\infty) = 1, \; \exists x \in K_\infty : \; \beta_F(x) = +\infty \tag{28}$$

is fulfilled then there exists a nonzero fixed point of the map F.

Proof. Denote $F_0(x) = (f_i^0(x))$, $F_\infty(x) = (f_i^\infty(x))$. By Lemma 4, the conditions (26), (28) are sufficient for the existence of a nonzero fixed point.

Let us now prove the sufficiency of (27). Let us show that condition $\lambda(F_0) > 1$ implies the existence of $\bar{x} \gneqq 0$ such that $F(\bar{x}) \ge \bar{x}$.

For any $x \in K_0$ we have: $0 \le F(x) \le F_0(x) = \lambda(F_0)x$, so from $x_i = 0$ follows $f_i(x) = 0$ and, by Lemma 2, $f_i(\alpha x) = 0$ for all $\alpha > 0$.

If $x_i > 0$, then $F_0(x) = \lambda(F_0)x$ implies $f_i^0(x) > 0$. Therefore, by (8), (9), for any arbitrarily small $\varepsilon > 0$ and sufficiently small $\alpha > 0$ (let for $\alpha < \alpha_0$) we have:

$$f_i^0(x) - \varepsilon \le \alpha^{-1}f_i(\alpha x) \le f_i^0(x) \quad (\forall i \in I^+(x))$$

$(I^+(x)$ is defined by (10)). Choosing any value ε satisfying the inequalities $0 < \varepsilon < \min_{i \in I^+(x)} (\lambda(F_0) - 1)x_i$, for $0 < \alpha < \alpha_0$ we get:

$$f_i(\alpha x) \ge \alpha(f_i^0(x) - \varepsilon) = \alpha(\lambda(F_0)x_i - \varepsilon) > \alpha x_i \quad (\forall i \in I^+(x)).$$

Taking into account the above equalities $f_i(\alpha x) = 0$ $(\forall i \in I^0(x))$ this means that $F(\bar{x}) \ge \bar{x}$, where $\bar{x} = \alpha x$ $(0 < \alpha < \alpha_0)$.

On the other hand, the condition $\lambda(F_\infty) < 1$ implies the existence of $\bar{y} > 0$ such that $F(\bar{y}) < \bar{y}$ [8, Theorem 2.13].

Thus, we have obtained the existence of $\bar{x} \gneqq 0$, $\bar{y} > 0$ such that $F(\bar{x}) \ge \bar{x}$, $F(\bar{y}) < \bar{y}$. Since $\bar{x} = \alpha x$ and $F(\bar{x}) \ge \bar{x}$ for all $0 < \alpha < \alpha_0$, one can choose $\bar{x}$ so that $\bar{x} < \bar{y}$. Thus, F has the invariant segment $[\bar{x}, \bar{y}]$. An application of Brouwer's theorem completes the proof.

As the next theorem shows, the conditions (26)–(27) are also necessary for the existence of a nonzero fixed point of concave map that is indecomposable at zero. Note that the map concave on $\mathbb{R}^q_+$ is monotone [1, Lemma 2.1.3].

Denote by N_F^+ the positive fixed points set of F.

Theorem 3. *Let the map F ($F(0) = 0$) be concave on $\mathbb{R}_+^q$ and indecomposable at zero. If there exists a nonzero fixed point $\bar{x}$ of F, then it is positive and one of the conditions (26), (27) is fulfilled.*

In case (26), $\operatorname{co}\{0, \bar{x}\} \subset N_F^+$ is true and all positive fixed points of F lie on the ray $R_{\bar{x}}$.

Proof. Let $\bar{x}$ be a nonzero fixed point of F; then $\bar{x} > 0$ because of indecomposability F at zero. By (9), $F_\infty(\bar{x}) \leq F(\bar{x}) = \bar{x} \leq F_0(\bar{x})$. It follows from $F_0(\bar{x}) \geq \bar{x}$ that $\lambda(F_0) \geq 1$ [6, Theorem 10.3]); in its turn, $F_\infty(\bar{x}) \leq \bar{x}$ results in $\lambda(F_\infty) \leq 1$ [7, Lemma 4]. So, we have $\lambda(F_\infty) \leq 1 \leq \lambda(F_0)$.

If $\lambda(F_0) = 1$, then $F_0(\bar{y}_0) = \bar{y}_0$, where $\bar{y}_0 \gneq 0$ is the eigenvector corresponding to the dominant eigenvalue $\lambda(F_0)$. The map F_0 is also indecomposable at zero [3, Theorem 3], so $\bar{y}_0 > 0$. Further, since $F(r\bar{y}_0) \leq F_0(r\bar{y}_0) = rF_0(\bar{y}_0) = r\bar{y}_0$ we have $F(r\bar{y}_0) \leq r\bar{y}_0$ ($\forall r > 0$). The sequence $\{F^t(r\bar{y}_0)\}_{t=0}^{+\infty}$ is monotonically decreasing and bounded, hence it converges; let its limit be $\bar{x}_r$. Choosing r such that $r\bar{y}_0 < \bar{x}$, we get $\bar{x}_r < \bar{x}$.

Because of the indecomposability of F at zero, all its positive fixed points lie on the same ray: $N_F^+ \subset R_{\bar{x}}$ [4, Corollary 3], so $\bar{x}_r = \alpha_r \bar{x}$ ($0 < \alpha_r < 1$). By Lemma 3, then $F(\alpha\bar{x}) = \alpha F(\bar{x})$ ($\forall \alpha \in [0,1]$) and $F_0(\bar{x}) = \lim\limits_{\alpha \to +0} \alpha^{-1} F(\alpha\bar{x}) = \bar{x}$, i.e. $F_0(\bar{x}) = \bar{x} = F(\bar{x})$. This means, by (23), (24), that $\alpha_F(\bar{x}) = 0$, i.e. situation (26) is realized. It follows from $F(\alpha\bar{x}) = \alpha F(\bar{x}) = \alpha\bar{x}$ ($\forall \alpha \in [0,1]$) that $\operatorname{co}\{0, \bar{x}\} \subset N_F^+$.

Case $\lambda(F_\infty) = 1$ is considered similarly: we have $F_\infty(\bar{y}_\infty) = \bar{y}_\infty$ for the eigenvector $\bar{y}_\infty \gneq 0$ corresponding to $\lambda(F_\infty)$, so, by (9), $F(s\bar{y}_\infty) \geq F_\infty(s\bar{y}_\infty) = s\bar{y}_\infty$, and $F^t(s\bar{y}_\infty) \geq (F_\infty)^t(s\bar{y}_\infty) \geq s\bar{y}_\infty$ ($\forall s > 0$, $t = 0, 1, 2, \ldots$). The sequence $\{F^t(s\bar{y}_\infty)\}_{t=0}^{+\infty}$ monotonically increases. Its boundedness this time follows from the existence of a sufficiently large $\beta > 1$ such that $s\bar{y}_\infty \leq \beta\bar{x}$. Indeed, by (6) we have $F^t(s\bar{y}_\infty) \leq F^t(\beta\bar{x}) \leq \beta F^t(\bar{x}) = \beta\bar{x}$, i.e. $F^t(s\bar{y}_\infty) \leq \beta\bar{x}$ ($t = 0, 1, 2, \ldots$).

The eigenvector $\bar{y}_\infty$ may have zero coordinates, but $I^+ = I^+(\bar{y}_\infty) \neq \varnothing$ (see (10)), where $\bar{y}_\infty = (\bar{y}_i)$. Let $i^+ \in I^+$, then there exists a sufficiently large s such that $s\bar{y}_{i^+} > \bar{x}_{i^+}$ (enough of $s > \bar{x}_{i^+}/\bar{y}_{i^+}$). Since $\bar{x}_s = \lim\limits_{t \to +\infty} F^t(s\bar{y}_\infty) \geq s\bar{y}_\infty$, we have $(\bar{x}_s)_{i^+} \geq s\bar{y}_{i^+} > \bar{x}_{i^+}$, i.e. $(\bar{x}_s)_{i^+} > \bar{x}_{i^+}$; so $\bar{x}_s \neq \bar{x}$. Thus, in this case, too, there are two distinct fixed points of F, so $\bar{x}_s = \beta_s\bar{x}$ ($\beta_s > 1$) [4, Corollary 3].

Hence, as in case $\lambda(F_0) = 1$, we get that $F_0(\bar{x}) = \bar{x} = F(\bar{x})$ and $\alpha_F(\bar{x}) = 0$, i.e. (26) is valid. The proof is complete.

Let us give an example demonstrating the essentiality of the indecomposability assumption for the validity of the conclusion of Theorem 3.

Example 3. Consider the map $F(x) = (f_1(x), f_2(x))$, where

$$f_1(x) = x_1/(1 + x_1) + x_2, \quad f_2(x) = \min\{x_1/2, x_2\}.$$

This map is concave on $\operatorname{int} \mathbb{R}_+^q$. From (8) we get:

$$F_0(x) = (x_1 + x_2, \min\{x_1/2, x_2\}), \quad \lambda(F_0) = 1,$$

$$F_\infty(x) = (x_2, \min\{x_1/2, x_2\}), \quad \lambda(F_\infty) = \sqrt{2}/2 < 1,$$

$$K_0 = \{(x_1, 0) \mid x_1 \in (0, +\infty)\}, \quad K_\infty = \{(x_1, \sqrt{2}\,x_1/2) \mid x_1 \in (0, +\infty)\}.$$

Let us find the sets of decomposability and indecomposability of this map.

On the set $\{x = (x_1, x_2) \mid 0 \le x_2 \le x_1/2\}$, the map F is decomposable, since on this set $f_2(x) = x_2$ and for $\bar{x} = (\bar{x}_1, \bar{x}_2)$, $\bar{x}' = (\bar{x}'_1, \bar{x}_2)$, with $\bar{x}_1 > \bar{x}'_1$ we have $f_2(\bar{x}) = f_2(\bar{x}') = \bar{x}_2$.

On the other hand, at the remaining points of $\mathbb{R}^2_+$, on the set $\{x = (x_1, x_2) \mid x_2 > x_1/2 \ge 0\}$, this map is indecomposable, since the function $f_1(x)$ is strong monotone, and the function $f_2(x) = x_1/2$ is strictly increasing in the variable x_1.

The set of positive fixed points has the form $N_F^+ = \{(x_1, \varphi(x_1)) \mid x_1 \in (0, 1]\}$, where $\varphi(x_1) = (x_1)^2/(1+x_1)$. As we can see, although $\lambda(F_0) = 1$, but $\alpha_F(x) \ne 0$ ($\forall x \in K_0$), and all these fixed points are on different rays.

Corollary 2. *The concave on $\mathbb{R}^q_+$ and indecomposable at zero map F ($F(0) = 0$) has a unique positive fixed point if and only if the condition (27) is satisfied.*

Note that situation (28), in the case of positive fixed point existence for concave indecomposable at zero map F, can also take place, but, as we have seen, it is a special case of situation (26). Let us show that the set of fixed points of F lying on the ray $R_{\bar{x}}$ in case $\lambda(F_\infty) = 1$ unbounded. Let $\bar{\beta}$ be any number, arbitrarily large, and $s > \max\{\bar{x}_{i+}/\bar{y}_{i+}, \bar{\beta}\bar{x}_{i+}/\bar{y}_{i+}\}$. From $\bar{x}_s = \beta_s\bar{x} \ge s\,\bar{y}_\infty$ we get $\beta_s\bar{x}_{i+} \ge s\,\bar{y}_{i+}$ and $\beta_s \ge s\,\bar{y}_{i+}/\bar{x}_{i+} > \bar{\beta}$, i.e. $\beta_s > \bar{\beta}$.

Because the fact $F(\beta_s\bar{x}) = \beta_s\bar{x} = \beta_s F(\bar{x})$ means that $\beta_s \in M_F(\bar{x})$, it follows (by Lemma 3) $(1, +\infty) \subseteq M_F(\bar{x})$ and $\beta_F(\bar{x}) = \sup M_F(\bar{x}) = +\infty$. Thus, by Lemma 4, $F_\infty(\bar{x}) = F(\bar{x}) = \bar{x}$, so we have the situanion (28).

Thus, situations (26), (28) are not alternative: they are realized simultaneously if the eigenrays of the maps F_0, F_∞ coincide with the positive ray from origin containing all positive fixed points of F.

This situation is realized, for example, for $F(x) = (f_1(x), f_2(x))$, where

$$f_1(x) = f_2(x) = \sqrt{x_1 x_2} + \sqrt{(x_1 - x_2)^2/(1 + (x_1 - x_2)^2)}.$$

Here to eigenvalues $\lambda(F_0) = \lambda(F_\infty) = 1$ corresponds the common eigenray $R = \{x = (t, t) \mid t \in (0, +\infty)\}$ containing the fixed points of the map F.

We also note that conditions (26)–(28) are not only sufficient, but also necessary conditions for the existence of a nonzero fixed point for a subhomogeneous map in the case of its indecomposability on the set of fixed points.

4   Conclusion

We have introduced here a weakening of the classical indecomposability concept—the concept of local indecomposability. In the case of a linear map, the concepts introduced coincide with the classical concept of map indecomposability. The coincidence of these properties was also proved for some classes of nonlinear maps. The connection between the introduced concepts and other well-known generalizations of map indecomposability concepts was shown. We have demonstrated that some assertions of the spectral theory for positively homogeneous maps remain valid when the property of global indecomposability is replaced by the property of local indecomposability (Theorem 1). Some properties of subhomogeneous local indecomposable maps were studied (Lemmas 2, 3, Corollary 1).

Sufficient conditions were obtained for the existence of a nonzero fixed point of a subhomogeneous map, which is not necessarily indecomposable (Theorem 2). The usefulness of the introduced local indecomposability concept for a more subtle analysis of the fixed points set structure was shown (Theorems 2, 3 and Corollary 2). For a map concave on $\mathbb{R}^q_+$, the obtained sufficient conditions are necessary under the assumption of its indecomposability at zero (Theorem 3), and the requirement of indecomposability at zero is essential (Example 3).

Finally, examples of maps illustrating the concept of local indecomposability were given.

References

1. Krause, U.: Dynamical Systems in Discrete Time: Theory, Models, and Applications. Walter de Gruyter GmbH, Berlin-Munich-Boston (2015)
2. Lemmens, B., Nussbaum, R.D.: Nonlinear Perron-Frobenius theory. In: Cambridge Tracts in Mathematics, vol. 189. Cambridge University Press, Cambridge (2012)
3. Mazurov, V.D., Smirnov, A.I.: The conditions of irreducibility and primitivity monotone subhomogeneous mappings. Trudy Inst. Mate. Mekh. UrO RAN **22**(3), 169–177 (2016) (in Russian). https://doi.org/10.12783/dtcse/optim2018/27933
4. Mazurov, V.D., Smirnov, A.I.: On the structure of the set of fixed points of decomposable monotone subhomogeneous mappings. Trudy Inst. Mate. Mekhan. UrO RAN **23**(4), 222–231 (2017) (in Russian). https://doi.org/10.21538/0134-4889-2017-23-4-222-231
5. Morishima, M.: Equilibrium, Stability, and Growth. Oxford University Press, Oxford (1964)
6. Nikaido, H.: Convex Structures and Economic Theory. Academic Press, New York (1968)
7. Oshime, Y.: On some weakenings of the concept of irreducibility. J. Math. Kyoto Univ. **23**, 803–830 (1983)
8. Oshime, Y.: Perron-Frobenius problem for weakly sublinear maps in a Euclidean positive orthant. J. Indust. Appl. Math. **9**, 313–350 (1992)
9. Piotrowski, T., Cavalcante, R.L.G.: The fixed point iteration of positive concave mappings converges geometrically if a fixed point exists (2021). https://arxiv.org/pdf/2110.11055.pdf
10. Schubert, M., Boche, H.: Interference calculus – a general framework for interference management and network utility optimization. In: Foundations in Signal Processing, Communications and Networking, vol. 7, pp. 1–252. Springer, Heidelberg (2012). https://doi.org/10.1007/978-3-642-24621-0
11. Mazurov, V.D., Smirnov, A.I.: A criterion for the existence of nondestructive controls in the problem of optimal exploitation of a binary-structured system. Proc. Steklov Inst. Math. **315**(1), S203–S218 (2021). https://doi.org/10.1134/S008154382106016X
12. Stetsenko, V.Y., Semiletov, V.A.: Indecomposable nonlinear operators. Nauka. Innov. Technol. **34**, 12–16 (2003). (in Russian)

One Relaxed Variant of the Proximal Level Method

Rashid Yarullin$^{(\boxtimes)}$ and Igor Zabotin

Kazan (Volga region) Federal University, Kazan, Russia
yarullinrs@gmail.com, iyazabotin@mail.ru

Abstract. We propose a method belonged to the class of bundle methods for solving convex programming problems. The points of the main sequence are constructed in the developed method with the relaxation condition in two stages. The point and parameters are fixed for estimating the approximation quality at the first stage. Further, at the second stage a piecewise affine model of the objective function is constructed on the basis of auxiliary points, and a "record" one is selected among these points. If the objective function is approximated well enough in the neighborhood of this "record" point by the constructed model, then the selected "record" point is fixed as the next point of the main sequence, and the first stage is performed with "cleaning" the model. Otherwise, the model of the objective function is refined, and the second stage of the proposed method is restarted. Various implementations of the developed method are discussed. The convergence of the proposed method is proved, and the main sequence of the objective function values converges to the optimal value at a linear rate. A complexity estimate of finding an approximate solution is obtained for the proposed method under certain conditions for constructing auxiliary points at the second stage.

Keywords: Nondifferentiable optimization · Convex function · Subgradient · Proximal level method · Convergence rate estimate · Relaxed sequence

1 Introduction

Information about subgradients which are obtained at previous iterations is quite often used in many nondifferentiable optimization methods while constructing iteration points (see, for example, [1–4]). The Kelley method [5] belongs to such methods, where some piecewise affine model of the objective function is minimized for constructing an iteration point. Note that this model is formed by the previously obtained subgradients. However, the Kelly method is characterized in [6] by instability, since the distance between neighboring points converges to zero very slowly, and the iteration points move in a "zigzag" pattern and can

This paper has been supported by the Kazan Federal University Strategic Academic Leadership Program ("PRIORITY-2030").

skip the solution neighborhood even in the case when the objective function is approximated well enough in this neighborhood by the piecewise affine model.

Various approaches have been proposed to speed up the convergence of the Kelly method (see, for example, [7–10]). In particular, the iteration points trajectory is stabilized in the proximal level method [7] as follows. The record value of the objective function and the record point named as the proximal point are fixed at each step. Then an approximation point is constructed by projecting the proximal point onto an auxiliary levels set of the objective function model. If the value of the objective function is less than the record value in this approximation point, then the record value and the proximal point are updated. Otherwise, the affine function, obtained on the basis of this approximation point, is added to the objective function model, and the next approximation one is found. This approach makes it possible to construct the relaxed sequence of proximal points and solves the above specific problems of the Kelly method.

In this paper we propose a new proximal level method that uses the above idea of constructing relaxed sequences. In the developed method each iteration point is found in two stages. At the first stage the point of the main sequence and the parameters that estimate the approximation quality are fixed. Further, at the second stage the objective function model is constrcuted on the basis of auxiliary points and the "record" point is chosen from these auxiliary points with the relaxation condition. If the objective function is sufficiently well approximated by the constructed model in the neighborhood of the "record" point, then the "record" point is selected as the next point of the main sequence, the model is "cleaned", and the first stage of the optimization process is restarted with the redefined estimation parameters of the approximation quality. The proposed method differs from the aforementioned proximal method in that the transition from the current iteration point of the main sequence to the next one is performed when relaxation occurs and a good approximation of the objective function is obtained by the piecewise affine model. In addition, the principal feature of the developed method is that each auxiliary point is chosen arbitrarily from the levels set of the objective function model, i.e. projection operation can be omitted. This opportunity of selection auxiliary points provides certain conveniences for the numerical implementation of the proposed method in practice.

For the well-known proximal level methods [6,7,10] an complexity estimate of finding an E-feasible solution is equal to $O(1/E^2)$, where $E > 0$. But this complexity estimate is valid under using the projection operation for constructing auxiliary points in the aforementioned proximal level methods. Note the proposed method is characterized by the fact that the main sequence of iteration points converges to the solution set of the optimization problem with the linear rate (the geometric progression rate) regardless of the involved approaches for finding the auxiliary points in the second stage. In addition, if the test points are constructed by the projection of the proximal points onto the auxiliary levels sets at the second stage of the proposed method, then the estimate of the complexity $O(\log_c E/E^2)$ is obtained, $c > 0$ is some parameter of the developed method.

2   Problem Settings

Let D be a convex bounded closed set from an n-dimensional Euclidian space R_n, $f(x)$ be a convex function in R_n. Solve the following problem:

$$\min\{f(x) : x \in D\}.$$

Suppose $f^* = \min\{f(x) : x \in D\}$, $X^* = \{x \in D : f(x) = f^*\} \neq \emptyset$, $x^* \in X^*$, $X^*(\varepsilon) = \{x \in D : f(x) \leq f^* + \varepsilon\}$, where $\varepsilon > 0$, $\partial f(x)$ be a subdifferential of the function $f(x)$ at the point $x \in R_n$, $K = \{0, 1, \dots\}$. Denote by $\lceil N \rceil$ the least integer no less than $N \in R_1$. Let $|D|$ be a diameter of the set D, $\mathrm{Pr}(z, A)$ be an projection operator of the point $z \in R_n$ on the set $A \subset R_n$.

Let us immediately note that there exists a positive number L (see, for example, Lemma 8 [11, p. 121]) in view of boundedness of the set D such that for any point $x \in D$ and any subgradient $s \in \partial f(x)$ the inequality $\|s\| \leq L$ is fulfilled.

3   Minimization method

The construction process of the points $\{x_k\}$, $k \in K$, by the proposed method with the following parameters:

$$\hat{x} \in R_n, \quad \hat{\alpha} \leq f^*, \quad \hat{\varepsilon} > 0, \quad \hat{\theta} \in (0, 1). \tag{1}$$

Step 0. Choose a subgradient $\hat{s} \in \partial f(\hat{x})$ and find a point $x_0 \in D$ such that

$$\max\{f(\hat{x}) + \langle \hat{s}, x_0 - \hat{x}\rangle; \hat{\alpha}\} \leq \max\{f(\hat{x}) + \langle \hat{s}, y - \hat{x}\rangle; \hat{\alpha}\} \quad \forall y \in D. \tag{2}$$

Assign
$$\alpha_0 = \max\{f(\hat{x}) + \langle \hat{s}, x_0 - \hat{x}\rangle; \hat{\alpha}\}, \tag{3}$$

$$\beta_0 = f(x_0), \tag{4}$$

$k = 0$.

Step 1. Choose a subgradient $s_k \in \partial f(x_k)$. Assign

$$\bar{x}_{k,0} = y_{k,0} = x_{k,0} = x_k, \tag{5}$$

$$\beta_{k,0} = \beta_k, \tag{6}$$

$$\alpha_{k,0} = \alpha_k, \tag{7}$$

$$s_{k,0} = s_k,$$

$i = 0$.

Step 2. If the equality

$$\beta_{k,i} = \alpha_{k,i}, \tag{8}$$

is fulfilled, then $\bar{x}_{k,i} \in X^*$, and the work of the proposed method is completed.

Step 3. If the inequality
$$\beta_{k,i} - \alpha_{k,i} \le \hat{\varepsilon}, \tag{9}$$
is fulfilled, then $\bar{x}_{k,i} \in X^*(\hat{\varepsilon})$, and if desired, the points finding process is terminated.

Step 4. Construct a model
$$m_{k,i}(y) = \max_{0 \le j \le i}\{f(x_{k,j}) + \langle s_{k,j}, y - x_{k,j}\rangle\}, \tag{10}$$
and assign a level value
$$l_{k,i} = (1 - \hat{\theta})\alpha_k + \hat{\theta}\beta_{k,i}. \tag{11}$$

Step 5. Find a point
$$y_{k,i+1} = \arg\min\{\max\{m_{k,i}(y); \alpha_{k,i}\} : y \in D\}, \tag{12}$$
and calculate a parameter
$$\alpha_{k,i+1} = \max\{m_{k,i}(y_{k,i+1}); \alpha_{k,i}\}. \tag{13}$$

Step 6. If the inequality
$$\alpha_{k,i+1} \ge l_{k,i}, \tag{14}$$
is fulfilled, then assign $i_k = i$,
$$\alpha_{k+1} = \alpha_{k,i_k+1}, \quad \beta_{k+1} = \min\{\beta_{k,i_k}; f(y_{k,i_k+1})\}. \tag{15}$$
$$x_{k+1} = \arg\min\{f(\bar{x}_{k,i_k}); f(y_{k,i_k+1})\}, \tag{16}$$
the value of k is increased by one, and go to Step 1.

Step 7. Construct a level set
$$U_{k,i} = \{y \in D : m_{k,i}(y) \le l_{k,i}\}, \tag{17}$$
choose a point $x_{k,i+1} \in U_{k,i}$, a subgradient $s_{k,i+1} \in \partial f(x_{k,i+1})$, and assign
$$\beta_{k,i+1} = \min\{\beta_{k,i}; f(x_{k,i+1}); f(y_{k,i+1})\}, \tag{18}$$
$$\bar{x}_{k,i+1} = \begin{cases} \bar{x}_{k,i}, \text{if } \beta_{k,i} \le f(x_{k,i+1}), \beta_{k,i} \le f(y_{k,i+1}); \\ x_{k,i+1}, \text{if } f(x_{k,i+1}) < \beta_{k,i}, f(x_{k,i+1}) \le f(y_{k,i+1}); \\ y_{k,i+1}, \text{if } f(y_{k,i+1}) < \beta_{k,i}, f(y_{k,i+1}) < f(x_{k,i+1}). \end{cases} \tag{19}$$
Go to Step 2 with incremented i.

Let us make some remarks concerning the developed method.

Remark 1. The function $m_{k,i}(y)$ is called a model of the function $f(x)$. The model $m_{k,i}(y)$ approximates the objective function $f(x)$ on the basis of the previously obtained data $x_{k,0}, \ldots, x_{k,i}, s_{k,0}, \ldots, s_{k,i}, f(x_{k,0}), \ldots, f(x_{k,i})$. Since the model $m_{k,i}(y)$ is formed by affine functions, then the function $m_{k,i}(y)$ is convex.

In the proposed method the sequence of the main points $\{x_k\}$, $k \in K$, is constructed at the first stage of the optimization process, and the sequence of the auxiliary points $\{x_{k,i}\}$, $i \in K$ is generated for implementing the transition from the point x_k to x_{k+1} at the second stage. The second stage is finished when the model $m_{k,i}(y)$, constructed on the basis of $x_{k,0}, \ldots, x_{k,i}$, approximates the objective function $f(x)$ well enough in a neighborhood of the point $\bar{x}_{k,i}$. The point $y_{k,i}$ is constructed to evaluate this approximation quality at Step 5 of the proposed method, and another record point $\bar{x}_{k,i+1}$ claiming the position x_{k+1} is fixed according to (19) in case of obtaining an unsatisfactory approximation, as well as a new affine function is added into the model $m_{k,i}(y)$.

Remark 2. At Step 5 the point $y_{k,i+1}$ is found by minimizing a nondifferentiable function, and the number $\alpha_{k,i+1}$ is calculated according to (13). However, the point $y_{k,i+1}$ and the number $\alpha_{k,i+1}$ can be found by solving the problem

$$\alpha \to \min_{\alpha \in R_1, y \in R_n},$$

$$f(x_{k,j}) + \langle s_{k,j}, y - x_{k,j} \rangle \le \alpha, \quad j = 0, \ldots, i,$$

$$\alpha \ge \alpha_{k,i},$$

$$y \in D.$$

If D is a polyhedron, then the above problem is a linear programming problem, and, in particular, the simplex method can be used to solve it.

Known cutting methods (see, for example, [5, 12, 13]) favorably differ from many optimization methods in that it is not difficult to obtain a lower estimate of the optimal value of the objective function at each iteration in practical calculations. This feature is also characteristic of the proposed method. Namely, in the proposed method we can estimate the value of f^* from below by constructing the minimum point $y_{k,i+1}$ of the model $m_{k,i+1}(y)$. This result is obtained in the following assertion.

Lemma 1. *Suppose the parameter $\alpha_{k,i}$ is obtained for some $k \ge 0$, $i \ge 0$ by the proposed method. Then the inequality*

$$\alpha_{k,i} \le f^* \tag{20}$$

is fulfilled.

Proof. Assume $k = 0$, $i = 0$. We have

$$\max\{f(\hat{x}) + \langle \hat{s}, x_0 - \hat{x} \rangle; \hat{\alpha}\} \le \max\{f(\hat{x}) + \langle \hat{s}, x^* - \hat{x} \rangle; f^*\} \le f^*$$

according to (2) and the second condition from (1) for x^*. Hence and from (3), (7) it follows $\alpha_{0,0} = \alpha_0 \le f^*$.

Now let estimate (20) be valid for $k = 0$ and any $i \ge 0$. If (8) or (9) occurs, then the method stops working, and $y_{k,i+1}$, $\alpha_{k,i+1}$ will not be constructed according to (12), (13). In this regard we assume that the inequality

$$\beta_{k,i} - \alpha_{k,i} > \hat{\varepsilon} \tag{21}$$

is fulfilled. Lets show that $\alpha_{k,i+1} \leq f^*$. Then from (12), (13), (20) it follows that

$$\alpha_{k,i+1} = \max\{m_{k,i}(y_{k,i+1}); \alpha_{k,i}\} \leq \max\{m_{k,i}(x^*); f^*\}. \tag{22}$$

Moreover, by definition of the subgradient we have

$$m_{k,i}(x^*) \leq f(x_{k,j_i}) + \langle s_{k,j_i}, x^* - x_{k,j_i} \rangle \leq f^*, \tag{23}$$

where the index $j_i \leq i$ is chosen according to the condition

$$f(x_{k,j_i}) + \langle s_{k,j_i}, x^* - x_{k,j_i} \rangle \geq f(x_{k,j}) + \langle s_{k,j}, x^* - x_{k,j} \rangle, \quad j = 0, \ldots, i.$$

Hence and from (22) we obtain $\alpha_{k,i+1} \leq f^*$.

Suppose that the inequalities (20), (21) hold for any $k \geq 0$, $i \geq 0$. Then the validity $\alpha_{k,i+1} \leq f^*$ can be proved by obtaining (22), (23) on the basis of (12), (13), (20). In addition, if there is a transition to the $(k+1)$-th iteration while executing inequality (14) at Step 6 of the proposed method, then the parameter $\alpha_{k+1} = \alpha_{k,i+1} \leq f^*$ is fixed according to (15), and in view of (7) we get $\alpha_{k+1,0} \leq f^*$. The lemma is proved.

Corollary 1. *Suppose the parameter α_k is computed for some $k \geq 0$ by the proposed method. Then $\alpha_k \leq f^*$.*

Remark 3. At the second stage the approximation progress is tracked by the model $m_{k,i}(y)$ based on the point $\bar{x}_{k,i}$ claiming the position x_{k+1}, and the record value of the objective function is entered into the parameter $\beta_{k,i}$.

Lemma 2. *Suppose the point $\bar{x}_{k,i}$ is constructed and the number $\beta_{k,i}$ is computed for some $k \geq 0$, $i \geq 0$ by the proposed method. Then*

$$f(\bar{x}_{k,i}) = \beta_{k,i}. \tag{24}$$

Proof. Suppose $k = 0$, $i = 0$. Then according to (4), (5), (6) we get

$$f(\bar{x}_{0,0}) = \beta_{0,0}.$$

Now let (24) holds under $k = 0$ and any $i \geq 0$. Lets obtain $\beta_{0,i+1} = f(\bar{x}_{0,i+1})$. Then in view of (19) and $\beta_{0,i} = f(\bar{x}_{0,i})$ we have

$$\bar{x}_{0,i+1} = \begin{cases} \bar{x}_{0,i}, & \text{if } f(\bar{x}_{0,i}) \leq f(x_{0,i+1}), f(\bar{x}_{0,i}) \leq f(y_{0,i+1}); \\ x_{0,i+1}, & \text{if } f(x_{0,i+1}) < f(\bar{x}_{0,i}), f(x_{0,i+1}) \leq f(y_{0,i+1}); \\ y_{0,i+1}, & \text{if } f(y_{0,i+1}) < f(\bar{x}_{0,i}), f(y_{0,i+1}) < f(x_{0,i+1}). \end{cases}$$

Moreover, in accordance with (18) the expression

$$\beta_{0,i+1} = \min\{f(\bar{x}_{0,i}); f(x_{0,i+1}); f(y_{0,i+1})\}$$

is fulfilled. Further, lets consider three cases.

1) If $f(\bar{x}_{0,i}) \le f(x_{0,i+1})$ and $f(\bar{x}_{0,i}) \le f(y_{0,i+1})$, then $\beta_{0,i+1} = f(\bar{x}_{0,i})$, $\bar{x}_{0,i+1} = \bar{x}_{0,i}$.

2) If $f(x_{0,i+1}) < f(\bar{x}_{0,i})$ and $f(x_{0,i+1}) \le f(y_{0,i+1})$, then $\beta_{0,i+1} = f(x_{0,i+1})$, $\bar{x}_{0,i+1} = x_{0,i+1}$.

3) If $f(y_{0,i+1}) < f(\bar{x}_{0,i})$ and $f(y_{0,i+1}) < f(x_{0,i+1})$, then $\beta_{0,i+1} = f(y_{0,i+1})$, $\bar{x}_{0,i+1} = y_{0,i+1}$. Thus, $\beta_{0,i+1} = f(\bar{x}_{0,i+1})$.

Now lets consider the case of transition from k-th to the $(k+1)$-th iteration. Assume that equality (24) is fulfilled for any $k \ge 0$ and $i = i_k$. Then according to (15) and (16) we get

$$\beta_{k+1} = \min\{f(\bar{x}_{k,i_k}); f(y_{k,i_k+1})\},$$

$$x_{k+1} = \arg\min\{f(\bar{x}_{k,i_k}); f(y_{k,i_k+1})\}.$$

If $f(\bar{x}_{k,i_k}) \le f(y_{k,i_k+1})$, then $\beta_{k+1} = f(\bar{x}_{k,i_k})$, $x_{k+1} = x_{k,i_k}$. Otherwise, $\beta_{k+1} = f(y_{k,i_k+1})$, $x_{k+1} = y_{k,i_k+1}$. Consequently, $\beta_{k+1} = f(x_{k+1})$, and in view of (5), (6) we get

$$\beta_{k+1,0} = \beta_{k+1} = f(x_{k+1}) = f(\bar{x}_{k+1,0}).$$

The lemma is proved.

Lemma 3. *Suppose the proposed method obtains $\alpha_{k,i}$, $\alpha_{k,i+1}$, $\beta_{k,i}$, $\beta_{k,i+1}$ for some $k \ge 0$, $i \ge 0$. Then*

$$\alpha_{k,i} \le \alpha_{k,i+1} \le f^* \le \beta_{k,i+1} \le \beta_{k,i}. \tag{25}$$

Proof. In view of (13) the inequality $\alpha_{k,i+1} \ge \alpha_{k,i}$ holds. Hence and from Lemma 1 we get

$$\alpha_{k,i} \le \alpha_{k,i+1} \le f^*. \tag{26}$$

Moreover, according to (18) we have $\beta_{k,i+1} \le \beta_{k,i}$, and $\bar{x}_{k,i} \in D$ by construction. Then taking into account Lemma 2 the expression

$$f^* \le \beta_{k,i+1} \le \beta_{k,i}$$

holds. Combining the last expression with (26), the assertion of the lemma is proved.

Corollary 2. *Suppose the parameters α_k, β_k are computed for some $k \ge 0$ by the proposed method. Then $\alpha_k \le f^* \le \beta_k$.*

On the basis of Lemmas 2, 3 it is not difficult to prove the following stopping criteria for the proposed method.

Theorem 1. *Suppose equality (8) is fulfilled for some $k \ge 0$, $i \ge 0$ at Step 2 of the proposed method. Then $\bar{x}_{k,i} \in X^*$.*

Theorem 2. *Suppose inequality (9) is fulfilled for some $k \ge 0$, $i \ge 0$ at Step 3 of the proposed method. Then $\bar{x}_{k,i} \in X^*(\hat{\varepsilon})$.*

Lemma 4. *Suppose the set $U_{k,i}$ and the point $y_{k,i+1}$ are constructed for some $k \geq 0$, $i \geq 0$ by the proposed method. Then*

$$y_{k,i+1} \in U_{k,i}. \tag{27}$$

Proof. The set $U_{k,i}$ is constructed according to the condition of the lemma. This means that there was a transition from Step 6 to Step 7 in the proposed method at the i-th iteration, i.e. the inequality $\alpha_{k,i+1} < l_{k,i}$ holds. Moreover, the point $y_{k,i+1} \in D$ is constructed according to (12), and the inequality $m_{k,i}(y_{k,i+1}) \leq \alpha_{k,i+1}$ is fulfilled in view of (13). Consequently, $m_{k,i}(y_{k,i+1}) < l_{k,i}$, i.e. $y_{k,i+1} \in U_{k,i}$. The lemma is proved.

Remark 4. The point $x_{k,i+1}$ can be chosen in various ways at Step 7 of the proposed method. In particular, if the constraint set D is given in a general form, then it is permissible to put $x_{k,i+1} = y_{k,i+1}$ in view of Lemma 4. If the bounded feasible set D is given by linear equalities and inequalities, then $U_{k,i}$ is a polyhedron. Then the approximation point $x_{k,i+1}$ can be found by projecting the point x_k onto the set $U_{k,i}$, i.e. $x_{k,i+1} = \Pr(x_k, U_{k,i})$, and such modification of the proposed method is conceptually close to some variants of the proximal level method [7,10,14]. If the set D is not a polyhedron and has a fairly simple structure, then the point $x_{k,i+1}$ can be found by any method for solving the system of equalities and inequalities.

In the proposed method all cutting planes that form the objective function model are discarded when the first stage of the optimization process is restarted as follows. The point x_k and the parameters α_k, β_k are fixed at the first stage. Then the model $m_{k,i}(y)$ is constructed based on the auxiliary points $x_{k,0}, \ldots, x_{k,i}$ at each step of the second stage and taking into account α_k the parameter $l_{k,i}$ is set to estimate the approximation quality of $f(x)$ by the model $m_{k,i}(y)$ in the neighborhood of the point $\bar{x}_{k,i}$. If inequality (14) is not fulfilled after finding the point $y_{k,i+1}$ and calculating the parameter $\alpha_{k,i+1}$, then the approximation quality is considered unsatisfactory, so the auxiliary point $x_{k,i+1}$ is selected for the next model $m_{k,i+1}(y)$, and the record point $\bar{x}_{k,i+1}$ is updated according to (19). Otherwise, firstly, x_{k+1} is fixed as the point of the main sequence, and the approximation progress is saved in form (15). Secondly, all earlier constructed cutting planes are discarded by adding one cutting plane formed on the basis of the point $x_{k+1,0} = x_{k+1}$ into the model $m_{k+1,0}(y)$, and the first stage of the optimization process is restarted.

Lemma 5. *Suppose for some $k \geq 0$, $i \geq 0$ the parameters $l_{k,i}$, $l_{k,i+1}$ are computed and the sets $U_{k,i}$, $U_{k,i+1}$ are constructed. Then the expressions*

$$l_{k,i+1} \leq l_{k,i}, \tag{28}$$

$$U_{k,i+1} \subset U_{k,i}. \tag{29}$$

hold.

Proof. Taking into account (11) we get $l_{k,i+1} - l_{k,i} = \hat{\theta}(\beta_{k,i+1} - \beta_{k,i})$. Hence and from (25) it follows that inequalitiy (28) is fulfilled.

Note that in accordance with Lemma 4 we have $U_{k,i} \neq \emptyset$, $U_{k,i+1} \neq \emptyset$. Lets fix any point $z' \in U_{k,i+1}$. Then according to (17), (28) we get

$$m_{k,i+1}(z') \leq l_{k,i}. \tag{30}$$

On the other hand, in view of (10) the expression

$$m_{k,i+1}(z') \geq \max_{0 \leq j \leq i} \{ f(x_{k,j}) + \langle s_{k,j}, z' - x_{k,j} \rangle \} = m_{k,i}(z')$$

holds. Combining the last expression with inequality (30) we obtain $m_{k,i}(z') \leq l_{k,i}$. Therefore, inclusion (29) holds true. The lemma is proved.

Corollary 3. *Suppose the sets*

$$U_{k,0}, \ldots, U_{k,i}$$

are constructed for some $k \geq 0$, $i \geq 0$ by the proposed method. Then

$$U_{k,i} \subset \cdots \subset U_{k,0}. \tag{31}$$

4   Convergence Research

Lemma 6. *Suppose for some $k \geq 0$ the point x_k is constructed and the number β_k is computed. Then $f(x_k) = \beta_k$.*

Proof. The validity of the lemma statement follows from equality (4) for $k = 0$.

Now suppose that the assertion of the lemma holds true for some $k \geq 0$. Lets prove the validity of the lemma assertion for $k+1$. Then according to Step 6 of the proposed method we have

$$\beta_{k+1} = \min\{\beta_{k,i_k}; f(y_{k,i_k+1})\},$$

$$x_{k+1} = \arg\min\{f(\bar{x}_{k,i_k}); f(y_{k,i_k+1})\}.$$

In addition, the equality $f(\bar{x}_{k,i_k}) = \beta_{k,i_k}$ is fulfilled in view of Lemma 2. Consequently, $f(x_{k+1}) = \beta_{k+1}$. The lemma is proved.

Lemma 7. *Suppose the points $x_{k,i}$, x_{k,p_i} are constructed for some $k \geq 0$, $p_i > i \geq 0$ by the proposed method. Then the inequality*

$$\|x_{k,p_i} - x_{k,i}\| \geq \frac{(1 - \hat{\theta})(\beta_{k,i} - \alpha_{k,i})}{L} \tag{32}$$

is fulfilled.

Proof. We have $x_{k,p_i} \in U_{k,p_i-1} \subset U_{k,i}$ according to Step 7 of the proposed method and condition (31), i.e.

$$m_{k,i}(x_{k,p_i}) = \max_{0 \leq j \leq i} \{f(x_{k,j}) + \langle s_{k,j}, x_{k,p_i} - x_{k,j} \rangle\} \leq l_{k,i}.$$

Therefore, the inequality

$$f(x_{k,i}) + \langle s_{k,i}, x_{k,p_i} - x_{k,i} \rangle \leq l_{k,i} \tag{33}$$

is fulfilled. Moreover, taking into account Lemma 6, condition (18) and expression (25) we have

$$f(x_{k,i}) \geq \beta_{k,i}, \quad \alpha_{k,i} \geq \alpha_k.$$

Then in view of (11) we obtain

$$f(x_{k,i}) - (1 - \hat{\theta})(\beta_{k,i} - \alpha_{k,i}) \geq \beta_{k,i} + (1 - \hat{\theta})(\alpha_k - \beta_{k,i}) = l_{k,i}.$$

Hence and from (33) it follows

$$f(x_{k,i}) - (1-\hat{\theta})(\beta_{k,i} - \alpha_{k,i}) \geq f(x_{k,i}) + \langle s_{k,i}, x_{k,p_i} - x_{k,i} \rangle \geq f(x_{k,i}) - L\|x_{k,p_i} - x_{k,i}\|.$$

Transforming the last expression we prove the validity of estimate (32). The lemma is proved.

Lemma 8. *For any $k \in K$ there exists a number i_k such that condition (14) is fulfilled.*

Proof. Suppose the point x_k be constructed for the index $k \in K$ by the proposed method. Let us prove the existence of the index $i_k \geq 0$ for which stop condition (14) will be held.

Assume that the sequences $\{\alpha_{k,i}\}$, $\{l_{k,i}\}$, $i \in K$ are constructed such that

$$l_{k,i} - \alpha_{k,i+1} > 0 \quad \forall i \in K.$$

Hence and from (11), (25) it follows

$$\alpha_k \leq \alpha_{k,i} \leq \alpha_{k,i+1} < l_{k,i} \leq \beta_{k,i} \leq \beta_k \quad \forall i \in K.$$

Consequently, for each $i \in K$ we have

$$0 < l_{k,i} - \alpha_{k,i+1} \leq l_{k,i} - \alpha_{k,i} \leq (1 - \hat{\theta})\alpha_{k,i} + \hat{\theta}\beta_{k,i} - \alpha_{k,i} \leq \beta_{k,i} - \alpha_{k,i}, \tag{34}$$

and the sequences $\{\alpha_{k,i}\}$, $\{l_{k,i}\}$, $i \in K$, are bounded.

Let $\{l_{k,i}\}$, $i \in K_1 \subset K$, $\{\alpha_{k,i+1}\}$, $i \in K_2 \subset K_1$, be any convergent sequences of the sequences $\{\alpha_{k,i}\}$, $\{l_{k,i}\}$, $i \in K$, respectively, and $\bar{\alpha}$, $\bar{l}$ be corresponding limit points. Further, the sequences $\{\alpha_{k,i+1}\}$, $\{l_{k,i}\}$, $i \in K_2$, correspond to the sequence $\{x_{k,i}\}$, $i \in K_2$. Select any convergent subsequence $\{x_{k,i}\}$, $i \in K_3 \subset K_2$, from the bounded sequence $\{x_{k,i}\}$, $i \in K_2$. Let $\bar{x}$ be a limit point of the sequence

$\{x_{k,i}\}$, $i \in K_3$. Now fix an index $p_i \in K_3$ for each $i \in K_3$ such that $p_i > i$. Then according to Lemma 7 and expression (34) for each $i \in K_3$ we have

$$0 < l_{k,i} - \alpha_{k,i+1} \leq \beta_{k,i} - \alpha_{k,i} \leq \frac{L}{1 - \hat{\theta}} \|x_{k,p_i} - x_{k,i}\|, \quad p_i > i, p_i \in K_3.$$

Since $x_{k,i} \to \bar{x}$, $x_{k,p_i} \to \bar{x}$, $l_{k,i} \to \bar{l}$, $\alpha_{k,i+1} \to \bar{\alpha}$ as $i \to \infty$, $i \in K_3$, then from the last expression we obtain the contradictory inequality $0 < \bar{l} - \bar{\alpha} \leq 0$. The lemma is proved.

Lemma 9. *Suppose the numbers* α_k, α_{k+1}, β_k, β_{k+1} *are constructed for some* $k \geq 0$ *by the proposed method. Then*

$$\beta_{k+1} - \alpha_{k+1} \leq (1 - \hat{\theta})(\beta_k - \alpha_k). \tag{35}$$

Proof. According to Lemma 8 there exists an index i_k such that expressions (14), (15) are fulfilled. Hence and from (6), (7), (25) it follows

$$\alpha_k = \alpha_{k,0} \leq \alpha_{k,i_k} \leq \alpha_{k,i_k+1} = \alpha_{k+1} \leq f^* \leq \beta_{k+1} \leq \beta_{k,i_k} \leq \beta_{k,0} = \beta_k.$$

Moreover, in view of (11) we have

$$l_{k,i_k} = (1 - \hat{\theta})\alpha_k + \hat{\theta}\beta_{k,i_k} \pm \beta_{k+1} \geq \beta_{k+1} + (1 - \hat{\theta})(\alpha_k - \beta_{k,i_k}).$$

Consequently, we obtain

$$0 \leq \alpha_{k,i_k+1} - l_{k,i_k} \leq \alpha_{k+1} - \beta_{k+1} - (1 - \hat{\theta})(\alpha_k - \beta_{k,i_k}) \leq \alpha_{k+1} - \beta_{k+1} + (1 - \hat{\theta})(\beta_k - \alpha_k)$$

from which inequality (35) holds true. The lemma is proved.

Theorem 3. *Suppose the sequence* $\{x_k\}$, $k \in K$, *is constructed by the proposed method. Then*

$$f(x_k) - \alpha_k \leq (1 - \hat{\theta})^k (\beta_0 - \alpha_0), \quad k \geq 0, \tag{36}$$

$$\lim_{k \in K} f(x_k) = f^*.$$

Proof. The sequence $\{x_k\}$, $k \in K$, corresponds to the sequence of numbers $\{\alpha_k\}$, $\{\beta_k\}$, $k \in K$, and according to Lemma 9 for each $k \in K$ inequality (35) holds. Consequently, the estimate

$$\beta_k - \alpha_k \leq (1 - \hat{\theta})^k (\beta_0 - \alpha_0), \quad k \geq 0 \tag{37}$$

is fulfilled. Moreover, in accordance with Lemma 6 we have $f(x_k) = \beta_k$, $k \in K$. Hence and from (37) we get (36).

Since the inequality $\alpha_k \leq f^*$ is fulfilled according to (25), then taking into account (36) we get

$$0 \leq f(x_k) - f^* \leq (1 - \hat{\theta})^k (\beta_0 - \alpha_0), \quad k \in K.$$

Hence and from $\hat{\theta} \in (0, 1)$ it follows that the expression

$$\lim_{k \in K} (1 - \hat{\theta})^k (\beta_0 - \alpha_0) = 0$$

is fulfilled. Consequently, $\lim_{k \in K} f(x_k) = f^*$. The theorem is proved.

Theorem 4. *Suppose the inequality $\beta_k - \alpha_k \leq \hat{\varepsilon}$ holds for some $k \geq 0$. Then the inclusion $x_k \in X^*(\hat{\varepsilon})$ is fulfilled.*

Proof. According to (25) and Lemma 6 we have $f(x_k) - f^* \leq \beta_k - \alpha_k$. Hence and from the condition $\beta_k - \alpha_k \leq \hat{\varepsilon}$ we obtain $x_k \in X^*(\hat{\varepsilon})$.

Now we are going to estimate the number of iterations of the first stage required to find an $\hat{\varepsilon}$-approximate solution by the proposed method.

Theorem 5. *Suppose the sequence $\{x_k\}$, $k \in K$, is constructed by the proposed method. Finding an $\hat{\varepsilon}$-approximate solution the number of k-iterations does not exceed*

$$\begin{cases} 1, & \text{if } \beta_0 - \alpha_0 \leq \hat{\varepsilon} \\ \lceil \log_{1-\hat{\theta}} \frac{\hat{\varepsilon}}{\beta_0 - \alpha_0} \rceil, & \text{if } \beta_0 - \alpha_0 > \hat{\varepsilon}. \end{cases} \tag{38}$$

Proof. If the inequality $\beta_0 - \alpha_0 \leq \hat{\varepsilon}$ is fulfilled, then in view of (6), (7) we get $\beta_{0,0} - \alpha_{0,0} \leq \hat{\varepsilon}$, and the optimization process is finished according to Step 3 of the proposed method. Then from Theorem 4 it follows that $x_0 \in X^*(\hat{\varepsilon})$, and estimation (38) holds true.

Assume that $\beta_0 - \alpha_0 > \hat{\varepsilon}$. Then according to Lemma 6 and estimation (36) the inequality

$$\beta_k - \alpha_k \leq (1 - \hat{\theta})^k (\beta_0 - \alpha_0)$$

is fulfilled for each $k \in K$. Hence and from $(1 - \hat{\theta})^k (\beta_0 - \alpha_0) \to 0$ as $k \to \infty$ there exists an index $k' \in K$ such that

$$\beta_{k'} - \alpha_{k'} \leq (1 - \hat{\theta})^{k'} (\beta_0 - \alpha_0) \leq \hat{\varepsilon}.$$

Consequently, the second part of estimation (38) is proved under $\beta_0 - \alpha_0 > \hat{\varepsilon}$ by taking the logarithm of the last expression to the base $1 - \hat{\theta}$. The theorem is proved.

Lemma 8 shows the finiteness of the second stage of the optimization process, but the complexity of this stage has not been studied. In this regard, we obtain an estimate for the number of iterations of the second stage taking into account additional conditions.

Lemma 10. *Suppose the points x_k, $x_{k,0}$, $\ldots$, x_{k,i_k} are constructed for some $k \in K$ by the proposed method, where i_k is an index fixed according to condition (14). In addition, put $x_{k,i} = \Pr(x_k, U_{k,i-1})$, $1 \leq i \leq i_k$ at Step 7 of the proposed method. Then*

$$i_k \leq 1 + \left\lceil \frac{|D|^2 L^2}{(1 - \hat{\theta})^2 \hat{\varepsilon}^2} \right\rceil. \tag{39}$$

Proof. If $i_k = 0$ or $i_k = 1$, then estimation (39) holds true.

Suppose $i_k > 1$. In this case (8), (9), (14) are not fulfilled for all $i = 0, \ldots, i_k - 1$, consequently, we have

$$\beta_{k,i} - \alpha_{k,i} > \hat{\varepsilon}, \quad i = 0, \ldots, i_k - 1. \tag{40}$$

The points $x_{k,0}, \ldots, x_{k,i_k-1}, x_{k,i_k}$ correspond to the sets

$$U_{k,0}, \ldots, U_{k,i_k-1}$$

such that in accordance with Lemma 5 $x_{k,i+1} \in U_{k,i} \subset U_{k,i-1}$, $i = 1, \ldots, i_k - 1$. Hence and from the generalized triangle inequality for the projection we have the relation

$$\|x_{k,i+1} - x_k\|^2 \geq \|x_{k,i+1} - \Pr(x_k, U_{k,i-1})\|^2 + \|\Pr(x_k, U_{k,i-1}) - x_k\|^2$$
$$\geq \|x_{k,i+1} - x_{k,i}\|^2 + \|x_{k,i} - x_k\|^2, \quad i = 1, \ldots, i_k - 1.$$

Consequently, taking into account Lemma 7 we obtain

$$\left(\frac{1-\hat{\theta}}{L}\right)^2 (\beta_{k,i} - \alpha_{k,i})^2 \leq \|x_k - x_{k,i+1}\|^2 - \|x_k - x_{k,i}\|^2, \quad i = 1, \ldots, i_k - 1.$$

Further, summing the last inequality over i from 1 to $i_k - 1$ we get

$$\sum_{i=1}^{i_k-1} \left(\frac{1-\hat{\theta}}{L}\right)^2 (\beta_{k,i} - \alpha_{k,i})^2 \leq \|x_k - x_{k,i_k}\|^2 \leq |D|^2.$$

Hence and from condition (40) it follows that estimation (39) holds true. The lemma is proved.

On the basis of the previous lemma and Theorem 5 we prove

Theorem 6. *Suppose the sequence* $\{x_k\}$, $k \in K$, *is constructed, the parameters* α_0, β_0 *are computed,* $x_{k,i+1} = \Pr(x_k, U_{k,i})$, $i \in K$, *at Step 6 by the proposed method. Then the complexity of finding an* $\hat{\varepsilon}$-*approximation point does not exceed*

$$\left(1 + \left\lceil \frac{|D|^2 L^2}{(1-\hat{\theta})^2 \hat{\varepsilon}^2} \right\rceil\right) * \begin{cases} 1, & \text{if } \beta_0 - \alpha_0 \leq \hat{\varepsilon} \\ \lceil \log_{1-\hat{\theta}} \frac{\hat{\varepsilon}}{\beta_0 - \alpha_0} \rceil, & \text{if } \beta_0 - \alpha_0 > \hat{\varepsilon}. \end{cases}$$

References

1. Konnov, I.V.: Conjugate subgradient type method for minimizing functionals. Issled. po prikl. matem. **12**, 59–62 (1984). (in Russian)
2. Mifflin, R., Sagastizabal, C.: A VU-algorithm for convex minimization. Math. Program. **104**(2)-(3), 583–608 (2005)
3. Shulgina, O.N., Yarullin, R.S., Zabotin, I.Y.: A cutting method with approximation of a constraint region and an epigraph for solving conditional minimization problems. Lobachevskii J. Math. **39**(6), 847–854 (2018). https://doi.org/10.1134/S1995080218060197
4. Yarullin, R.: Constructing mixed algorithms on the basis of some bundle method. CCIS **1275**, 150–163 (2020)
5. Kelley, J.E.: The cutting plane method for solving convex programs. J. SIAM **8**, 703–712 (1960)

6. Nesterov, Yu.: Introductory Lectures on Convex Optimization. Kluwer Academic Publishers, Boston (2004)
7. Lemarechal, C., Nemirovskii, A., Nesterov, Yu.: New variants of bundle methods. Math. Program. **69**(1), 111–147 (1995)
8. Kiwiel, K.C.: An ellipsoid trust region bundle method for nonsmooth convex minimization. SIAM J. Control Optim. **27**, 737–757 (1989)
9. Kiwiel, K.C.: Efficiency of proximal bundle methods. J. Optim. Theory Appl. **104**, 589–603 (2000)
10. Kiwiel, K.C.: Proximal level bundle methods for convex nondifferentiable optimization, saddle-point problems and variational inequalities. Math. Program. **69**, 89–109 (1995)
11. Polyak, B.T.: Introduction to Optimization. Nauka, Moscow (1983).[in Russian]
12. Bulatov, V.P.: Embedding Methods in Optimization Problems. Nauka, Novosibirsk (1977).[in Russian]
13. Zabotin, I.Ya., Kazaeva, K.E.: A version of the penalty method with approximation of the epigraphs of auxiliary functions. Uchen. Zap. Kazansk. Univ. Ser. Fiz.-Matem. Nauki. **161**(2), 263–273 (2019). (in Russian)
14. Yarullin, R.S.: A proximal level set method. In: Mesh Methods for Boundary-value Problems and Applications. Proceedings of 11th International Conference, Kazan, Kazan university, pp. 343–347 (2016). (in Russian)

Procedures for Updating Immersion Sets in the Cutting Method and Estimating the Solution Accuracy

Igor Zabotin⬥, Oksana Shulgina, and Rashid Yarullin[✉]⬥

Kazan (Volga Region) Federal University, Kazan, Russia
yarullinrs@gmail.com

Abstract. We propose a cutting method for solving a mathematical programming problem. This method approximates a feasible set by polyhedral sets. The procedures for updating immersion sets which are included to the proposed method allow to drop accumulated cutting planes at some iterations. The convergence of the cutting method is proved. We obtain estimation proximity of the constructed approximations to the solution set at each iteration under some additional propositions for the initial optimization problem.

Keywords: Cutting methods · Convex programming · Nonsmooth optimization · Epigraph · Cutting plane · Convergence · Solution accuracy

1 Introduction

Cutting methods using immersion of the feasible set into some polyhedral sets are often applied to solve practical convex programming problems with linear or convex quadratic objective functions. Convenience application of these methods is explained in the mentioned case by the fact that iteration points are constructed in these methods by solving linear or quadratic programming problems. For example, the methods from [1–4] belong to the group of such cutting methods.

Practical realization of cutting methods is enough difficult because cutting planes which form approximating sets are accumulated in the optimization process. Thus, problems of finding iteration points become more and more difficult as the number of iterations increases. Earlier in [5,6] the authors proposed and realized one approach of updating approximating sets in cutting methods due to dropping cutting planes. This approach based on the introduced criterion of quality approximating sets in the neighborhood of current iteration points. Here another criterion of estimating quality approximation is proposed. By using this

This paper has been supported by the Kazan Federal University Strategic Academic Leadership Program ("PRIORITY-2030").

criterion we develop a cutting method for solving mathematical programming problems in the general form, where there is opportunities to perform updating immersion sets. The proposed method has some advantages over the methods [5,6]. Several specific ways of such updates are discussed after describing the method.

The proposed method, in addition to the aforementioned approach of periodically updating the approximating sets, has some advantages over the known cutting methods [1,3,4]. Firstly, auxiliary problems of finding iteration points can be solved approximately. Secondly, the proposed method presents a convenient approach for constructing cutting planes based on estimating the proximity of auxiliary points to the corresponding sets that form the feasible set. Thirdly, the developed cutting method can solve convex programming problems with an empty interior of the admissible set.

In case, when the objective function is convex, and restrictions are defined by strongly convex functions, we obtain estimations of proximity iteration points and values of the objective function at this points to the solution set and, respectively, to the optimal value of the objective function.

2   Problem Setting

We solve the problem

$$\min\{f(x) : x \in D\}, \tag{1}$$

where $D = D' \cap D''$, $D' = \bigcap_{j \in J} D_j$, the sets D_j, $j \in J = \{1, \ldots, m\}$, and D'' are closed and convex from an n-dimensional Euclidian space R_n. Suppose the interior $\text{int} D_j$ of the set D_j is nonempty for each $j \in J$, and the function $f(x)$ is convex in R_n.

Note that we can put $D'' = R_n$ in (1). We also emphasize that the interior of the set D can be empty, in particular, according to emptiness of $\text{int} D''$ or $\text{int} D'$.

Suppose $f^* = \min\{f(x) : x \in D\}$, $X^* = \{x \in D : f(x) = f^*\}$. Let $W(z, D_j) = \{a \in R_n : \langle a, x - z \rangle \leq 0 \,\forall x \in D_j\}$ be a cone of the generalized support vectors for the set D_j at the point $z \in R_n$, $W^1(z, D_j) = \{a \in R_n : a \in W(z, D_j), \|a\| = 1\}$, $E(f, \gamma) = \{x \in R_n : f(x) \leq \gamma\}$, $\gamma \in R_1$, $K = \{0, 1, \ldots\}$.

3   Problem Solving Method

The proposed method constructs an approximation sequence $\{x_k\}$, $k \in K$, for solving problem (1) as follows.

Construct a convex closed set $M_0 \subset R_n$ containing any solution $x^* \in X^*$. Choose points $v^j \in \text{int} D_j$ for all $j \in J$. Set a number $\varepsilon_0 \geq 0$. Assign $k = 0$, $i = 0$.

1. Find a point

$$y_i \in M_i \cap D'' \cap E(f, f^*). \tag{2}$$

2. Form a set

$$J_i = \{j \in J : y_i \notin D_j\}.$$

If $J_i = \emptyset$, then $y_i \in X^*$, and the process is finished.
3. Choose a point $z_i^j \notin \text{int} D_j$ from the interval (v^j, y_i) for each $j \in J_i$ so that the inclusion $\bar{y}_i^j \in D_j$ is fulfilled for the point

$$\bar{y}_i^j = y_i + q_i^j(z_i^j - y_i), \qquad (3)$$

where $q_i^j \in [1, q]$, $1 \le q < +\infty$. Put $z_i^j = \bar{y}_i^j = y_i$ for all $j \in J \setminus J_i$.
4. Find an index $j_i \in J_i$ according to the condition

$$\|y_i - z_i^{j_i}\| = \max_{j \in J_i} \|y_i - z_i^j\|. \qquad (4)$$

5. If

$$\|y_i - z_i^{j_i}\| > \varepsilon_k, \qquad (5)$$

then

$$Q_i = M_i, \qquad (6)$$

and go to Step 6. If

$$\|y_i - z_i^{j_i}\| \le \varepsilon_k, \qquad (7)$$

then put $i_k = i$, $z_k = z_i^{j_i}$,

$$x_k = y_{i_k}, \qquad (8)$$

choose a convex closed set $Q_i = Q_{i_k} \subset \mathrm{R}_n$ such that

$$Q_i \subset M_0, \quad x^* \in Q_i, \qquad (9)$$

select $\varepsilon_{k+1} \ge 0$, increase the value of k by one, and go to Step 6.
6. Choose a set $H_i \subset J_i$ such that $j_i \in H_i$.
7. Select a finite subset $A_i^j \subset W^1(z_i^j, D_j)$ for each $j \in H_i$, assign

$$M_{i+1} = Q_i \bigcap_{j \in H_i} \{x \in \mathrm{R}_n : \langle a, x - z_i^j \rangle \ \forall a \in A_i^j\}, \qquad (10)$$

and go Step 1 with incremented i.

4   Method Discussion

Perform some remarks for the proposed method.

First of all, note that since the set $M_i \cap D'' \cap E(f, f^*)$ is nonempty, then it is possible to choose the point y_i in accordance with condition (2) for all $i \in K$. Indeed, the set M_0 contains the point x^* by construction, consequently, in view of (6) (9) (10) $x^* \in M_i$ for all $i \in K$, and, moreover, since $x^* \in D''$ and $x^* \in E(f, f^*)$, then

$$x^* \in M_i \cap D'' \cap E(f, f^*), \quad i \in K.$$

The point y_i which is satisfied (2) can be given by

$$y_i = \arg\min\{f(x) : x \in M_i \cap D''\}.$$

However, condition (2) allows to find y_i as an approximate minimum point of $f(x)$ on the set $M_i \cap D''$.

The optimality criterion of the point y_i given at Step 2 holds true, because $y_i \in D'$ under $J_i = \emptyset$, and, therefore, taking into account (2) the inclusion $y_i \in D \cap E(f, f^*)$ is fulfilled, i.e. $y_i \in X^*$.

There are many ways to choose the initial approximating set M_0. When the function $f(x)$ is linear or convex quadratic and the set D'' matches R_n or is defined by linear equalities or inequalities, then it is naturally to determine M_0 as polyhedron. In addition, if the intersection of the sets D_j, $j \in J' \subset J$, is a convex polyhedron, then it is convenient to put

$$M_0 = \bigcap_{j \in J'} D_j. \tag{11}$$

If $D'' = R_n$ and the function $f(x)$ reaches its minimal value on R_n, then it is possible to put $M_0 = R_n$, $y_0 = \arg\min\{f(x) : x \in R_n\}$. By selecting $M_0 = D$ the solution process of problem (1) is finished at the initial step.

The choice of the auxiliary points v^j is also wide at the preliminary step. If $\operatorname{int}D \neq \emptyset$ and there is the known point $v \in \operatorname{int}D$, then it is convenient to put $v^j = v$ for all $j \in J$. In case of giving the set M_0 by (11) there is no necessary to find the points $v^j \in \operatorname{int}D_j$ for $j \in J'$, because the inclusions $y_i \in D_j$, $j \in J'$, are fulfilled for all $i \in K$, and the points v^j, $j \in J'$, do not take a part in constructing cutting planes.

The points z_i^j, $j \in J_i$, which take a part in constructing cutting planes at the i-th step can be found as intersection points of the segments $[v^j, y_i]$ with boundaries of the sets D_j, $j \in J_i$, by putting $q_i^j = 1$ in (3). But such approach of setting the points z_i^j is difficult to realize. Therefore, in the method z_i^j can be selected other than mentioned intersection points by way of choosing the numbers q_i^j, $j \in J_i$.

Let us also pay attention that the index $j \in J_i$ which satisfies condition (4) must be included in the set H_i for each $i \in K$. In addition to j_i the indexes from J_i can be included in the H_i or no more.

There are many possibilities for choosing the set Q_i which takes a part in constructing M_{i+1} under $i = i_k$. In particular, it is permissible to put $Q_i = Q_{i_k} = M_i$. If the sets Q_i are given by (16) independently of conditions (5), (7) for all $i \in K$, then equality (10) has a form

$$M_{i+1} = M_i \bigcap_{j \in H_i} \{x \in R_n : \langle a, x - z_i^j \rangle \leq 0 \quad \forall a \in A_i^j\}$$

for all $i \in K$. In this case the cutting planes are accumulated from step to step, and the approximating sets are not updated. Lets show how to update the set M_i by choosing the sets Q_i at the iterations $i = i_k$.

Suppose inequality (17) is defined for some $i \in K$. Assign, for example,

$$Q_i = Q_{i_k} = M_{r_i},$$

where $0 \leq r_i \leq i$. Then condition (9) is fulfilled, and Q_i can be chosen from the sets $M_0, \ldots, M_i$ constructed to the i-th step. If we put $Q_i = M_0$, then full update occurs by dropping all cutting planes accumulated to the i-th step.

It is possible to save any earlier constructed cutting planes by $Q_i = Q_{i_k}$. For example, we can put

$$Q_i = M_0 \bigcap_{j \in H_{i-1}} \{x \in \mathrm{R_n} : \langle a, x - z_{i-1}^j \rangle \leq 0 \quad \forall a \in A_{i-1}^j\}$$

for $i \geq 1$, and we leave the cutting planes from the last step. It is clear that condition (9) allows to use another approaches of updating approximating sets by choosing Q_i. Below we prove that for ε_k there exists a number $i = i_k$ such that inequality (7) is fulfilled, consequently, there will be an opportunity to update the sets M_i.

5   Prove Convergence

Lets prove convergence of the proposed method. Below it is assumed that the sequence $\{y_i\}$, $i \in K$, constructed by the method is bounded. This assumption can be made, for example, by choosing the initial set M_0.

Lemma 1. *Let $U \subset \mathrm{R_n}$ be a convex set, L be its carrier subspace, Q be a bounded set from the affine shell of U and not contained in $\mathrm{ri}\,U$ (relative interior of the set U). If $v \in \mathrm{R_n}$ such that the inclusion $v \in \mathrm{ri}\,U$ is fulfilled, then there exists a number $\delta > 0$ such that the inequality $\langle a, v - z \rangle \leq -\delta$ for all $z \in Q \setminus \mathrm{ri}\,U$ and for all $a \in L \cap W^1(z, U)$.*

The proof of the assertion is represented in [2].

Lets give the following auxiliary assertion taking into account the mentioned above remark about possibility of selecting the sets Q_i by form (6) independently of conditions (5), (7).

Lemma 2. *Suppose the sequence $\{y_i\}$, $i \in K$, is constructed by the proposed method such that the sets Q_i are chosen according to (6) for all $i \in K$, $i \geq i' \geq 0$. Let $\{y_i\}$, $i \in K' \subset K$, be a convergence subsequence of the sequence $\{y_i\}$, $i \in K$, $i \geq i'$, and $l \in J$ be an index such that the set*

$$K_l = \{i \in K' : l = j_i\} \tag{12}$$

has an infinite number of elements, where j_i is defined in (4). Then

$$\lim_{i \in K_l} \|y_i - z_i^l\| = 0. \tag{13}$$

Proof. For any $i \in K_l$ fix an index $p_i \in K_l$ such that $p_i > i$. In view of (6), (10) $M_{p_i} \subset M_i$. Moreover, $y_{p_i} \in M_{p_i}$, and the inclusion $a \in W^1(z_i^l, M_{p_i})$ is fulfilled for any $a \in A_i^l$. Then

$$\langle a, y_{p_i} - z_i^l \rangle \leq 0 \quad \forall a \in A_i^l.$$

Hence and taking into account

$$z_i^l = y_i + \gamma_i^l(v^l - y_i), \tag{14}$$

where $\gamma_i^l \in [0, 1)$, we have the inequality

$$\langle a, y_i - y_{p_i} \rangle \geq \gamma_i^l \langle a, y_i - v^l \rangle \tag{15}$$

for all $a \in A_i^l$.

According to Lemma 1 there exists a number $\delta_l > 0$ such that $\langle a, y_i - v^l \rangle \geq \delta_l$ $\forall i \in K_l$, $a \in A_i^l$. Then in accordance with (15) $\langle a, y_i - y_{p_i} \rangle \geq \gamma_i^l \delta_l \ \forall a \in A_i^l$, and since $\|a\| = 1 \ \forall a \in A_i^l$, then

$$\|y_i - y_{p_i}\| \geq \gamma_i^l \delta_l \quad \forall i, p_i \in K_l, \quad p_i > i. \tag{16}$$

Since $\{y_i\}$, $i \in K_l$, convergences, then $\|y_i - y_{p_i}\| \to 0$, $i \in K_l$, and in view of (16) $\gamma_i^l \to 0$, $i \in K_l$. Consequently, from (14) taking into account boundedness of $\{\|v^l - y_i\|\}$, $i \in K_l$, we get (13). The lemma is proved.

Lemma 3. *Suppose the sequence $\{y_i\}$, $i \in K$, be constructed by the proposed method, and, in addition, the numbers ε_k are given by*

$$\varepsilon_k > 0 \quad \forall k \in K. \tag{17}$$

Then there exists an index $i = i_k$ for each $k \in K$ such that equality (8) is fulfilled.

Proof. 1) Suppose $k = 0$. If $\|y_0 - z_0^{j_0}\| \leq \varepsilon_0$, then according to Step 5 of the method we have $i_0 = 0$, $x_0 = y_0$, and equality (8) is determined under $k = 0$. Therefore, assume that $\|y_0 - z_0^{j_0}\| > \varepsilon_0$. Lets obtain existence of the index $i = i_0 > 0$ such that inequality (7) is fulfilled under $i = i_0$, $k = 0$. Then (8) will be proved.

Suppose the inverse, i.e.

$$\|y_i - z_i^{j_i}\| > \varepsilon_0 \quad \forall i \in K, i > 0. \tag{18}$$

Choose a convergent subsequence $\{y_i\}$, $i \in K'$, from the bounded sequence $\{y_i\}$, $i \in K$, $i > 0$. It is clear that there exists $l \in J$ such that the index j_i is equal to l for infinitely many indexes $i \in K'$. Choose the subset K_l from the set K' according to (12). Note that taking into account the assumptions the set Q_i has form (6) for all $i > 0$. Then by Lemma 2 we get equality (13). On the other hand, in view of (18) $\|y_i - z_i^l\| > \varepsilon_0 \ \forall i \in K_l$ that contradicts condition (13). Thus, equality (8) is proved for $k = 0$.

2) Now suppose equality (8) is fulfilled for any fixed $k \geq 0$, i.e. the point $x_k = y_{i_k}$, $k > 0$, is constructed. Lets obtain existence of the index $i = i_{k+1} > i_k$ such that

$$x_{k+1} = y_{i_{k+1}},$$

then the lemma will be proved.

Suppose the inverse. In this case

$$\|y_i - z_i^{j_i}\| > \varepsilon_{k+1} \quad \forall i > i_k. \tag{19}$$

Choose a convergent subsequence $\{y_i\}$, $i \in K''$, from the points y_i, $i > i_k$. As in the first part of the proof lets fix an index $l \in J$ such that $j_i = l$ is determined for infinitely many indexes $i \in K''$, and define $K_l = \{i \in K'' : l = j_i\}$. In accordance with (19) equality (6) is fulfilled for all $i \in K$, $i > i_k$, consequently, by Lemma 2 we get (13). But according to the same assumption (19) $\|y_i - z_i^l\| > \varepsilon_{k+1} \forall i \in K_l$. The last inequality contradicts equality (13) and the lemma is proved.

By Lemma 3 under executing condition (17) the sequence $\{x_k\}$, $k \in K$, is constructed together with $\{y_i\}$, $i \in K$.

Lemma 4. *Suppose the sequence $\{x_k\}$, $k \in K$, is constructed by the proposed method such that the numbers ε_k, $k \in K$, are chosen according to (17) and*

$$\varepsilon \to 0, \quad k \to \infty. \tag{20}$$

Let $\{x_k\}$, $k \in K' \subset K$, be a convergent subsequence of the sequence $\{x_k\}$, $k \in K$, and $\bar{x}$ be its limit point. Then

$$\bar{x} \in D. \tag{21}$$

Proof. Since according to (2) $x_k = y_{i_k} \in D'' \ \forall k \in K$, and the set D'' is closed, then $\bar{x} \in D''$. Therefore, lets obtain the inclusion

$$\bar{x} \in D' \tag{22}$$

to prove assertion (21).

Note that the sequences $\{z_{i_k}^j\}$, $\{\bar{y}_{i_k}^j\}$, $k \in K'$, are fixed for each $j \in J$ together with the sequence $\{x_k\}$, $k \in K'$. Suppose $l \in J$ such that the index $j_{i_k} \in J_{i_k}$ found in accordance with (4) under $i = i_k$ is equals to l for infinitely many indexes $k \in K'$. Put

$$K_l = \{k \in K' : j_{i_k} = l\}.$$

In view of (7), (8) for all $k \in K_l$ we have

$$0 \le \|y_{i_k} - z_{i_k}^l\| = \|x_k - z_{i_k}^l\| \le \varepsilon_k.$$

Hence and from (20) it follows that

$$\lim_{k \in K_l} \|x_k - z_{i_k}^l\| = 0. \tag{23}$$

In view of (4) for all $k \in K_l$ inequalities $\|y_{i_k} - z_{i_k}^l\| \ge \|y_{i_k} - z_{i_k}^j\|$, $j \in J_k$, do hold true. Moreover, according to Step 3 of the method we get $\|y_{i_k} - z_{i_k}^j\| = 0$ $\forall j \in J \setminus J_k$, $k \in K_l$. Therefore, taking into account (8)

$$\|x_k - z_{i_k}^l\| \ge \|x_k - z_{i_k}^j\| \quad \forall k \in K_l, j \in J.$$

Hence and from (23) it follows that

$$\lim_{k \in K_l} \|x_k - z_{i_k}^j\| = 0 \quad \forall j \in J. \tag{24}$$

Further, according to Step 3 for all $k \in K_l$ and $j \in J$ the point $\bar{y}_{i_k}^j$ either is equal to x_k or has the form

$$\bar{y}_{i_k}^j = x_k + q_{i_k}^j (z_{i_k}^j - x_k). \tag{25}$$

Then from the boundedness of the sequence $\{x_k\}$, $k \in K_l$, it follows that the sequences $\{\bar{y}_{i_k}^j\}$, $k \in K_l$, are bounded for all $j \in J$. Now for each $j \in J$ choose a convergent subsequence $\{\bar{y}_{i_k}^j\}$, $k \in K_l^j \subset K_l$, from $\{\bar{y}_{i_k}^j\}$, $k \in K_l$, and let u_j be its limit point. Note that

$$u_j \in D_j \quad \forall j \in J \tag{26}$$

because of closedness of the set D_j.

For all $j \in J$ put

$$P_1^j = \{k \in K_l^j : j \in J_i\}, \quad P_2^j = K_l^j \setminus P_1^j.$$

At least one of the sets P_1^j, P_2^j is infinite for each $j \in J$. Now for each fixed $j \in J$ go to the limit in equalities (25) under $k \in P_1^j$ taking into account (24) if the set P_1^j is infinite or in the equalities $\bar{y}_{i_k}^j = x_k$ under $k \in P_2^j$ if the set P_2^j is infinite. Then we get equalities $u_j = \bar{x}$ for all $j \in J$ from which in view of (26) it follows (22). The lemma is proved.

Theorem 1. *Suppose the sequence $\{x_k\}$, $k \in K$, be constructed by the proposed method in accordance with conditions (17), (20). Then its any limit point belongs to the set X^*.*

Proof. Let $\{x_k\}$, $k \in K' \subset K$, be a convergent subsequence of the sequence $\{x_k\}$, $k \in K$, and $\bar{x}$ be its limit point. Lets show that

$$f(\bar{x}) = f^*, \tag{27}$$

then the theorem will be proved.

By Lemma 4 inclusion (21) is fulfilled, consequently,

$$f(\bar{x}) \geq f^*. \tag{28}$$

On the other hand, according to (2) we have $f(x_k) \leq f^* \ \forall k \in K'$, and, therefore, $f(\bar{x}) \leq f^*$. From this inequality and inequality (28) it follows (27). The theorem is proved.

Note that for choosing the minimizing subsequence $\{x_{k_l}\}$, $l \in K$, from the sequence $\{x_k\}$, $k \in K$, it is enough to get indexes k_l according to the condition

$$f(x_{k_{l+1}}) > f(x_{k_l}).$$

If the indexes $i = i_k \geq 1$ is fixed at Step 5 such that the inequality $f(y_{i_k}) \geq f(y_{i_{k_1}})$ is fulfilled together with condition (7), then for the corresponding sequence $\{x_k\}$ it is possible to prove the following assertion.

Theorem 2. *Suppose the conditions of Theorem 1 are fulfilled, and, moreover, the inequalities hold*

$$f(x_{k+1}) \geq f(x_k) \quad \forall k \in K. \tag{29}$$

Then the whole sequence $\{x_k\}$ converges to the set X^*.

Proof. In view of (29) and the boundedness of $\{x_k\}$ the sequence $\{f(x_k)\}$ is convergent. Then according to Theorem 1 we have $\lim\limits_{k \in K} f(x_k) = f^*$. Hence and from the famous theorem (e.g., [7, p. 62]) it follows that the assertion is proved.

Lets discuss opportunities of selecting the numbers ε_k in the method.

Note that it is possible to put $\varepsilon_k = 0$ for any $k \in K$ in the method. For example, suppose $\varepsilon_0 = 0$. Then while constructing the sequence $\{y_i\}$ the condition $\|y_i - z_i^{j_i}\| > 0$ is fulfilled for all $i \in K$. This means that none of the points x_k will be fixed, and at the same time the sets Q_i will be given by (6) for all $i \in K$.

Let $\{y_i\}$, $i \in K' \subset K$, be a convergent subsequence of the sequence $\{y_i\}$, $i \in K$, which is constructed in accordance with $\varepsilon_0 = 0$, and $\bar{y}$ be its limit point. It is not difficult to prove $\bar{y} \in D$ by the technique of the Lemma 4, i.e. $f(\bar{y}) \geq f^*$. Otherwise, according to (2) $f(y_i) \leq f^*$, $i \in K'$, and, consequently, $f(\bar{y}) \leq f^*$. Thus, $\bar{y} \in X^*$. However note that the process of constructing such sequence $\{y_i\}$, $i \in K$, will be stripped of any updates of approximating sets. Similar reasoning can be carried out in the case $\varepsilon_k = 0$ under $k > 0$.

Now pay attention to the case, when ε_k is chosen according to (17). Note that the numbers ε_k, $k \geq 1$, can be chosen at the initial step of the method. But then $\{\varepsilon_k\}$ is not adapted to the minimization procedure. Therefore, the method provides for the possibility of specifying $\{\varepsilon_k\}$ in the process of constructing the sequence $\{x_k\}$. Namely, the numbers ε_{k+1}, $k \geq 0$ are chosen at Step 5 of the method only after finding the next approximation x_k. Lets give an example of such constructing $\{\varepsilon_k\}$.

Lets consider that ε_0 is an infinitely large positive number. Then $\|y_0 - z_0^{j_0}\| < \varepsilon_0$, and according to condition (7) we have $x_0 = y_0$, $z_0 = z_0^j$. For all $k \geq 0$ assign

$$\varepsilon_{k+1} = \gamma_k \|x_k - z_k\|,$$

where $\gamma_k \in (0, 1)$. Note that this sequence satisfies (17), and condition (20) is fulfilled in case $\gamma_k \to 0$, $k \in K$.

6   Solution Accuracy Estimations

Further, lets obtain estimates of the proximity of the values $f(y_i)$ and the points y_i to the value f^* and to the set X^* respectively for each $i \in K$ (including each $i = i_k$, $k \in K$) under some additional conditions for the set D and the function $f(x)$.

Suppose we have $D'' = \mathbb{R}_n$, int $D \neq \emptyset$ in (1), and the points $v^j = v$ are fixed, where $v \in \text{int } D$, for all $j \in J$. The points $\bar{y}_i^j \in D_j$ are found according to (3) for all $j \in J_i$ at the i-th iteration. In addition, suppose the inclusion $\bar{y}_i^{p_i} \in D$ is fulfilled for the index $p_i \in J_i$. Then taking into account (2) we have estimates

$$f(y_i) \leq f^* \leq f(\bar{y}_i^{p_i}).$$

Below we will consider that the function $f(x)$ is convex, $D'' = R_n$ and

$$D_j = \{x \in R_n : f_j(x) \leq 0\} \quad \forall j \in J,$$

where $f_j(x)$, $j \in J$, are strongly convex functions defined in R_n with strongly convex constants μ_j respectively. Moreover, suppose the set D satisfies Slater's condition, and any absolute minimum point of the function $f(x)$ (in case of existence) does not belong to the set int D.

Put

$$F(x) = \max_{j \in J} f_j(x).$$

Note immediately that the function $F(x)$ is strongly convex with the strongly convex constant

$$\mu = \min_{j \in J} \mu_j.$$

It is not also difficult to check the fact that problem (1) has a unique solution because of introducing above additional assumptions concerning this problem. Therefore, lets consider that $X^* = \{x^*\}$.

Lemma 5. *Suppose $y \notin D$ and the point $z = x^* + \alpha(y - x^*)$ satisfies the inequality $F(z) > 0$ for all $\alpha \in (0, 1]$. Then the estimation*

$$\|y - x^*\| \leq \delta \tag{30}$$

holds, where $\delta = \sqrt{F(y)/\mu}$.

Proof. Taking into account the conditions of the lemma and the equality $F(x^*) = 0$ it is easy to prove the inequality

$$F'(x^*, y - x^*) \geq 0, \tag{31}$$

where $F'(x^*, y - x^*)$ is a derivative of the function $F(x)$ at the point x^* in the direction $y - x^*$. Let $\partial F(x^*)$ be a subdifferential of the function $F(x)$ at the point x^*. Since (see, for example, [8, p. 74])

$$F'(x^*, y - x^*) = \max\{\langle c, y - x^* \rangle : c \in \partial F(x^*)\},$$

then from (31) it follows existence $c^* \in \partial F(x^*)$ such that

$$\langle c^*, y - x^* \rangle \geq 0. \tag{32}$$

Since $F(x)$ is strongly convex, then (see [7]) for the vector c^* we get the inequality

$$F(y) - F(x^*) \geq \langle c^*, y - x^* \rangle + \mu \|y - x^*\|^2,$$

from which taking into account (32) and the equality $F(x^*) = 0$ it follows estimation (30). The lemma is proved.

Lemma 6. *Suppose the conditions of Lemma 5 are fulfilled, and, moreover, the function $F(x)$ satisfies the Lipschitz condition with the constant L. Then for the point y the inequality*

$$|f(y) - f^*| \le \gamma \tag{33}$$

is fulfilled together with inequality (30), where $\gamma = L\delta$.

The proof of the assertion follows from the inequality

$$|f(y) - f(x^*)| \le L\|y - x^*\| \tag{34}$$

and estimation (30).

Lemma 7. *Let the point $y \in R_n$ be such that $y \in E(f, f^*)$ and $y \ne x^*$. Then inequality (30) is fulfilled for this point. If the function $f(x)$ satisfies the Lipschitz condition with the constant L, then estimation (33) also holds for the point y.*

Proof. Lets show that the inequality $F(z) > 0$ is determined for the point $z = \alpha y + (1 - \alpha)x^*$ for any $\alpha \in (0, 1)$.

Assume the inverse, i.e. $F(z) \le 0$ for some $\alpha \in (0, 1)$. Then $z \in D$ and $f(z) \ge f^*$. On the other hand, from the inclusions $y, x^* \in E(f, f^*)$ and the convexity of the set $E(f, f^*)$ it follows that $z \in E(f, f^*)$, i.e. $f(z) \le f^*$. Consequently,

$$f(z) = f^*. \tag{35}$$

As noted above the set X^* consists of the unique point x^*. Since according to the conditions $y \ne x^*$ and $\alpha > 0$, then $z \ne x^*$ that contradicts equality (35). Thus, we prove that the point y satisfies conditions of Lemma 5. Consequently, inequality (30) holds true. The second assertion of the lemma follows from (30), (34). The lemma is proved.

Further, according to Steps 1, 2 of the method the inclusion $y_i \in E(f, f^*)$ is fulfilled for all points of constructed sequence $\{y_i\}$, $i \in K$, and, moreover, $y_i \ne x^*$. Then in view of Lemma 7 for all $i \in K$ we get estimations

$$\|y_i - x^*\| \le \delta_i,$$

where $\delta_i = \sqrt{F(y_i)/\mu}$, and if the function $f(x)$ satisfies the Lipschitz condition, then the inequalities

$$|f(y_i) - f^*| \le L\delta_i$$

hold true.

Since $y_{i_k} = x_k$ for all $i = i_k \in K$, $k \in K$, and wherein $x_k \in E(f, f^*)$, $x_k \ne x^*$, $k \in K$, then, in view of the remark just made, we have

Theorem 3. *Suppose the sequence $\{x_k\}$, $k \in K$, is constructed by the proposed method. Then for each $k \in K$ the inequality*

$$\|x_k - x^*\| \le \Delta_k$$

is fulfilled, where $\Delta_k = \sqrt{F(x_k)/\mu}$, and if the function $f(x)$ satisfies the Lipschitz condition with the constant L, then the estimation

$$\|f(x_k) - f^*\| \le L\Delta_k$$

also holds true.

References

1. Bulatov, V.P.: Embedding Methods in Optimization Problems. Nauka, Novosibirsk (1977) (in Russian)
2. Zabotin, I.Ya.: On the several algorithms of immersion-severances for the problem of mathematical programming. Bull. Irkutsk State Univ. Ser. Math. **4**(2), 91–101 (2011) (in Russian)
3. Levitin, E.C., Polyak, B.T.: Minimization methods for feasible set. Zhurn. Vychisl. Matem. i Matem. Fiz. **6**(5), 878–823 (1966) (in Russian)
4. Nesterov, Y.E.: Introductory to convex optimization. MCCME, Moscow (2010) (in Russian)
5. Zabotin, I. Ya., Yarullin, R.S.: One approach to constructing cutting algorithms with dropping of cutting planes. Russian Math. (Iz. VUZ) **57**(3), 60–64 (2013)
6. Zabotin, I.Ya., Yarullin, R.S.: A cutting method with updating approximating sets and its combination with other algorithms. Bull. Irkutsk State Univ. Ser. Math. **10**, 13–26 (2014) (in Russian)
7. Vasil'ev, F.P.: Optimization Methods, vol. 1. MCCME, Moscow (2011) (in Russian)
8. Pshenichnyy, B.N.: Convex analysis and extremal problems. Nauka, Moscow (1980) (in Russian)

Game Theory and Optimal Control

Synthesis of Motion Control of Rod Heating Sources with Optimization of Places for Measuring States

Kamil R. Aida-Zade$^{(\boxtimes)}$ (ID) and Vugar A. Hashimov (ID)

Institute of Control Systems of ANAS, B. Vahabzade 9, AZ1141 Baku, Azerbaijan
kamil_aydazade@rambler.ru
http://www.isi.az

Abstract. The problem of control synthesis for the heating process of a rod by lumped sources moving along the rod is studied. The problem of feedback control of moving heat sources during rod heating is considered. The speed of point-wise sources are assigned depending on the state of the processes at the measurement points. The formulas for the gradient components of the objective functional allowing for the numerical solution of the problem using of the first-order optimization methods are obtained.

Keywords: Rod heating · Feedback control · Moving sources · Temperature measurement points · Feedback parameters

1 Introduction

We study the problem of synthesis of control of the rod heating process by point-wise sources moving along the rod. The current values of the speeds of movement of the sources are determined depending on the temperature at the points of measurement, the location of which is being optimized. The paper proposes to use the linear dependence of the control actions by the motion of the sources on the measured temperature values. Constant coefficients involved in these dependencies are the desired feedback parameters.

Note that the problems of synthesis of control of objects described by both ordinary and partial differential equations are the most difficult both in the theory of optimal control and in the practice of their application [1–7].

For the problems of synthesis of control of objects with lumped parameters, there are certain, fairly general approaches to their solution, in particular, for linear systems [2–4,8]. There are no such approaches for objects with distributed parameters [1,3,5,6]. Firstly, this is due, to a wide variety of both mathematical models of such objects and possible variants of the corresponding formulations of control problems [1,3]. Secondly, the implementation of currently known methods for controlling objects with feedback in real time requires the use of expensive telemechanics, measuring and computing equipment [3,5,6].

Y. Kochetov et al. (Eds.): MOTOR 2022, CCIS 1661, pp. 233–246, 2022.
https://doi.org/10.1007/978-3-031-16224-4_16

In this paper, the determination of the feedback parameters is reduced to the problem of parametric optimal control. For the numerical solution of the problem, first-order optimization methods were used. Formulas for the gradient of the objective functional with respect to the feedback parameters are obtained.

The proposed approach to the synthesis of the control of moving sources can be easily extended to other processes described by other types of differential equations and initial-boundary conditions.

2   Formulation of the Problem

Consider the following process of heating a rod by moving heat sources [5]:

$$u_t(x,t) = a^2 u_{xx}(x,t) - \lambda_0 [u(x,t) - \theta] + \sum_{i=1}^{N_c} q_i \delta(x - z_i(t)), \qquad (1)$$

$$x \in (0, l), \quad t \in (0, T],$$

$$u_x(0,t) = \lambda_1(u(0,t) - \theta), \quad t \in (0,T], \qquad (2)$$
$$u_x(l,t) = -\lambda_2(u(l,t) - \theta), \quad t \in (0,T],$$

Here $u(x,t)$ is the temperature of the rod at the point x at the moment of time t; $\delta(\cdot)$ is a Dirac function; l – rod length; T – heating process duration; N_c – number of point sources; a, λ_0, λ_1, λ_2 – given coefficients; q_i piece-wise continuous functions by t, and $z_i(t)$ that determine the i-th rule of motion of source on the rod; moreover

$$0 \le z_i(t) \le l, \quad t \in [0, T], \quad i = 1, 2, \dots, N_c. \qquad (3)$$

θ – constant temperature of the external environment with a known set of possible values Θ and the density function $\rho_\Theta(\theta)$ such that

$$\rho_\Theta(\theta) \ge 0, \quad \theta \in \Theta, \quad \int_\Theta \rho_\Theta(\theta)d\theta = 1.$$

With respect to the initial temperature of the rod the set of its possible values are known. This set is determined by a parametrically defined function:

$$u(x,0) = \varphi(x; p), \quad x \in [0, l], \quad p \in P \subset R^s. \qquad (4)$$

where P a given set of parameter values with a density function $\rho_P(p) \ge 0$

$$\rho_P(p) \ge 0, \quad p \in P, \quad \int_P \rho_P(p)dp = 1.$$

The motions of the sources are determined by the equations

$$\dot{z}_i(t) = a_i z_i(t) + \vartheta_i(t), \quad t \in (0, T], \tag{5}$$

$$z_i(0) = z_i^0, \quad i = 1, 2, \ldots, N_c. \tag{6}$$

Here a_i are the given parameters of the source motion; z_i^0 are given initial positions of sources; $\vartheta_i(t)$ is a piece-wise continuous control, satisfying the conditions:

$$V = \left\{ \vartheta_i(t) : \quad \underline{\vartheta_i} \leq \vartheta_i(t) \leq \overline{\vartheta_i}, \quad t \in [0, T], \quad i = 1, 2, \ldots, N_c \right\}. \tag{7}$$

The problem of control is to determine the vector-function $\vartheta = \vartheta(t) = (\vartheta_1(t), \vartheta_2(t), \ldots, \vartheta_{N_c}(t))$, minimizing the given functional:

$$J(\vartheta) = \int\limits_P \int\limits_\Theta I(\vartheta; p, \theta) \rho_\Theta(\theta) \rho_P(p) d\theta dp, \tag{8}$$

$$I(\vartheta; p, \theta) = \int\limits_0^l \mu(x)[u(x, T) - U(x)]^2 dx + \varepsilon \left\| \vartheta(t) - \hat{\vartheta} \right\|_{L_2^{N_c}[0,T]}^2. \tag{9}$$

Here $U(x)$, $\mu(x) \geq 0$, $x \in [0, l]$ are a given piece-wise continuous functions; $u(x, t) = u(x, t; \vartheta, p, \theta)$ is a solution to the initial-boundary value problem (1), (2), (3) under given control $\vartheta(t) \in V$, parameters $p \in P$ of initial condition $\varphi(x; p)$ and ambient temperature $\theta \in \Theta$. $\varepsilon > 0$, $\hat{\vartheta}$ – are given parameters of regularization of the functional of the problem.

The objective functional in the problem under consideration estimates the control vector function $\vartheta(t)$ on the behavior of the heating process on average over all possible values of the parameters of the initial conditions $p \in P$ and ambient temperature $\theta \in \Theta$.

Let at the N_o optimized points $\xi_j \in [0, l]$, $j = 1, 2, \ldots, N_o$ continuously temperature is measured:

$$u_j(t) = u(\xi_j, t), \quad t \in [0, T], \quad \xi_j \in [0, l], \quad j = 1, 2, \ldots, N_o.$$

Based on the measurement results, it is required to determine the current optimal control values $\vartheta(t)$. To do this, we use the following linear dependences of control on the measured temperature values

$$\vartheta_i(t, y) = \sum_{j=1}^{N_o} \beta_i^j [u(\xi_j, t) - \tilde{\gamma}_i^j], \quad t \in [0, T], \quad i = 1, 2, \ldots, N_c. \tag{10}$$

Here $y = \left\{ \beta_i^j, \tilde{\gamma}_i^j, \xi^j, i = 1, 2, \ldots, N_c, j = 1, 2, \ldots, N_o \right\}$, $N = (2N_c + 1)N_o$ dimensional vector determined by synthesized constant feedback parameters. The parameters β_i^j by analogy with synthesis problems for objects with lumped parameters, we will call the amplification factors.

Introducing the notation

$$\gamma_i^\vartheta = \sum_{j=1}^{N_o} \beta_i^j \tilde{\gamma}_i^j,$$

for dependence (10) we obtain

$$\vartheta_i(t, y) = \sum_{j=1}^{N_o} \beta_i^j u(\xi_j, t) - \gamma_i^\vartheta, \quad t \in [0, T], \quad i = 1, 2, \ldots, N_c. \tag{11}$$

The objective functional in this case can be written as follows:

$$J(y) = \int_P \int_\Theta I(y; p, \theta) \rho_\Theta(\theta) \rho_P(p) d\theta dp, \tag{12}$$

$$I(y; p, \theta) = \int_0^l \mu(x)[u(x, T) - U(x)]^2 dx + \varepsilon \|y - \hat{y}\|_{R^N}^2. \tag{13}$$

Substituting dependencies (11) into Eq. (5), we obtain

$$\dot{z}_i(t) = a^i z_i(t) + \sum_{j=1}^{N_o} \beta_i^j u(\xi_j, t) - \gamma_i^\vartheta, \quad t \in (0, T]. \tag{14}$$

From technological considerations, the range of possible temperature values at the points of the rod during its heating can be considered known:

$$\underline{u} \leq u(x, t) \leq \overline{u}, \quad x \in [0, l], \quad t \in [0, T]. \tag{15}$$

Then, taking into account the linearity of dependences (11), from constraints (7) we obtain linear constraints on the feedback parameters:

$$\underline{\vartheta_i} \leq \sum_{j=1}^{N_o} \beta_i^j \tilde{u}_i^k - \gamma_i \leq \overline{\vartheta_i}, \quad i = 1, 2, \ldots, N_c, \quad k = 1, 2, \ldots, \tilde{N}. \tag{16}$$

$$0 \leq \xi_j \leq l, \quad j = 1, 2, \ldots, N_o. \tag{17}$$

Here $\tilde{u}^k$ is the jth column of the matrix $\tilde{U}$ of size $N_o \times \tilde{N}$, $\tilde{N} = 2^{N_o}$. The columns of the matrix $\tilde{U}$ are N_o dimensional vectors consisting of various combinations of values $\underline{u}, \overline{u}$.

The conditions (17) are natural restrictions on the location of measurement points.

Thus, the source control problem for moving sources (1)–(9) with feedback (11) is reduced to the parametric optimal control problem (12), (13), (1), (2), (3) [7,9].

We note the following specific features of the investigated parametric optimal control problem.

First, the original problem of control of moving sources (1)–(9) is generally not convex. This is due to the non-linear third term in Eq. (1) $\delta(x - z_i(t; \vartheta_i(t)))$. The resulting problem of parametric optimal control is also not convex in the feedback parameters. This can be seen from the differential equations (14) and dependences (11).

Secondly, the problem is specific because of the objective functional (12), (13), which estimates the behavior of a beam of phase trajectories with initial conditions from a parametrically given set.

In general, the obtained problem can also be classified as a class of finite-dimensional optimization problems with respect to the vector $y \in R^N$. In this problem, to calculate the objective functional at any point, it is required to solve initial-boundary value problems for differential equations with partial and ordinary derivatives.

3   Determination of Feedback Parameters

To minimize the functional (12), (13), taking into account the linearity of constraints (17), we use the gradient projection method [9]:

$$y^{n+1} = \mathcal{P}_{(16,17)} \left[y^n - \alpha_n \mathbf{grad} J\left(y^n\right) \right], \tag{18}$$

$$\alpha_n = \arg \min_{\alpha \geq 0} J\left(\mathcal{P}_{(16,17)} \left[y^n - \alpha \mathbf{grad} J\left(y^n\right) \right]\right), \quad n = 0, 1, \ldots$$

Here α_n is the one-dimensional minimization step, y^0 is an arbitrary starting point of the search from R^N; $\mathcal{P}_{(16,17)}[\cdot]$ – is the operator of projecting an arbitrary point $y \in R^N$ onto the admissible domain defined by constraints (16), (17). Taking into account the linearity of constraints (16), (17), the operator $\mathcal{P}_{(16,17)}[\cdot]$ is easy to construct constructively [9]. It is known that iterative procedure (18) allows one to find only the local minimum of the objective functional closest to the point y^0. Therefore, for procedure (18), it is proposed to use the multistart method from different starting points. From the obtained local minimum points, the best functional is selected.

In the implementation of procedure (18), analytical formulas for the components of the gradient of the objective functional play an important role. Therefore, below we will prove the differentiability of the functional with respect to the optimized parameters and obtain formulas for its gradient, which make it possible to formulate the necessary optimal conditions for the synthesized feedback parameters y.

Theorem 1. *Under conditions on the functions and parameters involved in the problem (1), (2), (4), (6), (12)–(14), the functional (12), (13) is differentiable* *and components of its gradient with respect to the feedback parameters are determined by the formulas:*

$$\frac{\partial J(y)}{\partial \beta_i^j} = \int\limits_P \int\limits_\Theta \left\{ -\int\limits_0^T \varphi_i(t) u\left(\xi_j, t\right) dt + 2(\beta_i^j - \hat{\beta}_i^j) \right\} \rho_\Theta(\theta) \rho_P(p) d\theta dp,$$

$$\frac{\partial J(y)}{\partial \xi_j} = \int\limits_P \int\limits_\Theta \left\{ -\sum_{i=1}^{N_c} \int\limits_0^T \beta_i^j \varphi^i(t) u_x(\xi^j, t) dt + 2\left(\xi_j - \hat{\xi}_j\right) \right\} \times \qquad (19)$$

$$\times \rho_\Theta(\theta) \rho_P(p) d\theta dp,$$

$$\frac{\partial J(y)}{\partial \gamma_i} = \int\limits_P \int\limits_\Theta \left\{ \int\limits_0^T \varphi_i(t) dt + 2\varepsilon(\gamma_i - \hat{\gamma}_i) \right\} \rho_\Theta(\theta) \rho_P(p) d\theta dp.$$

$i = 1, 2, \ldots, N_c$, $j = 1, 2, \ldots, N_o$. *Functions* $\psi(x, t)$ *and* $\varphi_i(t)$, $i = 1, 2, \ldots, N_c$ *for every given parameters* $\theta \in \theta$ *and* $p \in P$ *are solutions of the following conjugate initial-boundary value problems:*

$$\psi_t(x, t) = -a^2 \psi_{xx}(x, t) + \lambda_0 \psi(x, t) - \sum_{i=1}^{N_c} \sum_{j=1}^{N_o} \left(\beta_i^j \varphi_i(t)\right) \delta\left(x - \xi_j\right),$$

$$x \in \Omega, \quad t \in [0, T),$$

$$\psi(x, T) = -2\mu(x)\left(u(x, T) - U(x)\right), \quad x \in \Omega, \qquad (20)$$

$$\psi_x(0, t) = \lambda_1 \psi(0, t), \quad t \in [0, T),$$

$$\psi_x(l, t) = \lambda_2 \psi(l, t), \quad t \in [0, T),$$

$$\dot{\varphi}_i(t) = -a_i \varphi_i(t) - \psi_x(z_i(t), t) q_i, \quad t \in [0, T), \qquad (21)$$

$$\varphi_i(T) = 0, \quad i = 1, 2, \ldots, N_c.$$

Proof. To prove the differentiability of the functional $J(y)$ with respect to y, we use the increment method.

From the obvious dependence of all parameters of the initial conditions $p \in P$ and the temperature of the environment $\theta \in \theta$, the validity of the formula follows:

$$\mathbf{grad} J(y) = \mathbf{grad} \int\limits_P \int\limits_\Theta I(y; p, \theta) \rho_\Theta(\theta) \rho_P(p) d\theta dp = \qquad (22)$$

$$= \int\limits_P \int\limits_\Theta \mathbf{grad} I(y; p, \theta) \rho_\Theta(\theta) \rho_P(p) d\theta dp.$$

Therefore, it is sufficient to obtain formulas for $\mathbf{grad} I(y; p, \theta)$ for arbitrarily given admissible values of $p \in P$ and $\theta \in \theta$. On the right side of Eq. (1), we introduce the notation

$$W(t; y) = \sum_{i=1}^{N_c} q_i \delta\left(x - z_i(t)\right), \quad t \in [0, T].$$

Let $u(x,t) = u(x,t; y, p, \theta)$, $z(t) = z(t; y, p, \theta)$ are solutions, respectively, initial boundary-value problem (1), (2), (4) and Cauchy problem (14), (6) for given values of the parameters p and θ. Let the feedback parameters y get an increment Δy: $\tilde{y} = y + \Delta y$. It is clear that the corresponding solutions of problems (1), (2), (4) and (14), (6) also get increments, which we denote as follows:

$$\tilde{u}\left(x, t; \tilde{y}, p, \theta\right) = u(x, t; y, p, \theta) + \Delta u(x, t; y, p, \theta),$$

$$\tilde{z}\left(t; \tilde{y}, p, \theta\right) = z(t; y, p, \theta) + \Delta z(t; y, p, \theta).$$

The increments $\Delta u(x, t; y, p, \theta)$ and $\Delta z(t; y, p, \theta)$ are solutions of the following initial-boundary-value problems with an accuracy of small above the first order of with respect to $\|\Delta u(x,t)\|$, $\|\Delta y\|$:

$$\Delta u_t(x,t) = a^2 \Delta u_{xx}(x,t) - \lambda_0 \Delta u(x,t) + \Delta W(t; y), \quad x \in (0, l), \quad t \in (0, T], \quad (23)$$

$$\Delta u(x, 0) = 0, \quad x \in [0, l], \tag{24}$$

$$\Delta u_x(0, t) = \lambda_1 \Delta u(0, t), \quad t \in (0, T], \tag{25}$$

$$\Delta u_x(l, t) = -\lambda_2 \Delta u(l, t), \quad t \in (0, T]. $$

$$\Delta \dot{z}_i(t) = a_i \Delta z_i(t) + \Delta \vartheta_i(t), \quad t \in (0, T], \tag{26}$$

$$\Delta z_i(0) = 0, \quad i = 1, 2, \ldots, N_c. \tag{27}$$

The functional $\Delta I(y; p, \theta)$ will get an increment

$$\Delta I(y) = I(y + \Delta y; p, \theta) - I(y; p, \theta) = \tag{28}$$

$$= 2 \int_0^l \mu(x)\left(u(x, T) - U(x)\right) \Delta u(x, T) dx + 2\varepsilon \langle y - \hat{y}, \Delta y \rangle + \mathcal{R}_1,$$

$$\mathcal{R}_1 = o\left(\|\Delta u(x, t)\|, \|\Delta y\|\right).$$

Move the right-hand sides of differential Eqs. (23) and (26) to the left, multiply both sides of the obtained qualities by so far arbitrary functions $\psi(x, t)$ and $\varphi_i(t)$, respectively. We integrate over $t \in (0, T)$ and $x \in (0, l)$. The resulting left-hand sides equal to zero are added to (28). Will have:

$$\Delta I(y) = 2 \int_0^l \mu(x)\left(u(x, T) - U(x)\right) \Delta u(x, T) dx + 2\varepsilon_1 \langle y - \hat{y}, \Delta y \rangle +$$

$$+ \int_0^T \int_0^l \psi(x, t)\left(\Delta u_t(x, t) - a^2 \Delta u_{xx}(x, t) + \lambda_0 \Delta u(x, t) - \Delta W(t; y)\right) dx dt +$$

$$+ \sum_{i=1}^{N_c} \int_0^T \varphi_i(t) \left(\Delta \dot{z}_i(t) - a_i \Delta z_i(t) - \Delta \vartheta_i(t) \right) dt + \mathcal{R}_1.$$

Integrating by parts, grouping and taking into account conditions (24), (25), (27), we obtain the following expression for the increment of the functional:

$$\Delta I(y) = 2 \int_0^l \mu(x) \left(u(x,T) - U(x) \right) \Delta u(x,T) dx + \int_0^l \psi(x,T) \Delta u(x,T) dx + \quad (29)$$

$$+ \int_0^T \int_0^l \left(-\psi_t(x,t) - a^2 \psi_{xx}(x,t) + \lambda_0 \psi(x,t) \right) \Delta u(x,t) dx dt -$$

$$- \sum_{i=1}^{N_c} \sum_{j=1}^{N_o} \int_0^T \int_0^l \left\{ \beta_i^j \varphi_i(t) \right\} \delta \left(x - \xi_j \right) \Delta u(x,t) dx dt +$$

$$+ a^2 \int_0^T \left(\psi_x(l,t) + \lambda_2 \psi(l,t) \right) \Delta u(l,t) dt - a^2 \int_0^T \left(\psi_x(0,t) - \lambda_1 \psi(0,t) \right) \Delta u(0,t) dt +$$

$$+ \sum_{i=1}^{N_c} \sum_{j=1}^{N_o} \Delta \beta_i^j \left\{ - \int_0^T \varphi_i(t) u \left(\xi_j, t \right) dt + 2\varepsilon \left(\beta_i^j - \hat{\beta}_i^j \right) \right\} +$$

$$+ \sum_{i=1}^{N_c} \Delta \gamma_i \left\{ \int_0^T \varphi_i(t) dt + 2\varepsilon \left(\gamma_i - \hat{\gamma}_i \right) \right\} +$$

$$+ \sum_{j=1}^{N_o} \Delta \xi_j \left\{ - \sum_{i=1}^{N_c} \int_0^T \left(\beta_i^j \varphi_i(t) \right) u_x(\xi_j,t) dt + 2 \left(\xi_j - \hat{\xi}_j \right) \right\} + \sum_{i=1}^{N_c} \varphi_i(T) \Delta z_i(T) +$$

$$+ \sum_{i=1}^{N_c} \int_0^T \left\{ -\dot{\varphi}_i(t) - a_i \varphi_i(t) - \psi_x \left(z_i(t), t \right) q_i \right\} \Delta z_i(t) dt + \mathcal{R}_2,$$

$$\mathcal{R}_2 = o \left(\| \Delta u(x,t) \|, \| \Delta z(t) \|, \| \Delta y \| \right).$$

Using the well-known results on the solution of the boundary value problem (1), (2) and the Cauchy problem (5), (6), one can obtain estimates $\| \Delta u(x,t) \| < k_1 \| \Delta y \|$, $\| \Delta u(x,t) \| < k_2 \| \Delta y \|$. Then from (29) it follows that the functional of the problem is differentiable.

Using the arbitrariness of the choice of the functions $\psi(x,t)$ and $\varphi_i(t)$, we require them to satisfy conditions (20).

Then the components of the gradient of the functional $I(y; p, \theta)$, determined by the linear parts of the functional increment with the corresponding feedback parameters [9], are defined by the following formulas:

$$\frac{\partial I(y; p, \theta)}{\partial \beta_i^j} = - \int_0^T \varphi_i(t) u\left(\xi_j, t\right) dt + 2(\beta_i^j - \hat{\beta}_i^j),$$

$$\frac{\partial I(y; p, \theta)}{\partial \xi^j} = - \sum_{i=1}^{N_c} \int_0^T \left\{ \beta_i^j \varphi_i(t) \right\} u_x(\xi_j, t) dt + 2\left(\xi_j - \hat{\xi}_j \right), \qquad (30)$$

$$\frac{\partial I(y; p, \theta)}{\partial \gamma_i} = \int_0^T \varphi_i(t) dt + 2\varepsilon \left(\gamma_i - \hat{\gamma}_i \right).$$

Taking into account formula (22) from (30), we obtain the required formulas (19) given in Theorem 1.

4   Numerical Experiments

The purpose of the numerical experiments carried out to solve test problems of the form (1)–(9) with feedback (10) was as follows. We studied 1) the quality of the objective functional, namely the property of multi-extremality; 2) dependence of the solution of the control problem on the number of sources and points for measuring the state; 3) the influence of the state measurement error on the control of the heating process as a whole.

Numerical experiments were carried out on the example of test problem, in which the parameters and functions involved in (1)–(9) were as follows:

$$a^2 = 1, \quad \lambda_0 = 0.001, \quad \lambda_1 = \lambda_2 = 0.0001, \quad l = 1, \quad T = 1, \quad \underline{\vartheta} = -2, \quad \overline{\vartheta} = 2,$$

$$q_1 = q_2 = 5, \quad z_1(t) \in [0.05; 0.95], \quad z_2(t) \in [0.05; 0.95],$$

$$\xi_1 \in [0.05; 0.3], \quad \xi_2 \in [0.3; 0.5], \quad \xi_3 \in [0.5; 0.7], \quad \xi_4 \in [0.7; 0.95],$$

$$N_\varphi = N_\theta = 3, \quad \Phi = \{0; 1; 2\}, \quad \theta = \{4.9; 5; 5.1\},$$

$$U(x) = 10, \quad x \in [0, 1], \quad P(\varphi = \varphi_i) = P(\theta = \theta_j) = 1/3, \quad i, j = 1, \ldots, 3.$$

When minimizing the objective functional to take into account constraints (16) we used the gradient projection method (18) [9]. To find the step α at each iteration of (18), the golden section method was applied [9].

To solve the direct and conjugate boundary value problems of parabolic type, an implicit scheme of the grid method was used: with a step in the spatial variable $h_x = 0.01$, and for the time variable $h_t = 0.001$.

To solve the direct and conjugate Cauchy problems (5), (6) and (21), the fourth-order Runge-Kutta method was used with a step $h_t = 0.001$. To approximate the functionand Dirac, taking into account that the sources are moving, the scheme proposed in [12] was used.

First, we studied the case with the number of sources $N_o = 2$ and the number of measurement points $N_c = 4$. Tables 1 and 2 show the results of some intermediate iterations of the objective function gradient projection method.

As can be seen from the tables, different initial values of feedback parameters β, γ and locations of measurement points ξ were used for these experiments.

Table 1. The results of solution of the problem for $N_o = 2$, $N_c = 4$, $q_1 = q_2 = 5$, $T = 1$ from the first starting point.

n	β				γ				ξ		$J(y)$
0	-0.0118	-0.0311	-0.0213	-0.0216	10.027	10.006	9.9941	9.9918	0.2254	0.4114	0.6850
	0.0121	0.0217	-0.0214	-0.0119	10.052	10.007	9.9818	9.9853	0.6097	0.8167	
1	-0.0118	-0.0311	-0.0213	-0.0216	10.027	10.006	9.9941	9.9918	0.2254	0.4114	0.3886
	0.0315	0.0411	-0.0020	0.0075	10.052	10.007	9.9818	9.9853	0.6097	0.8167	
2	0.0054	-0.0146	-0.0053	-0.0046	10.027	10.006	9.9940	9.9917	0.2241	0.4105	0.2909
	0.0175	0.0268	-0.0148	-0.0062	10.049	10.006	9.9818	9.9848	0.6069	0.8171	
3	-0.0266	-0.0436	-0.0407	-0.0345	10.011	9.9953	9.9946	9.9922	0.1975	0.3909	0.0601
	0.0422	0.0465	0.0340	0.0233	9.9966	9.9906	9.9823	9.9742	0.5513	0.8270	
4	-0.0283	-0.0409	-0.0410	-0.0353	10.004	9.9908	9.9948	9.9924	0.1861	0.3825	0.0213
	0.0392	0.0397	0.0395	0.0244	9.9740	9.9839	9.9824	9.9697	0.5432	0.8310	
5	-0.0277	-0.0381	-0.0391	-0.0345	10.002	9.9895	9.9949	9.9925	0.1828	0.3801	0.0145
	0.0381	0.0368	0.0402	0.0252	9.9675	9.9820	9.9825	9.9685	0.5204	0.8321	
6	-0.0326	-0.0334	-0.0365	-0.0386	9.9982	9.9871	9.9954	9.9930	0.1750	0.3748	0.0085
	0.0376	0.0301	0.0411	0.0330	9.9535	9.9781	9.9830	9.9660	0.5051	0.8346	

Table 2. The results of solution the problem for $N_o = 2$, $N_c = 4$, $q_1 = q_2 = 5$, $T = 1$ from the second starting point.

n	β				γ				ξ		$J(y)$
0	0.0375	-0.1032	0.0260	0.0301	10.019	9.9960	9.9851	9.9834	0.0950	0.4981	0.4614
	0.0259	0.0641	0.0009	-0.0770	10.058	10.013	9.9873	9.9927	0.5294	0.9038	
1	0.0309	-0.1088	0.01864	0.0235	10.019	9.9960	9.9852	9.9835	0.0960	0.4964	0.4125
	0.0318	0.0702	0.0079	-0.0698	10.057	10.013	9.9873	9.9922	0.5291	0.9029	
2	-0.0036	-0.0410	-0.0131	-0.0119	10.003	9.9889	9.9930	9.9908	0.1561	0.4036	0.1019
	0.0191	0.0240	0.0178	-0.0101	9.9778	9.9861	9.9839	9.9723	0.5108	0.8516	
3	-0.0043	-0.0219	-0.0132	-0.0131	10.000	9.9874	9.9945	9.9921	0.1677	0.3856	0.0434
	0.0181	0.0165	0.0212	0.0030	9.9625	9.9809	9.9833	9.9685	0.5072	0.8417	
4	-0.0036	-0.0116	-0.0122	-0.0125	9.9986	9.9868	9.9952	9.9928	0.1733	0.3769	0.0139
	0.0121	0.0072	0.0170	0.0036	9.9551	9.9784	9.9830	9.9666	0.5055	0.8369	
5	-0.0079	-0.0080	-0.0105	-0.0141	9.9982	9.9870	9.9954	9.9930	0.1749	0.3747	0.0086
	0.0130	0.0043	0.0150	0.0083	9.9534	9.9781	9.9829	9.9660	0.5050	0.8347	

Note that as a result of solving the problem from different initial points for minimization, the obtained values of the measurement points differ little. The obtained values of the feedback parameters differ quite strongly. In this case, the obtained minimum values of the objective functional are the same.

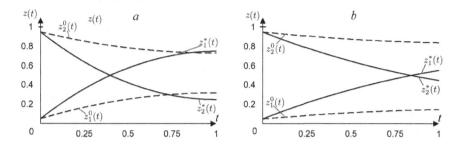

Fig. 1. Trajectories of sources for different initial minimization parameters (- - - initial trajectory, − optimal trajectory).

Figure 1 a), b) shows the initial and optimal source trajectories for different initial approximations y^0.

For comparison, the case was also considered when the power of the sources is equal to $q_1 = q_2 = 10$, and the duration of the control time is equal to $T = 0.5$.

Table 3. Results of solution of the problem for $N_o = 2$, $N_c = 4$, $q_1 = q_2 = 10$, $T = 0.5$ using the first starting point.

n	y										J(y)
	β				γ				ξ		
0	−0.0118	−0.0311	−0.0213	−0.0216	10.027	10.006	9.9941	9.9918	0.2254	0.4114	0.3445
	0.0121	0.0217	−0.0214	−0.0119	10.052	10.007	9.9818	9.9853	0.6097	0.8167	
1	0.0007	−0.0173	−0.0075	−0.0095	10.027	10.006	9.9941	9.9918	0.2254	0.4114	0.1689
	0.0237	0.0345	−0.0086	−0.0008	10.052	10.007	9.9818	9.9853	0.6099	0.8171	
2	−0.0113	−0.0307	−0.0208	−0.0208	10.027	10.006	9.9941	9.9918	0.2255	0.4115	0.0366
	0.0307	0.0423	−0.0009	0.0057	10.052	10.007	9.9817	9.9853	0.6099	0.8170	
3	−0.0100	−0.0293	−0.0194	−0.0195	10.027	10.006	9.9941	9.9918	0.2256	0.4115	0.0344
	0.0313	0.0430	−0.0002	0.0063	10.052	10.007	9.9817	9.9853	0.6099	0.8170	
4	−0.1758	0.0541	0.0459	−0.1746	10.041	10.008	9.9952	10.005	0.0599	0.4048	0.0311
	0.1929	−0.0772	−0.0120	0.1474	10.068	10.006	9.9882	9.9960	0.6465	0.9379	

Tables 3, 4 show the results of intermediate iterations from different initial procedure (18). Figure 2 shows the source motion trajectories for the initial and obtained values of the optimized parameters. As can be seen from the tables, despite the fact that the obtained minimum values of the objective functional are the same, the optimized parameters are different.

Table 4. The results of solution of the problem for $N_o = 2$, $N_c = 4$, $q_1 = q_2 = 10$, $T = 0.5$ using the second starting point.

n	y										J(y)
	β				γ				ξ		
0	0.0375	−0.1032	0.0260	0.0301	10.019	9.9960	9.9851	9.9834	0.0950	0.4981	0.2602
	0.0259	0.0641	0.0009	−0.0770	10.058	10.013	9.9873	9.9927	0.5294	0.9038	
1	0.0176	−0.1296	−0.0003	0.0103	10.019	9.9965	9.9850	9.9833	0.0948	0.4981	0.1004
	0.0427	0.0864	0.0231	−0.0604	10.058	10.013	9.9873	9.9924	0.5295	0.9050	
2	0.0279	−0.1187	0.0105	0.0203	10.019	9.9962	9.9850	9.9833	0.0948	0.4981	0.0578
	0.0407	0.0845	0.0212	−0.0623	10.058	10.013	9.9873	9.9924	0.5295	0.9051	
3	0.0255	−0.1215	0.0077	0.0178	10.019	9.9963	9.9850	9.9833	0.0948	0.4981	0.0345
	0.0374	0.0807	0.0174	−0.0657	10.058	10.013	9.9873	9.9925	0.5295	0.9051	
4	0.0175	−0.1805	0.0017	−0. 0586	9.849	10.396	9.7530	9.9385	0.1143	0.4825	0.0312
	0.0536	0.2107	0. 0747	−0.0177	9.928	10.018	9.9898	9.9956	0.5193	0.9175	

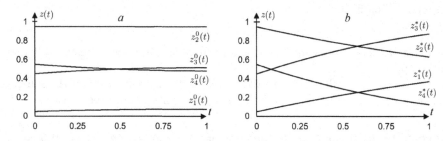

Fig. 2. Trajectories of motion of sources for different initial (a) and obtained feedback parameters.

Table 5 shows the results of solving the problem with the number of sources $N_c = 4$ and the number of measurement points $N_o = 4$. Duration of control process is $T = 1$, and the powers of the sources are $q_i = 2.5$, $i = 1, ..., 4$.

An increase in the number of sources made it possible to significantly reduce the value of the objective functional. Figure 2 shows the initial and optimal trajectories of the sources.

Numerous calculations of the heating process were carried out at optimal values of the feedback parameters, at which the state measurements had an error of 2%, 3%, 5%. The calculations showed that both the deviations of the trajectories of the sources and the course of the heating process itself do not differ significantly from the corresponding indicators with accurate measurements. Due to their small difference, it is not possible to present comparative results in a visual form.

Table 5. Results of solution the problem for $N_o = 4$, $N_c = 4$, $q_1 = q_2 = 10$, $T = 1$.

No	y									$J(y)$
	β		γ						ξ	
0	-0.1118	-0.1811	-0.0113	-0.0216	9.5275	9.7462	9.9241	9.1918	0.2542	0.5682
	0.2021	0.2179	-0.0140	-0.2189	9.9524	9.2570	9.9818	9.3853	0.4624	
	-0.8024	-0.0078	-0.0215	-0.3125	9.8274	9.9264	9.5941	9.5932	0.6897	
	0.1302	0.0409	0.1127	0.2016	9.9721	9.5171	9.9818	9.4253	0.9367	
1	-0.0018	-0.0011	-0.0013	-0.0016	9.2755	9.0524	9.0274	9.9021	0.1254	0.0061
	0.0021	0.0017	-0.0014	-0.0019	9.0062	9.0070	9.9264	9.5171	0.4914	
	-0.0024	—0.0078	-0.0015	-0.0025	9.9041	9.2818	9.5941	9.2818	0.6597	
	0.0092	0.0009	0.0027	0.0016	9.1918	9.3853	9.5932	9.4253	0.9167	
2	-0.0007	-0.0008	0.0002	-0.0009	9.5275	9.0062	9.9041	9.1918	0.1343	0.0031
	0.0042	0.0039	0.0005	-0.0001	9.0524	9.0070	9.2818	9.3853	0.4948	
	-0.0052	-0.0111	-0.0049	-0.0058	9.0274	9.9264	9.5941	9.5932	0.6672	
	0.0129	0.0041	0.0058	0.0047	9.9021	9.5171	9.2818	9.4253	0.9208	
3	-0.0042	-0.0024	-0.0031	-0.0025	9.5275	9.0062	9.9041	9.1918	0.1554	0.0009
	0.0056	0.0054	0.0019	0.0012	9.0524	9.0070	9.2818	9.3853	0.4642	
	-0.0067	-0.0129	-0.0067	-0.0076	9.0274	9.9264	9.5941	9.5932	0.6508	
	0.0149	0.0059	0.0075	0.0064	9.9021	9.5171	9.2818	9.4253	0.9209	
4	-0.0049	-0.0035	-0.0037	-0.0035	9.5275	9.0062	9.9041	9.1918	0.1254	0.0005
	0.0064	0.0062	0.0026	0.0017	9.0524	9.0070	9.2818	9.3853	0.4624	
	-0.0073	-0.0137	-0.0081	-0.0090	9.0274	9.9264	9.5941	9.5932	0.6897	
	0.0163	0.0065	0.0080	0.0070	9.9021	9.5171	9.2818	9.4253	0.9267	

5   Conclusion

In the article the problem of synthesis of motion control of pointwise sources of heating of a rod with piece-wise continuous specified powers in systems with distributed parameters is investigated. The rules of motion of pointwise sources are assigned depending on the state of the process at the measurement points. A formula for the linear dependence of the synthesized parameters on the measured temperature values is proposed. he differentiability of the functional with respect to the feedback parameters is shown, formulas for the gradient of the functional with respect to the synthesized parameters are obtained. The formulas allow for solving the source control synthesis problem to use efficient first-order numerical optimization methods and available standard software packages.

The proposed approach to the synthesis of lumped source control can be used in systems for automatic control and automatic control of lumped sources for many other technological processes and technical objects. The objects themselves can be described by other partial differential equations and types of boundary conditions.

References

1. Butkovskiy, A.G.: Methods of control of systems with distributed parameters. Nauka, Moscow (1984) (in Russian)
2. Utkin, V.I.: Sliding Modes in Control and Optimization. Springer, Berlin (1992). https://doi.org/10.1007/978-3-642-84379-2
3. Ray W.H.: Advanced Process Control. McGraw-Hill Book Company (2002)
4. Yegorov, A.I.: Bases of the control theory. Fizmatlit, Moscow (2004) (in Russian)
5. Butkovskiy, A.G., Pustylnikov, L.M.: The theory of mobile control of systems with distributed parameters. Nauka, Moscow (1980) (in Russian)
6. Sirazetdinov, T.K.: Optimization of systems with distributed parameters. Nauka, Moscow (1977) (in Russian)
7. Sergienko, I.V., Deineka, V.S.: Optimal control of distributed systems with conjugation conditions. Kluwer Acad. Publ., New York (2005)
8. Polyak, B.T., Khlebnikov, M.V., Rapoport, L.B.: Mathematical theory of automatic control. LENAND, Moscow (2019) (in Russian)
9. Vasilyev, F.P.: Optimization methods, 824. Faktorial Press, Moscow (2002) (in Russian)
10. Guliyev, S.Z.: Synthesis of zonal controls for a problem of heating with delay under non-separated boundary conditions. Cybern. Syst. Analysis. **54**(1), 110–121 (2018)
11. Aida-zade, K.R., Abdullaev, V.M.: On an approach to designing control of the distributed-parameter processes. Autom. Remote. Control. **73**(9), 1443–1455 (2012)
12. Aida-zade, K.R., Hashimov, V.A., Bagirov, A.H.: On a problem of synthesis of control of power of the moving sources on heating of a rod. Proc. Inst. Math. Mech. ANAS **47**(1), 183–196 (2021)
13. Nakhushev, A.M.: Loaded equations and their application. Nauka, Moscow (2012) (in Russian)
14. Abdullaev, V.M., Aida-Zade, K.R.: Numerical method of solution to loaded nonlocal boundary value problems for ordinary differential equations. Comput. Math. Math. Phys. **54**(7), 1096–1109 (2014). https://doi.org/10.1134/S0965542514070021
15. Abdullayev, V.M., Aida-zade, K.R.: Finite-Difference Methods for Solving Loaded Parabolic Equations. Comp. Math. Math. Phys. **56**(1), 93–105 (2016)

Stability of Persons' Pareto Behavior with Respect to Different Mechanisms in Resource Allocation Game Experiments

Nikolay A. Korgin, Vsevolod O. Korepanov$^{(\boxtimes)}$, and Anna I. Blokhina

V. A. Trapeznikov Institute of Control Sciences, Moscow, Russia
vkorepanov@ipu.ru

Abstract. In the present paper we analyze data obtained in experiments of resource allocation games with two mechanisms: the Yang-Hajek's mechanism (YH) and mechanism based on the alternating direction method of multipliers (ADMM). Previous research has shown that there is some correlation between consensus achievement and share of certain types of behavior. This paper considers behavior models based on best response and Pareto improvement. The aim of the research was to investigate individuals' behavior variation on mechanism change. A behavior based on Pareto improvement is demonstrated by players in both mechanisms slightly more often than only in one mechanism. At the same time, more players demonstrate behavior based on best response in either YH or in ADMM than in both mechanisms. Players had an opportunity to subjectively rank mechanisms; dependence between players behavior and subjective ranking is also studied. It is possible that players give higher rank to the mechanism where they behave according to the best-response and Pareto improvement models more often. There can be inverse relationship between the ranking of YH mechanism and the share of Pareto improvement behavior.

Keywords: Resource allocation problem · Pareto efficiency · Yang-Hajek's mechanism · ADMM mechanism · Mechanism design

1 Introduction

The Mechanism Design propose methodology to create "mechanisms" for solving some economic (socio-economic) problem in game-theory manner. Despite of good theoretical properties real people behavior can decrease efficiency of such mechanisms. For example, some mechanism constructs a game with unique Nash equilibrium which gives efficient outcome (solution of the problem), but people can end a game not in that Nash equilibrium situation and, as a consequence, the economic problem doesn't get solution.

Causes of such failures and their consequences may be different. We investigate best response based behaviors and Pareto Improvement behavior of people

© The Author(s), under exclusive license to Springer Nature Switzerland AG 2022
Y. Kochetov et al. (Eds.): MOTOR 2022, CCIS 1661, pp. 247–259, 2022.
https://doi.org/10.1007/978-3-031-16224-4_17

as cause of consensus end of games - can these models describe that people stop changing their actions (which we interpret as a consensus). But the main question we address is a change of one player behavior when she plays one mechanism after another.

In the present paper we analyze data obtained in experiments in the form of resource allocation games with two mechanisms: the Yang-Hajek's mechanism (YH) and mechanism based on the alternating direction method of multipliers (ADMM). We analyze the decisions made by players and their questionaries. The aim of the research was to investigate individuals' behavior variation due to mechanism change.

2   Resource Allocation Mechanisms

2.1   Resource Allocation Problem

An organizational system consists of a single Principal and a set $N = 1, ..., n$ of players. Principal allocates some infinitely divisible good in a limited positive amount $R \in \mathbb{R}^1_+$, which can be allotted among the players in any proportion. The utility of each player $i \in N$ in terms of the good allotted him is described by a function $u_i : \mathbb{R}^1_+ \to \mathbb{R}^1$ belonging to a certain set U_i of admissible utility functions. The set of admissible allocations:

$$A = \{x = (x_1, ..., x_n) : \sum_{i \in N} x_i \leqslant R, x \in \mathbb{R}^n_+\}. \tag{1}$$

The set of possible utility profiles:

$$U = \{u = (u_1, ..., u_n) : u_i \in U_i, i \in N\}. \tag{2}$$

The problem is to find the allocation $g : U \to A$ which maximizes the total utility of all players for any $u \in U$, i.e.:

$$g(u) \in \underset{x \in A}{\text{Argmax}} \sum_{i \in N} u_i(x_i) \tag{3}$$

In this paper we consider a special case of the problem described above with $N = 1, 2, 3$, utility function $u_i(x_i) = \sqrt{r_i + x_i}$, where $r = (1, 9, 25)$ is the profile of types of utility functions. For example, r_i can be seen as inner resource of player i. Amount of disposable resource is $R = 115$.

The games were conducted as follows. At each step every player makes a bid. The set of players' bids at step k is called 'situation' at this step and denoted as $s^k = (s^k_1, s^k_2, s^k_3)$. Then in according to a chosen allocation mechanism allocation x^k and transfers of players t^k_i are calculated, the result is reported to all players. The game stops, if none of the players changes his bid (the players achieve consensus), or if the maximum admissible number of steps (which is known to all players) have been made. The final payoff of each player is defined as the profit at the last step.

2.2 Yang-Hajek Mechanism

In the Yang-Hajek mechanism [4] the message of each player (the desired amount of resource for him) $s_i \in \mathbb{R}_+$ is not limited by the available amount. The resource allotted to the player is defined by

$$x_i = \frac{s_i}{S} R,$$

where $S = \sum_{i=1}^{n} s_i$.

Each player is assigned the additional transfer:

$$t_i = \beta S_{-i} s_i,$$

where $S_{-i} = S - s_i$, $\beta = 5 \times 10^{-4}$ - "penalty" (transfer) strictness.

The payoff at step k is defined as the utility at this step minus the assigned transfer: $\phi_i^k = u_i^k - t_i^k$. Where u_i^k, t_i^k, s_i^k is utility, transfer and bid of player i at step k: $u_i^k = u_i(Rs_i^k/(\sum_j s_j^k))$ and $t_i^k = \beta s_i^k \sum_{j \neq i} s_j^k$.

2.3 ADMM Mechanism

Originally in [1] it was a distributed optimization algorithm proposed by authors for the Sharing problem. But it is only technical method to solve (3), it is not address the incentive compatibility problem from the Mechanism design [2]. ADMM was adapted to resource allocation mechanism and its vulnerabilities are showed in [3]. The ADMM mechanism is as follows.

At step k the resource allotted to each player x_i^k coincides with his message s_i^k (i.e. the desired amount of resource), and the message is not limited by the available amount of resource. We point out that the mechanism allow non-admissible resource allocation (as an algorithm it converges to admissible allocation).

Each player is assigned the additional transfer:

$$t_i^k = \beta(s_i^k - s_i^{k-1} + \overline{x^{k-1}} - R/n + y^{k-1})^2,$$

where $\beta = 5 \times 10^{-4}$, $s_i^0 = R/n$, $\overline{x^{k-1}} = \frac{1}{n} \sum_{j=1}^{n} x_j^{k-1}$, $y^{k-1} = \overline{x^{k-1}} - R/n + y^{k-2}$, $y^0 = 0$.

The payoff at step k is defined as the utility at this step minus the assigned transfer: $\phi_i^k = u_i^k - t_i^k$.

3 Behavior Models

3.1 Rational Behavior

Rational behavior models are based on the player's best response (BR) to the situation at the previous step $k - 1$:

$$br_i(s^{k-1}) \in \underset{y_i \in \mathbb{R}_+}{\mathrm{Argmax}} \, \phi_i(y_i, s_{-i}^{k-1}),$$

where s_{-i}^{k-1} is the messages made by the opponents of player i at $k-1$ step.

Two rational behavior models are considered:

1) We treat s_i to be near best response decision with accuracy ε $(BR(\varepsilon))$ if:

$$|s_i - br_i(s^{k-1})| < \varepsilon;$$

2) We treat s_i to be Toward best response decision $(TwBR)$ if:

$$\begin{cases} s_i = s_i^{k-1}, & if\, br_i(s^{k-1}) = s_i^{k-1}, \\ (s_i - s_i^{k-1})/(br_i(s^{k-1}) - s_i^{k-1}) > 0, & \text{otherwise.} \end{cases}$$

3.2   Pareto Efficiency Behavior

Previous research [6] has shown that there is some correlation between consensus achievement and share of behavior that increases "local" Nash function which is positive when each player's payoff increased at step t comparing to step $t-1$. It is a strong Pareto improvement situation.

We treat s_i to be a strong Pareto Improvement (PI) *decision* at step k if:

$$\forall j : \phi_j(s_i, s_{-i}^{k-1}) > \phi_j(s^{k-1})$$

We can also speak about s^k as strong PI *situation* at step k:

$$\forall j : \phi_j(s^k) > \phi_j(s^{k-1})$$

Here we also consider a weak Pareto improvement. We treat s_i to be a weak Pareto improvement decision at step k if:

$$\forall j : \phi_j(s_i, s_{-i}^{k-1}) \geq \phi_j(s^{k-1})$$
$$\exists j : \phi_j(s_i, s_{-i}^{k-1}) > \phi_j(s^{k-1})$$

In what follows, we will not distinguish between strong and weak Pareto since there are no weak Pareto improvement decisions that are not strong Pareto improvement decisions in the experimental data we analyze; we will denote this type of behavior as PI, meaning strong Pareto improvement.

4   Experimental Data Analysis

We obtain experimental data from 14 games with YH mechanism and 14 games with ADMM mechanism. We chose only persons who participated as player in two games: with YH and ADMM mechanisms. It should be noted that each person participate firstly in game with YH mechanism, and then (perhaps after games with other mechanisms) in game with ADMM mechanism.

It should be noted that participants know all information about mechanisms and games excluding types of other players and their bids at history and current step. We also make presentation about mechanism and educational game before real games, so we think that players understand a mechanism rules. Game interface contains 'calculator' to calculate a mechanism's result from a player's action with non-changed actions of other players.

For each step of the games we have the set of players' bids, and calculated utilities, payoffs and best responses. 11 YH games, of which 3 are consensus-ended, and 7 ADMM games, of which 4 are consensus-ended, were selected for analysis. Not-consensus games ended by reaching the maximum admissible step.

Next after games with mechanisms it was questionnaires where persons ranked the mechanisms from 1 (best) to 5 (worst) according to their subjective evaluation of the mechanism applicability for resource allocation. We emphasize that persons rankings were made on their experience with mechanisms.

4.1   Pareto Efficient Behavior Analysis

For each person in YH and ADMM games the share of her Pareto improvement decisions was calculated. Figure 1 demonstrates the results.

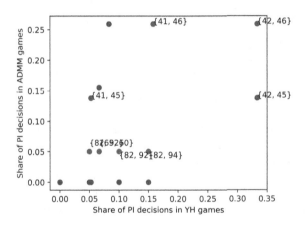

Fig. 1. Share of PI decisions for each person in YH and ADMM games

Each point in the diagram corresponds to one person. Its abscissa coordinate is the share of this player's PI decisions in the YH game; its ordinate coordinate is the corresponding value in the ADMM game. We use green points if both games where the person participated ended with consensus; red points – if none of two games ended with consensus; yellow points – if only the YH game ended with consensus (there are no such games); magenta points – if only the ADMM game ended with consensus. For some persons we put on the diagram games' ids in which they participated.

Figure 2 demonstrates data about each person's share of PI decisions in YH and ADMM games and consensus-state of games ends.

Person	YH game	ADMM game	Share of PI decisions in YH,	Share of PI decisions in ADMM, %	Consensus in YH	Consensus in ADMM
28	41	46	16	26	+	+
26	41	45	5	14	+	+
31	42	45	33	14	+	+
30	42	46	33	26	+	+
25	40	45	7	16	-	+
32	43	46	8	26	-	+
35	92	80	5	0	-	+
38	93	81	10	0	-	+
39	93	81	5	0	-	+
40	93	81	15	0	-	+
37	94	80	0	0	-	+
36	94	80	5	0	-	+
3	60	65	7	5	-	-
42	92	82	5	5	-	-
41	92	82	10	5	-	-
43	94	82	15	5	-	-

Fig. 2. Share of PI decisions for each person in YH and ADMM games, table

In YH games 4 persons didn't use PI at all, in ADMM – 6 persons without PI. But it's just some overall notes. We concentrate on persons decisions and Fig. 2 doesn't contains all data about decisions of all players in games; that is why we will not make analysis of decisions only in YH or only in ADMM games further. For example, person 25 participated in game 40, but Fig. 2 doesn't contain information about other persons of game 40 - each of them participated only in one of YH or ADMM games, and we are interesting in persons who participate in YH and in ADMM both. Nevertheless there is all decisions of game 40 in experiments' data.

There are 14 (of 16) persons who demonstrated PI decisions at least in one mechanism. 8 of them demonstrated PI decisions in both mechanisms, and 6 – only in one mechanism. A following hypothesis can be tested: if a person applies PI decisions in one game, then probability that he will apply such decisions in all games is higher than probability that he will do it only in one mechanism.

3 of 4 persons, who participated in consensus-ended games for both mechanisms, demonstrated high share of PI decisions in both games.

4.2   Rational Behavior Analysis

For each person in YH and ADMM games the share of near best response with accuracy 1 (BR(1)) decisions was calculated. Figure 3 demonstrates the results; Figure 4 demonstrates the complete data.

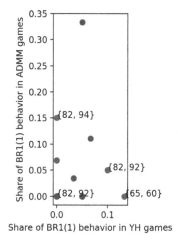

Fig. 3. Share of best response with accuracy 1 behavior

Person	YH game	ADMM game	Share of BR(1) decisions in YH, %	Share of BR(1) decisions in ADMM, %	Consensus in YH	Consensus in ADMM
28	41	46	0	0	+	+
26	41	45	0	7	+	+
31	42	45	0	7	+	+
30	42	46	0	0	+	+
25	40	45	3	3	-	+
32	43	46	7	11	-	+
35	92	80	5	0	-	+
38	93	81	5	0	-	+
39	93	81	0	0	-	+
40	93	81	5	0	-	+
37	94	80	0	0	-	+
36	94	80	5	33	-	+
3	60	65	13	0	-	-
42	92	82	10	5	-	-
41	92	82	0	0	-	-
43	94	82	0	15	-	-

Fig. 4. Share of best response with accuracy 1 behavior, table

There are 11 persons who demonstrated BR(1) behavior at least in one mechanism. 4 of them demonstrated BR(1) behavior in both mechanisms, and 7 – only in one mechanism. Therefore, a following hypothesis can be tested: if a person applies BR(1) behavior in one game, then probability that he will apply this behavior in all games is lower than probability that he will do it only in one mechanism.

Similar data were obtained for Toward best response behavior. Figures 5 and 6 show the results.

There are 16 persons who demonstrated TwBR behavior at least in one mechanism. 15 of them demonstrated TwBR behavior in both mechanisms, and 1 – only in one mechanism. However, TwBR is not a precise enough model. A hypothesis can be tested: share of TwBR decisions is in average larger than share of non-TwBR decisions for each person, i.e. players in average tend to increase their payoffs at each step.

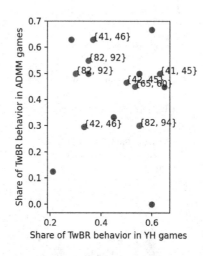

Fig. 5. Share of TwBR behavior in YH and ADMM games

5   Persons' Rankings Analysis

Persons were asked to assign some rank from 1 (best) to 5 (worst) to each mechanism. Figure 7 show association between the share of person's PI decisions in YH and ADMM games respectively and rank assigned to the mechanism by the person. Each point corresponds to one person; its abscissa coordinate is the share of this person's PI decisions in a game; its ordinate coordinate is the rank assigned to the mechanism by the person.

It is possible that there is inverse relationship between the ranking of YH mechanism and the share of PI decisions. May be it's because of persons figure out that players can decrease transfers if proportionally decrease their bids. Such

Person	YH game	ADMM game	Share of TwBR decisions in YH, %	Share of TwBR decisions in ADMM, %	Consensus in YH	Consensus in ADMM
28	41	46	37	63	+	+
26	41	45	63	50	+	+
31	42	45	50	47	+	+
30	42	46	33	30	+	+
25	40	45	65	45	-	+
32	43	46	28	63	-	+
35	92	80	60	67	-	+
38	93	81	60	0	-	+
39	93	81	21	13	-	+
40	93	81	55	50	-	+
37	94	80	45	33	-	+
36	94	80	35	50	-	+
3	60	65	53	45	-	-
42	92	82	35	55	-	-
41	92	82	30	50	-	-
43	94	82	55	30	-	-

Fig. 6. Share of TwBR behavior in YH and ADMM games, table

decisions are PI with high probability. For example, let players have *positive* bids $s = (s_1, s_2, s_3)$, $s_i > 0$ at some step of a game. Gain of player i:

$$\phi_i(s) = u_i\left(\frac{s_i}{\sum_j s_j}R\right) - \beta s_i \sum_{j \neq i} s_j$$

If each player decreases his bid by $a > 1$ times, then their bids $s(a) = (s_1/a, s_2/a, s_3/a)$ will be PI situation:

$$\phi_i(s_a) = u_i\left(\frac{s_i}{\sum_j s_j}R\right) - \beta/a^2 s_i \sum_{j \neq i} s_j > \phi_i(s)$$

In one YH game bids was $s^3 = (1, 99, 115)$, then player 2 decreases bid to $s_2^4 = 40$ at step 4. So increase of $\phi_1(s_1^3, s_2^4, s_3^3), \phi_3(s_1^3, s_2^4, s_3^3)$ it's almost obvious due to decrease of player 2 bid: x_1, x_3 increase and t_1, t_3 decrease. And gain of player 2 also increases:

$$\phi_2(s) = \sqrt{99 * 115/(215)} - 0.0005 * 99 * 116 \approx 1.53 < 3.11 = \phi_2(s_1^3, s_2^4, s_3^3)$$

So we can hypothesize that if a person figures out that possibility and actively uses it, then his rank of YH can decrease – he thinks that the game goal shifts from resource distribution to transfers decreasing.

For ADMM mechanism in Fig. 7 it is not likely that there is relationship between the ranking and the share of PI decisions. But it is interesting that players from consensus-ended games didn't use PI decisions at all (unlike YH games).

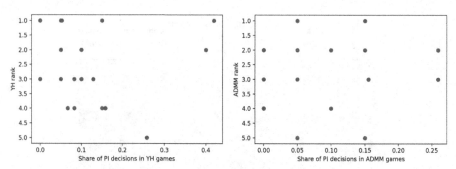

(a) Share of PI decisions and game rank in (b) Share of PI decisions and game rank in
YH games   ADMM games

Fig. 7. PI decisions

Figure 8 demonstrates ranking variation on change from YH (arrow start) to ADMM (arrow end) mechanism and share of PI decisions. Each arrow joints points corresponding to one person.

In 4 cases out of 7 decrease of PI decisions share coincides with decrease of rank; only in 1 case decrease of PI decisions share coincides with increase of rank. At the same time, in 3 cases out of 4 increase of PI decisions share coincides with increase of rank, and in 1 case out of 4 – with decrease of rank. A hypothesis, that there is correlation between a decrease of share of PI decisions and decrease of subjective ranking, can be tested.

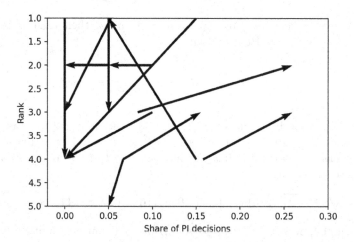

Fig. 8. Variation of share of PI decisions and ranks on change from YH to ADMM mechanism

Figures 9, 10, 11, and 12 demonstrate similar results for rational behavior. It is unlikely that there are any dependencies for BR(1) behavior and YH rank

(Fig. 9); but can be relation between BR(1) share and ADMM rank (Fig. 9). For TwBR there can be direct relationship between share of TwBR decisions and ranking of ADMM mechanism (Fig. 10).

In all cases, when both share of BR(1) behavior and ranks change on change from YH to ADMM, increase of share of BR(1) behavior coincides with increase of rank, and decrease of such behavior coincides with decrease of rank (Fig. 11). For TwBR, in 2 cases out of 7 when share of such decisions increases, rank increases too; in 2 cases rank decreases. In 5 cases out of 6 when share of TwBR decisions decreases, rank decreases and in 1 case out of 6 rank increases (Fig. 12).

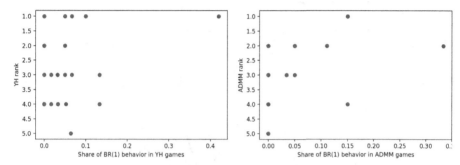

(a) Share of BR(1) behavior in YH games and rankings

(b) Share of BR(1) behavior in ADMM games and rankings

Fig. 9. BR(1) decisions

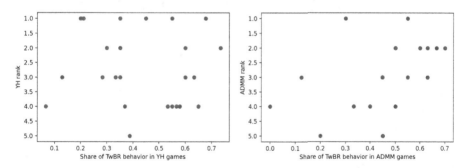

(a) Share of TwBR behavior in YH games and rankings

(b) Share of TwBR behavior in ADMM games and rankings

Fig. 10. TwBR decisions

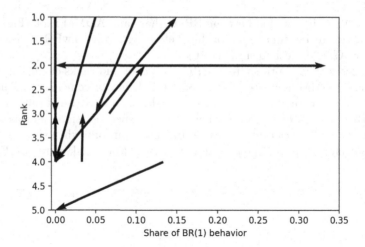

Fig. 11. Variation of share of BR(1) decisions and ranks on change from YH to ADMM mechanism

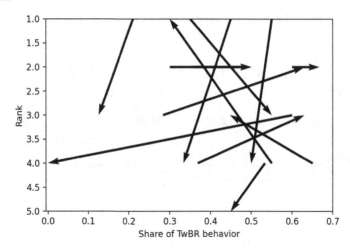

Fig. 12. Variation of share of TwBR decisions and ranks on change from YH to ADMM mechanism

6   Conclusion

Persons' decisions in games with two resource allocation mechanisms were pre-analyzed. Data of frequency of different types of behavior were obtained.

Comparison of persons' behavior in games with different mechanisms demon-strates that we can test a number of hypothesis further:

- some persons keep the same behavior models (PI or BR(1)) frequently enough in different mechanisms. In particular, there are persons who apply Pareto improvement in different mechanisms.

- persons decisions coincide with TwBR behavior in 50% cases
- for behavior models there can be dependence between their frequency and subjective rankings of mechanisms.

We plan to develop some experiments to formally test that hypothesis, due to small amount of data that we have now. May be we should have experiments with change of YH-ADMM playing order. More literature review is also needed to see other works with (Pareto improvement) behavior models stable to mechanism change, to consensus reaching.

References

1. Boyd, S., et al.: Distributed optimization and statistical learning via the alternating direction method of multipliers. Found. Trends Mach. Learn. **3**(1), 1–122 (2011)
2. Korgin, N.A., Korepanov, V.O.: An efficient solution of the resource allotment problem with the Groves-Ledyard mechanism under transferable utility. Automat. Remote Control **77**(5), 914–942 (2016)
3. Korgin, N.A., Korepanov, V.O.: Experimental gaming analysis of ADMM dynamic distributed optimization algorithm. IFAC-PapersOnLine **49**(12), 574–579 (2016)
4. Korgin, N.A., Korepanov, V.O.: Experimental gaming comparison of resource allocation rules in case of transferable utilities. Int. Game Theory Rev. **19**(02), 1750006 (2017)
5. Korgin, N.A., Korepanov, V.O.: Nash bargaining solution as negotiation concept for resource allocation problem: analysis of experimental data. Contrib. Game Theory Manag. **13**, 207–217 (2020)
6. Korepanov, V.O., Korgin, N.A., Blokhina, A.I.: Comparison of resource allocation mechanism by use of Nash bargaining solution approach (in Russian: Sravneniye mekhanosmov raspredeleniya resursa s pomoshyu resheniya Nesha dlya torgov. In: Large-Scale Systems Control, Papers of XVII All-Russian School Conference of Young Scientists. Moscow), pp. 605–616 (2021)

Network Centralities Based on Non-additive Measures

Natalia Nikitina[✉] and Vladimir Mazalov

Institute of Applied Mathematical Research, Karelian Research Center of the Russian Academy of Sciences, Pushkinskaya 11, 185910 Petrozavodsk, Russia
{nikitina,mazalov}@krc.karelia.ru

Abstract. Network models are widely employed in many areas of science and technology. Mathematical analysis of their properties includes various methods to characterize, rank and compare network nodes. A key concept here is centrality, a numerical value of node importance in the whole network. As the links in a network represent interactions between the nodes, non-additive measures can serve to evaluate characteristics of sets of nodes considering these interactions, and thus to define new centrality measures. In this work, we investigate variants of network centralities based on non-additive measures, calculated as the Choquet integral of a function of the distance between pairs of vertices. We illustrate the applications of the constructed centrality measures on three examples: a social network, a chemical space network and a transportation network. The proposed centrality measures can complement existing methods of node ranking in different applications and serve as a starting point for developing new algorithms.

Keywords: Network centrality · Vertex ranking · Non-additive measure · Choquet integral · Fuzzy measure · Fuzzy integral

1 Introduction

Non-additive measures (fuzzy measures, capacities) extend the scope of additive measures such as probabilities. Importantly, they allow to represent uncertainty under decision making or the worth of a group of agents in mathematical game theory. Non-additive measures have been first studied by Choquet in 1954 [1]. The Choquet integral is a generalization of the Lebesgue integral with respect to non-additive measures. It allows to extend the concept of the weighted average to the case with interaction among criteria.

Non-additive measures and the Choquet integral have been studied and reviewed by M. Grabisch, V. Torra, Y. Narukawa, L. Godo, G. Beliakov and other researchers, for instance, in [2,3,5–8]. Recent comprehensive reviews of

The study was funded through Russian Science Foundation grant # 22-11-20015 implemented in collaboration with Republic of Karelia authorities with funding from the Republic of Karelia Venture Capital Fund.

the field of research and applications of non-additive measures and fuzzy integrals are presented in [8,11].

An important connection with cooperative game theory has been studied in [4,9,10], as the set capacity represents interactions between the set elements and thus can express the value of a coalition of players. This includes games on networks, where coalition formation is restricted by the network topology. For instance, in [10], the authors describe the use of Choquet and Sugeno integrals with respect to non-additive measures for network analysis. They consider a non-additive measure corresponding to the connection degree of a set of nodes and study the properties of the Choquet integral characterizing either each separate node or the network in a whole. The results extend the use of game theory for network analysis and allow for computationally efficient algorithms.

In network research, characterising and ranking the nodes is one of the most important problems. A key concept here is centrality [12], a numerical value of node importance in the network. Depending on the mathematical model, node importance may be defined in different ways. To name the most popular ones, the degree centrality indicates the number of direct connections that the node has; the closeness centrality indicates the connectedness of the node with the rest of the network; the betweenness centrality indicates to what extent does the node serve as a bridge between all possible pairs of nodes.

As the links in a network represent interactions between the nodes, non-additive measures can serve to evaluate characteristics of sets of nodes, and thus to define new centrality measures. In this paper, we consider the application of non-additive measures to define several network centrality measures as the discrete Choquet integrals of a function of the distance between pairs of vertices. For a social network, we consider a non-additive measure based on the number of edges in a subgraph. For a chemical space network, we consider a non-additive measure based on the minimal node weight. Finally, for a public transport network, we consider a measure based on the summarized node weights.

All the proposed centrality measures can complement existing methods of node ranking in different applications and can serve as a starting point for developing new mathematical models, for example, game-theoretic ones.

2 Network Centralities Based on Non-additive Measures

2.1 Non-additive Measures and the Choquet Integral

Consider a finite set $X \neq \emptyset$ and a simple function $f : X \to \mathbb{R}$. The function takes values $\boldsymbol{ran} f = \{a_1, \ldots, a_n\}$, where $0 \leq a_1 < a_2 < \ldots < a_N$. A set function $\mu : 2^X \to [0, 1]$ is a non-additive measure (capacity or fuzzy measure) [5] if it satisfies the following axioms:

(1) $\mu(\emptyset) = 0$, $\mu(X) = 1$ (boundary conditions);
(2) $A \subseteq B$ implies $\mu(A) \leq \mu(B)$ (monotonicity).

Let us permute the indices of set elements so that $0 \leq f(x_{s(1)}) \leq \ldots \leq f(x_{s(N)}) \leq 1$; $f(x_{s(0)}) = 0$; $W_{s(i)} = \{x_{s(i)}, \ldots, x_{s(N)}\}$. The Choquet integral [1] of function f with respect to the non-additive measure μ is defined as

$$(C) \int f d\mu = \sum_{i=1}^{N} \left[f(x_{s(i)}) - f(x_{s(i-1)}) \right] \mu(W_{s(i)}). \tag{1}$$

2.2 Node Centralities Expressed via the Choquet Integral

In the calculation of the Choquet integral, there is a restriction that subsets of the players form a specific chain (a more general case is the concave integral). In network analysis, such a chain is defined by the network topology. In many applications, it is useful to partition network nodes into subsets basing on their distance from the target node.

Consider an undirected connected graph $G = (V, E)$ with diameter d, $|V| = N$. For a fixed vertex i, one can consider a family of functions

$$f_i(j) = f(l(i,j)), \tag{2}$$

where $l(i,j)$ is the distance (the length of a shortest path) between i and j. If f_i is non-increasing and the eccentricity of i is equal to e_i, the partitioning of graph vertices to integrate it by Choquet takes the following form (Table 1).

Table 1. Subsets of graph vertices to integrate function f_i (2) by Choquet (1).

Subset $W_{s(k)}$	Capacity $\mu(W_{s(k)})$	Value of $f_i(x_{s(k)})$
None	None	$f(x_{s(0)}) = 0$
$W_{s(1)} = \{j : l(i,j) \leq e_i\} = V$	1	$f(x_{s(1)}) = f(e_i)$
$W_{s(2)} = \{j : l(i,j) \leq e_i - 1\}$	$\mu(W_{s(2)})$	$f(x_{s(2)}) = f(e_i - 1)$
...	...	...
$W_{s(k)} = \{j : l(i,j) \leq e_i - k + 1\}$	$\mu(W_{s(k)})$	$f(x_{s(k)}) = f(e_i - k + 1)$
...	...	...
$W_{s(e_i)} = \{j : l(i,j) \leq 1\}$	$\mu(W_{s(e_i)})$	$f(x_{s(e_i)}) = f(1)$
$W_{s(e_i+1)} = \{j : l(i,j) \leq 0\} = \{i\}$	$\mu(W_{s(e_i+1)})$	$f(x_{s(e_i+1)}) = 1$

With specific f_i and $\mu(W)$, the Choquet integral (1) can express different characteristics of vertex i including several known centrality measures. For example, with

$$f_i(j) = \begin{cases} 1, & \text{if } j \text{ is a neighbour of } i \\ 0, & \text{otherwise} \end{cases} \tag{3}$$

and

$$\mu(W) = \frac{|W_E|}{|E|}, \tag{4}$$

the Choquet integral of the vertex i is the normalized *degree centrality*,

$$DC(i) = \frac{1}{|E|} \sum_j a_{ij}. \tag{5}$$

Here, a_{ij} is the element of the adjacency matrix indicating the existence of an edge between i and j in an undirected graph.

With

$$f_i(j) = r^{l(i,j)}, \ 0 < r < 1 \tag{6}$$

and an additive measure

$$\mu(W) = \frac{|W \setminus \{i\}|}{N - 1}, \tag{7}$$

the Choquet integral of the vertex i is the normalized *closeness centrality*,

$$CC(i) = \sum_{j \neq i} r^{l(i,j)}. \tag{8}$$

With

$$f_i(j) = \frac{1}{r^{l(i,j)}} \tag{9}$$

and set capacity defined by (7), the Choquet integral of the vertex i is the normalized *harmonic centrality*,

$$HC(i) = \sum_{j \neq i} \frac{1}{r^{l(i,j)}}. \tag{10}$$

In the next paragraph, we construct and investigate three centrality measures based on the Choquet integral of function $f_i(j)$ (6).

3   Applications

3.1   Node Centrality Based on the Relative Edge Density

Let us consider function (6) and set capacity (4). Capacities in Table 1 take form

$$\mu(W_m) = \mu(W_{m+1}) + \delta(i, m),$$

where $\delta(i, m)$ is the number of edges added when constructing vertex subsets

$$W_m = W_{m+1} \bigcup \{j : l(i,j) = d - m + 1\}.$$

These are all edges between i's neighbours of orders $(d - m)$ and $(d - m + 1)$ and edges between i's neighbours of order $(d - m + 1)$.

$$\delta(i, m) = \sum_j \left(b_{ij}^{(d-m)} + \frac{1}{2} b_{ij}^{(d-m+1)} \right) \sum_k a_{jk} b_{ik}^{(d-m+1)}.$$

$$\delta(i, d) = \sum_j a_{ij} \left(1 + \frac{1}{2} \sum_k a_{jk} a_{ik} \right).$$

Here, matrix $B^{(m)}$ enumerates i's neighbours of order m,

$$b_{ij}^{(m)} = \begin{cases} 1, \text{if } l(i, j) = m \\ 0, \text{otherwise.} \end{cases}$$

Note that $B^{(0)} = E$. For $m \geq 1$, these matrices can be calculated as

$$B^{(m)} = A^{(m)} \odot \bar{A}^{(m-1)} \odot \ldots \odot \bar{A}^{(0)}, \tag{11}$$

where matrix $A^{(m)}$ enumerates i'th neighbours up to order m and can be obtained from the m-th power of the adjacency matrix A as

$$a_{ij}^{(m)} = \begin{cases} 1, \text{if } a_{ij}^m > 0 \\ 0, \text{otherwise.} \end{cases}$$

$\bar{X}$ is the complement of the logical matrix X, $\odot$ is the Hadamard product. For an isolated vertex, $(C) \int f_i d\mu = 0$. For a full graph, $(C) \int f_i d\mu = r$. For $d \geq 2$, the Choquet integral takes form

$$(C) \int f_i d\mu = r^d + \sum_{m=2}^{d} (r^{d-m+1} - r^{d-m+2}) \mu(W_m) =$$

$$= r^d + \sum_{m=2}^{d} (r^{d-m+1} - r^{d-m+2})(\mu(W_{m+1}) + \delta(i, m)) =$$

$$= r^d + \sum_{m=2}^{d} (r^{d-m+1} - r^d) \delta(i, m) \tag{12}$$

Or, in a matrix form,

$$(C) \int f_i d\mu = r^d + \sum_{m=2}^{d} (r^{d-m+1} - r^d) \left\{ \left(\left(B^{(d-m)} + \frac{1}{2} B^{(d-m+1)} \right) \odot \left(B^{(d-m+1)} \cdot A \right) \right) e \right\}_i,$$

where e is a unit vector, and $\{x\}_i$ is the i-th component of vector x.

As $r \to 1$, $(C) \int f_i d\mu$ counts all edges of the graph with the same factor and tends to 1. As $r \to 0$, $(C) \int f_i d\mu$ is proportional to the vertex degree plus the number of edges between its neighbours, $\sum_j a_{ij} (1 + \frac{1}{2} \sum_k a_{ik})$.

Let us illustrate the considered centrality in application to the known Zachary's karate club network [15, 16] presented in Fig. 1, with $r = 0.5$. The nodes with min-max normalized centralities at least 0.75 are depicted in green.

In Fig. 2, we present the min-max normalized centralities of all nodes of network (Fig. 1) calculated as closeness centrality (Algorithm 1), based on the

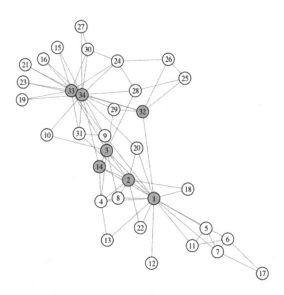

Fig. 1. Zachary's karate club network. Nodes 1 and 34 have the maximal centralities (12), $r = 0.5$. Nodes depicted in green have min-max normalized centralities at least 0.75.

Choquet integral (12) (Algorithm 2) and degree centrality (Algorithm 3). One observes some differences in vertex ranking by the considered algorithms.

Each algorithm assigns the highest rank to vertices 1 and 34. However, algorithm 2 allows to distinguish between vertices with equal degrees if they are connected to vertex subsets that accumulate different numbers of edges. This is the case of vertex 17 ranked lower than vertices 15, 16, 18 and 19 of degree 2.

Also, Algorithm 2 allows to better distinguish between a community center and its neighbours that are well connected with other parts of the network but have small degrees. This is the case of vertices 1 and 3, 34 and 32.

To summarize, the proposed centrality measure based on the Choquet integral (12) can be used to rank network nodes or to complement existing centrality measures, for example, closeness and degree centrality, to make the differences between certain vertices more specific. A possible application of this centrality measure is characterization of community structure in social network analysis.

3.2   Ranking of Nodes in a Chemical Space Network

Network representation has been recently proposed to effectively model and visualize the chemical space of small molecules [14]. The nodes of a chemical space network are individual small molecules. The link weights represent chemical similarities between pairs of molecules (generally, undirected).

For a given set of molecules, one can obtain the full graph by calculating the similarity coefficients between each two nodes. In order to reduce the complexity of investigating such a dense network, an appropriate threshold τ is set to

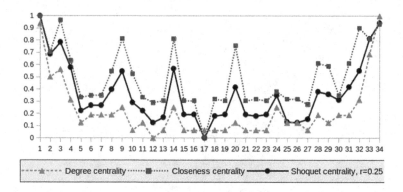

Fig. 2. Normalized vertex centralities of the Zachary's karate club network.

remove links with weights lower than τ. The resulting network can be studied by mathematical methods to obtain vertex ranking, community structure and other characteristics related to biochemical properties of the set of molecules.

In particular, mathematical methods can aid the selection of the most representative subset of molecules from a large set. Such a problem arises in drug discovery, when it is necessary to rank the results of a virtual drug screening and filter the compounds to pass onto the next stage of research.

Let us illustrate the possible application of the Choquet integral for ranking of nodes in a chemical space network. In Fig. 3, we present a network of 15 selected unique known inhibitors of the SARS coronavirus 3C-like proteinase obtained from the ChEMBL database [18] via the Web interface [19].

Molecular structures are represented as binary strings, and we apply the commonly used Tanimoto coefficient [20] to calculate similarities between pairs of the molecules and link the nodes correspondingly.

The choice of $f_i(j)$ and set capacity $\mu(W)$ depends on the problem being solved and can incorporate different aspects of nodes interaction and properties. For example, consider the problem of selecting a subset of molecules with a simple shape, strongly connected to other molecules with simple shapes. Such a subset might represent important structural properties characterizing the whole set of molecules. In order to select such a subset, let us consider molecules which have a low molecular weight mw and are strongly connected to other molecules with low molecular weights. We use function (6) and set capacity

$$\mu(W) = \frac{\min\limits_{i \in V} mw(i)}{\min\limits_{i \in W} mw(i)}. \tag{13}$$

Obviously, this measure is non-additive, monotonic and satisfies the boundary conditions $\mu(\emptyset) = 0$ and $\mu(V) = 1$.

The Choquet integral (1) allows to calculate vertex centralities and rank the nodes accordingly. In Fig. 3, we provide the resulting network. The size of a node is proportional to its molecular weight. Five nodes with the highest centralities are depicted in green.

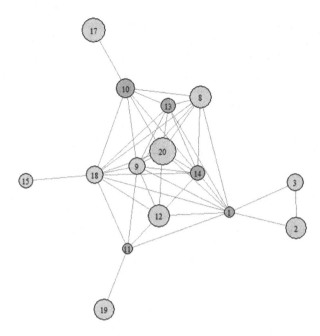

Fig. 3. Network of 15 of the known inhibitors of the SARS coronavirus 3C-like proteinase, similarity threshold $\tau = 0.6$, function parameter $r = 0.1$. Nodes depicted in green have the highest calculated centralities. (Color figure online)

3.3   Ranking of Nodes in a Transportation Network

An important application of network analysis is research, design and optimization of transportation networks. Centrality plays an important role in such models [21,22] as it expresses to what extent is the node involved in the process of passenger or vehicle flows going through the network. Various node ranking methods can be developed to evaluate the local traffic intensity, detect bottlenecks, select optimal points for new traffic lights etc.

In particular, we can consider a network model of a public transport. The nodes of the network are bus stops. The links represent direct road connections. Each node is associated with a non-negative number indicating the number of people living nearby, and thus, contributes to the whole passenger flow depending on its location and population.

In such a model, the Choquet integral can express a variant of node centrality to rank bus stops according to the number of people using each of them daily. Such ranking can help, for example, to determine the most profitable points for newspaper stands or recycling containers. Let us consider measure μ on a set of network nodes expressing the total number of people who start their daily routes from bus stops of set W:

$$\mu(W) = \frac{\sum_{i \in W} w_i}{W_s}. \tag{14}$$

In this case, the measure is additive which reduces the complexity of its computation. The Choquet integral reduces to the weighted sum.

In this model, function $f_i(j)$ (6) expresses the fraction of the number of passengers who travel from node j to node i. The node centrality based on the Choquet integral expresses the relative number of passengers using bus stop i as start or end point of their daily route:

$$R(i) = (C) \int f_i d\mu = \sum_{m=0}^{e_i} r^m \sum_{\substack{j: \\ l(i,j)=m}} w_j = \sum_{m=0}^{e_i} r^m \sum_j b_{ij}^{(m)} w_j. \quad (15)$$

Let us illustrate the application of the proposed centrality measure (15) to determine the bus stops used by the highest number of people in the public transport network of Petrozavodsk. In Fig. 4, we provide the network with ten highest ranked nodes highlighted in green.

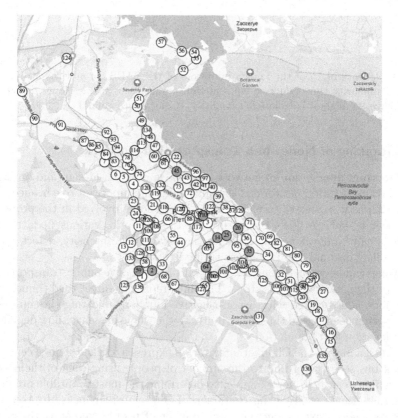

Fig. 4. Public transport network of Petrozavodsk. Ten nodes ranked the highest by measure (15) ($r = 0.5$) are depicted in green. (Color figure online)

In Fig. 5, we compare top-10 nodes selected by different centrality measures. One observes that the selected subsets of nodes differ, and so do the characteristics of the corresponding groups of bus stops. The differences should be taken into account when changing bus routes, introducing new elements such as new traffic lights, building new houses or supermarkets. On the other hand, some of the nodes (for example, 26, 45 and 110) are ranked the highest by four or three of the algorithms. It emphasizes the importance of these nodes in a transportation network. They may be potential bottlenecks, especially during the rush hour. Such nodes may be subject to optimization to decrease the load on them.

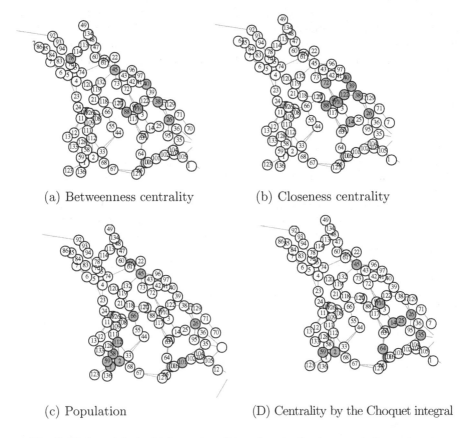

(a) Betweenness centrality   (b) Closeness centrality

(c) Population   (D) Centrality by the Choquet integral

Fig. 5. Nodes with the highest centrality values in the transportation network.

4   Conclusion

In this paper, we have considered the application of non-additive measures to calculate node centralities in a network. With the proposed non-additive measures, we integrated the function $f_i(j) = f(l(i,j))$ by Choquet to obtain node centralities. Here, $l(i, j)$ is the distance between nodes i and j, or the length of a shortest path between them. The partition of network nodes depending on their distance from the target node makes sense in multiple applications. For illustrations, we calculated and commented on node centralities in a social network, a chemical space network and a transportation network.

A similar approach can be used to define group centrality measures, in particular, for transportation networks where nodes can be grouped in accordance with the geographical location or a bus route.

Non-additive measures on sets can represent interactions between the set elements and thus can express the value of a coalition of players. In future work, we plan to develop and research network game models based on application of non-additive measures and the Choquet integral to define the characteristic function. Design of the characteristic function in a cooperative game is a challenging task [23]. Non-additive measures can help in case of monotonic games.

References

1. Choquet, G.: Theory of capacities. Annales de l'institut Fourier **5**, 131–295 (1954). https://doi.org/10.5802/aif.53
2. Murofushi, T., Sugeno, M.: A theory of fuzzy measures: representations, the Choquet integral, and null sets. J. Math. Anal. Appl. **159**(2), 532–549 (1991). https://doi.org/10.1016/0022-247x(91)90213-j
3. Grabisch, M.: Fuzzy integral in multicriteria decision making. Fuzzy Sets Syst. **69**(1995), 279–298 (1995). https://doi.org/10.1016/0165-0114(94)00174-6
4. Bortot, S., Marques Pereira, R.A., Stamatopoulou, A.: Consensus dynamics, network interaction, and Shapley indices in the Choquet framework. Soft. Comput. **24**(18), 13757–13768 (2019). https://doi.org/10.1007/s00500-019-04512-3
5. Torra, V., Narukawa, Y.: Modeling Decisions. 1st edn. Springer, Heidelberg (2007). https://doi.org/10.1007/978-3-540-68791-7
6. Beliakov, G., Pradera, A., Calvo, T.: Aggregation functions: A guide for practitioners. STUDFUZZ, vol. 221. Springer (2007). https://doi.org/10.1007/978-3-540-73721-6
7. Grabisch, M., Labreuche, C.: A decade of application of the Choquet and Sugeno integrals in multi-criteria decision aid. Ann. Oper. Res. **175**(1), 247–290 (2010). https://doi.org/10.1007/s10288-007-0064-2
8. Torra, V., Narukawa, Y., Sugeno, M. (eds.) Non-Additive Measures: Theory and Applications. STUDFUZZ, vol. 310. Springer, Cham (2014). https://doi.org/10.1007/978-3-319-03155-2
9. Fujimoto, K.: Cooperative game as non-additive measure. In: Non-Additive Measures, vol. 131–171 (2014). https://doi.org/10.1007/978-3-319-03155-2_6
10. Torra, V., Narukawa, Y.: On network analysis using non-additive integrals: extending the game-theoretic network centrality. Soft. Comput. **23**(7), 2321–2329 (2018). https://doi.org/10.1007/s00500-018-03710-9
11. Grabisch, M.: Set Functions, Games and Capacities in Decision Making, vol. 46. Springer, Berlin (2016). https://doi.org/10.1007/978-3-319-30690-2_2
12. Freeman, L.C.: Centrality in social networks conceptual clarification. Soc. Netw. **1**(3), 215–239 (1978–1979) https://doi.org/10.1016/0378-8733(78)90021-7

13. Suri, N.R., Narahari, Y.: Determining the top-k nodes in social networks using the Shapley value. In: Proceedings of the 7th International Joint Conference on Autonomous Agents and Multiagent Systems, vol. 3, pp. 1509–1512 (2008). https://doi.org/10.1109/tase.2010.2052042

14. Maggiora, G.M., Bajorath, J.: Chemical space networks: a powerful new paradigm for the description of chemical space. J. Comput. Aided Mol. Des. **28**(8), 795–802 (2014). https://doi.org/10.1007/s10822-014-9760-0

15. Kunegis, J.: KONECT – the Koblenz network collection. In: Proceedings of the 22nd International Conference on World Wide Web, pp. 1343–1350. Association for Computing Machinery, New York (2013). https://doi.org/10.1145/2487788. 2488173

16. Zachary, W.: An information flow model for conflict and fission in small groups. J. Anthropol. Res. **33**(4), 452–473 (1977). https://doi.org/10.1086/jar.33.4.3629752

17. Lusseau, D., Schneider, K., Boisseau, O.J., et al.: The bottlenose dolphin community of Doubtful Sound features a large proportion of long-lasting associations. Behav. Ecol. Sociobiol. **54**, 396–405 (2003). https://doi.org/10.1007/s00265-003-0651-y

18. Gaulton, A., Hersey, A., Nowotka, M., et al.: The ChEMBL database in 2017. Nucleic Acids Res. **45**(D1), D945–D954 (2017). https://doi.org/10.1093/nar/gkw1074

19. Davies, M., Nowotka, M., Papadatos, G., et al.: ChEMBL web services: streamlining access to drug discovery data and utilities. Nucleic Acids Res. **43**(W1), W612–W620 (2015). https://doi.org/10.1093/nar/gkv352

20. Tanimoto, T.T.: An elementary mathematical theory of classification and prediction. IBM Report, Nov 1958

21. Wang, K. and Fu, X.: Research on centrality of urban transport network nodes. In: AIP Conference Proceedings, vol. 1839(1), p. 020181. AIP Publishing LLC. (2017). https://doi.org/10.1063/1.4982546

22. Cheng, Y.-Y., Lee, R.K.-W., Lim, E.-P., Zhu, F.: Measuring centralities for transportation networks beyond structures. In: Kazienko, P., Chawla, N. (eds.) Applications of Social Media and Social Network Analysis. LNSN, pp. 23–39. Springer, Cham (2015). https://doi.org/10.1007/978-3-319-19003-7_2

23. Mazalov, V.V., Avrachenkov, K.E., Trukhina, L.I., Tsynguev, B.T.: Game-theoretic centrality measures for weighted graphs. Fund. Inform. **145**(3), 341–358 (2016). https://doi.org/10.3233/FI-2016-1364

Optimal Control of Output Variables Within a Given Range Based on a Predictive Model

Margarita Sotnikova$^{(\boxtimes)}$ and Ruslan Sevostyanov

Saint-Petersburg State University, Universitetskii Prospekt 35, Petergof, Saint Petersburg, Russia
m.sotnikova@spbu.ru

Abstract. This paper is devoted to the problem of digital control design to keep the output variables of the controlled process in a given range. Such a problem is of particular importance in control practice if it is only necessary to maintain some variables of a dynamic process in a stated range, but not to track the reference signal. The control design approach is developed for a nonlinear digital model. This approach is based on the predictive model and a special introduced cost functional that is minimized over the prediction horizon. This functional include two terms: the first term represent intensity of the control actions, the second term is a penalty for violating the specified range. It is shown that the optimal control design at each instant of discrete time is reduced to nonlinear programming problem. This problem always has a solution due to the addmissible set, which includes additional variables that guarantee the existence of the solution even in the case of violating the given range. The application of the proposed approach for oil refining in the distillation column is given. The examples of process simulation are presented and discussed.

Keywords: Optimal control · Predictive model · Digital system · Nonlinear programming · Distillation column · Given range

1 Introduction

This paper is devoted to the problem of designing optimal control to keep controlled variables of a dynamic process in a given range. Unlike the traditional approach, the control objective in the considered problem is to maintain output variables in a certain range, and not to track some reference signal. The importance of such a problem is determined by practical applications, when it is sufficient to maintain some specified limits for controlled variables in order to ensure acceptable quality of the resulting product, for example, in the oil refining industry [1].

The formulation of the problem is discussed in relation to a general nonlinear discrete-time model with input constraints and specified limits on output variables. The proposed control design approach is based on the model predictive

The reported study was funded by RFBR, project number 20-07-00531.

control (MPC) scheme [2–5]. The MPC control is widely used in practic due to its ability to deal with the nonlinear models, constraints, and multiple inputs and outputs [6–8].

The main idea of the proposed approach is to introduce a functional representing the intensity of control actions and the penalty for violating the stated limits by output variables. This allows to adjust the intensity of the control and minimize the control actions if the output variables are within the range. It is shown that the optimization of this functional on the prediction horizon leads to a nonlinear programming problem, which is solved repeatedly at each instant of discrete time in accordance with the MPC strategy.

The application of the proposed approach to the particular case of a linear input-output model with a time delay is presented. The initial model in the form of transfer functions, which is convenient for engineering applications, is transformed into the state-space form by expanding the state vector with additional components corresponding to the time delay. It is shown that in this particular case the problem of choosing the optimal control strategy on the prediction horizon leads to the quadratic programming. The derivation of the corresponding optimization problem is given. The practical example of the oil refining in a distillation column is considered and discussed.

This paper develops the results [9, 10] that were received earlier. In particular, the control design approach is given here for a general case of a nonlinear discrete-time model. The corresponding results for a linear model are derived. Also, the proposed approach is distinguished by the vector of additional variables that allow to regulate the violation of the admissible range at each instant of discrete time on the prediction horizon.

2   Problem Formulation

Let consider nonlinear mathematical model of a controlled object, represented in discrete time

$$
\begin{aligned}
\mathbf{x}[k+1] &= \mathbf{f}(\mathbf{x}[k], \mathbf{u}[k], \mathbf{w}[k]), \\
\mathbf{y}[k] &= \mathbf{g}(\mathbf{x}[k], \mathbf{u}[k], \mathbf{v}[k]).
\end{aligned}
\tag{1}
$$

Here the following notations are used: k is the instant of discrete time, $\mathbf{x} \in E^n$ is the state vector, $\mathbf{u} \in E^m$ is the control input, $\mathbf{y} \in E^r$ is the measured output, $\mathbf{w} \in E^{n_w}$ is the external disturbances, $\mathbf{v} \in E^{n_v}$ is the measurement noise.

Unlike the traditional reference control approach, we consider the problem of keeping the output variables $\mathbf{y}$ in the given range, taking into account limited control resources. In order to formalize description of the problem, let us introduce admissible sets $U \subseteq E^m$ and $Y \subseteq E^r$ for control input $\mathbf{u}$ and model output $\mathbf{y}$ respectively. Then the control objective is to keep the constraints

$$
\begin{aligned}
\mathbf{y}[k] &\in Y, \ \forall k \ge k_0, \\
\mathbf{u}[k] &\in U, \forall k \ge 0,
\end{aligned}
\tag{2}
$$

where $k_0 \ge 0$ is some integer number. It is important to note that the constraints imposed on output variables can be violated at the initial time instant $k = 0$.

So, the control problem is to bring output variables $\mathbf{y}[k]$ into the given range as soon as possible and then keep them inside the range with minimal control actions. The contstraints on the input variables $\mathbf{u}$ are rigid, and the constraints on the output variables $\mathbf{y}$ are softer and can sometimes be violated, for example, due to external disturbances.

In the following, we use the representetion of the admissible set Y using the upper and lower boundaries, i.e.

$$\mathbf{y}_{min} \leq \mathbf{y}[k] \leq \mathbf{y}_{max}, \ \forall k \geq k_0, \tag{3}$$

where $\mathbf{y}_{min}, \mathbf{y}_{max}$ are the given vectors. Similarly, we introduce boundaries for the control input $\mathbf{u}$, which define the admissible set U:

$$\begin{aligned} \mathbf{u}_{min} \leq \mathbf{u}[k] \leq \mathbf{u}_{max}, \\ \Delta\mathbf{u}_{min} \leq \Delta\mathbf{u}[k] \leq \Delta\mathbf{u}_{max}, \forall k \geq 0, \end{aligned} \tag{4}$$

where $\Delta\mathbf{u}[k] = \mathbf{u}[k] - \mathbf{u}[k-1]$ is the control action at time instant k.

As a result, let us consider a problem of optimal feedback control design, which ensures boundary constraints (3) and (4) on input and output variables relative to the mathematical model (1) of the object dynamics.

3   Optimization Approach to Control Design

The stated problem requires the development of a specific control design approach, which differs from the traditional method of reference control design, since there is no need to provide some reference signal, but only to keep controlled variables in the given range (3) subject to input constraints (4).

The proposed control design approach is based on ideology of Model Predictive Control (MPC). To present the main idea of the approach, let us introduce the predictive model taking into account equations (1) of object dynamics:

$$\begin{aligned} \mathbf{x}[i+1] &= \mathbf{f}(\mathbf{x}[i], \mathbf{u}[i], \bar{\mathbf{w}}[i]), \ i = k+j, \ j = 0, 1, 2, ..., \ \mathbf{x}[k] = \tilde{\mathbf{x}}[k], \\ \mathbf{y}[i] &= \mathbf{g}(\mathbf{x}[i], \mathbf{u}[i]). \end{aligned} \tag{5}$$

Here $\bar{\mathbf{w}}[i]$ are the estimated or measured external disturbances, $\tilde{\mathbf{x}}[k]$ is the estimated state of the controlled object (1) at time instant k. Measurement noise $\mathbf{v}$ is omitted in predictive model (5) for simplicity. This model is initialized with the current state of the object $\tilde{\mathbf{x}}[k]$ and allows to predict its future behavior. This means that for any given control sequence $\bar{\mathbf{u}} = \{\mathbf{u}[i]\}, i = \overline{k, k+P-1}$ we can get corresponding output sequence $\bar{\mathbf{y}}(\bar{\mathbf{u}}) = \{\mathbf{y}[i]\}, i = \overline{k+1, k+P}$ according to the model (5), where P is the prediction horizon.

The key idea of MPC is to choose a control sequence $\bar{\mathbf{u}}$ that optimizes the dynamics of a closed-loop system with respect to some cost function, taking into account the constraints imposed on input and output variables. This optimal control sequence is applied to the object only at the current instant of time k. At the next instant of time $k+1$ the entire optimization process repeats again, providing feedback control.

In order to formulate the optimization problem on the prediction horizon, we first construct the admissible set of control actions. To this end, the given range (3) is represented on the prediction horizon P by the inequalities:

$$y_j^{min} \leq y_j[i] \leq y_j^{max}, \ i = \overline{k+1, k+P}, \ j = \overline{1, r}, \tag{6}$$

where y_j, y_j^{max}, y_j^{min} are the j-th components of the vectors $\mathbf{y}$, $\mathbf{y}_{max}$, $\mathbf{y}_{min}$ respectively. As mentioned above, these constraints can be violated, for example, at the initial instant of time or during transient processes. In this connection, let us introduce additional slack variables, that allow such violations, and represent inequalities (6) as follows:

$$y_j^{min} - \varepsilon_{ij} \leq y_j[k+i] \leq y_j^{max} + \varepsilon_{ij}, \ i = \overline{1, P}, \ j = \overline{1, r},$$
$$\varepsilon_{ij} \geq 0. \tag{7}$$

Here, the variable ε_{ij} determines the violation of the given range for the output variable y_j at the i-th step of the prediction horizon. If $\varepsilon_{ij} = 0$, then the output variable y_j is inside the range (6) at the instant of time $k + i$, and if $\varepsilon_{ij} > 0$, then the variable y_j is out of the admissible range (6).

Constraints (7) considered together over the entire prediction horizon P can be rewritten in vector form:

$$\overline{\mathbf{y}}_{min} - \varepsilon \leq \overline{\mathbf{y}}(\overline{\mathbf{u}}) \leq \overline{\mathbf{y}}_{max} + \varepsilon,$$
$$\varepsilon \geq \mathbf{0}, \tag{8}$$

where $\overline{\mathbf{y}}_{min} = (\mathbf{y}_{min}, ..., \mathbf{y}_{min})^T \in E^{rP}$, $\overline{\mathbf{y}}_{max} = (\mathbf{y}_{max}, ..., \mathbf{y}_{max})^T \in E^{rP}$ are auxiliary vectors, $\overline{\mathbf{y}} = (\mathbf{y}[k+1], ..., \mathbf{y}[k+P])^T \in E^{rP}$ is the output variables, $\overline{\mathbf{u}} = (\mathbf{u}[k], ..., \mathbf{u}[k+P-1])^T \in E^{mP}$ is the control input on the prediction horizon, $\varepsilon = (\varepsilon_{11}, \varepsilon_{12}, ..., \varepsilon_{1r}, ..., \varepsilon_{P1}, \varepsilon_{P2}, ..., \varepsilon_{Pr})^T \in E^{rP}$ is the vector of additional slack variables. It can be noted that the constraints (8) are nonlinear.

In accordance with (8), the admissible set of control actions on the prediction horizon takes the form:

$$\Omega = \{(\overline{\mathbf{u}}, \varepsilon) \in E^{(m+r)P} : \mathbf{u}[k+i] \in U, \ i = \overline{0, P-1},$$
$$\overline{\mathbf{y}}_{min} - \varepsilon \leq \overline{\mathbf{y}}(\overline{\mathbf{u}}) \leq \overline{\mathbf{y}}_{max} + \varepsilon, \varepsilon \geq \mathbf{0}\}. \tag{9}$$

The admissible set (9) is generally non-convex.

Let us introduce cost functional that determines the quality of control processes on the prediction horizon:

$$J = J(\Delta \overline{\mathbf{u}}, \varepsilon) = \sum_{j=0}^{P-1} \Delta \mathbf{u}[k+j]^T \mathbf{Q} \Delta \mathbf{u}[k+j] + \rho \varepsilon^T \mathbf{R} \varepsilon. \tag{10}$$

Here we use the notation: $\Delta \mathbf{u}[k] = \mathbf{u}[k] - \mathbf{u}[k-1]$ is a vector of control action at time instant k, $\Delta \overline{\mathbf{u}} = (\Delta \mathbf{u}[k], \Delta \mathbf{u}[k+1], ..., \Delta \mathbf{u}[k+P-1]) \in E^{mP}$ is an auxiliary vector of control actions on the prediction horizon, $\mathbf{Q}$ and $\mathbf{R}$ are positive definite diagonal weight matrices, $\rho > 0$ is a weight multiplier.

It is important to note that the functional (10) achieves its minimum zero value inside the range (6) if the control input is fixed. At the same time if the controlled output $\mathbf{y}$ is outside the range (6), then the desire to minimize the functional (10) leads to the retraction of these variables inside the range (6). On the other hand, if the controlled output $\mathbf{y}$ is already inside the range (6), then minimizing the functional (10) leads to minimal control actions that allow to keep variables inside the range.

Using the notation introduced earlier, functional (10) takes the form:

$$J = J(\Delta\bar{\mathbf{u}}, \varepsilon) = \Delta\bar{\mathbf{u}}^T\bar{\mathbf{Q}}\Delta\bar{\mathbf{u}} + \rho\varepsilon^T\mathbf{R}\varepsilon, \tag{11}$$

where $\bar{\mathbf{Q}} = diag(\mathbf{Q}, ..., \mathbf{Q})$ is a diagonal weight matrix with size $mP \times mP$.

Before proceeding to the optimal problem formulation, let us transform the constraints imposed on the control input. It can be seen that there is the following linear relationship between the vectors $\bar{\mathbf{u}}$ and $\Delta\bar{\mathbf{u}}$:

$$\bar{\mathbf{u}} = \mathbf{M}_0\mathbf{u}[k-1] + \mathbf{M}_1\Delta\bar{\mathbf{u}}, \tag{12}$$

where

$$\mathbf{M}_0 = \begin{pmatrix} \mathbf{E}_{n\times n} \\ \mathbf{E}_{n\times n} \\ \vdots \\ \mathbf{E}_{n\times n} \end{pmatrix}, \mathbf{M}_1 = \begin{pmatrix} \mathbf{E}_{n\times n} & 0 & \cdots & 0 \\ \mathbf{E}_{n\times n} & \mathbf{E}_{n\times n} & \cdots & 0 \\ \cdots & \cdots & \ddots & 0 \\ \mathbf{E}_{n\times n} & \mathbf{E}_{n\times n} & \cdots & \mathbf{E}_{n\times n} \end{pmatrix}.$$

Hence, we get that constraints (4), if they are considered over the entire prediction horizon, are equivalent to the system of linear inequalities

$$\mathbf{A}\Delta\bar{\mathbf{u}} \le \mathbf{b} + \mathbf{b}_u\mathbf{u}[k-1]. \tag{13}$$

Here the following notations are used:

$$\mathbf{A} = \begin{pmatrix} \mathbf{M}_1 \\ -\mathbf{M}_1 \\ \mathbf{E}_{mP\times mP} \\ -\mathbf{E}_{mP\times mP} \end{pmatrix}, \mathbf{b} = \begin{pmatrix} \bar{\mathbf{u}}_{max} \\ -\bar{\mathbf{u}}_{min} \\ \Delta\bar{\mathbf{u}}_{max} \\ -\Delta\bar{\mathbf{u}}_{min} \end{pmatrix}, \mathbf{b}_u = \begin{pmatrix} -\mathbf{M}_0 \\ \mathbf{M}_0 \\ \mathbf{0}_{mP\times m} \\ \mathbf{0}_{mP\times m} \end{pmatrix}.$$

Taking into account inequalities (13), the admissible set (9) can be represented in the form:

$$\Omega = \{(\Delta\bar{\mathbf{u}}, \varepsilon) \in E^{(m+r)P} : \mathbf{A}\Delta\bar{\mathbf{u}} \le \mathbf{b} + \mathbf{b}_u\mathbf{u}[k-1], $$
$$\bar{\mathbf{y}}_{min} - \varepsilon \le \bar{\mathbf{y}}(\Delta\bar{\mathbf{u}}) \le \bar{\mathbf{y}}_{max} + \varepsilon, \varepsilon \ge 0\}. \tag{14}$$

As a result, using equations (11) and (14), we can formulate the optimization problem of finding the best control strategy over the prediction horizon:

$$J_k = J_k(\Delta\bar{\mathbf{u}}, \varepsilon) = \Delta\bar{\mathbf{u}}^T\bar{\mathbf{Q}}\Delta\bar{\mathbf{u}} + \rho\varepsilon^T\mathbf{R}\varepsilon \rightarrow \min_{(\Delta\bar{\mathbf{u}}\ \varepsilon)\in\Omega\subseteq E^{(m+r)P}}. \tag{15}$$

The optimization problem (15) in the general case is a nonlinear programming problem, which can be solved, for example, using sequential quadratic programming method [11]. The solution of the problem (15) defines the optimal control sequnce $\Delta u^*[k]$, $\Delta u^*[k+1], ..., \Delta u^*[k+P-1]$ and the vector of slack variables ε^*. In accordance with the basic MPC idea, this optimal control sequence is used only at the current time instant k, that is, only the vector $\Delta u^*[k]$ is applied as the control input. At the next time instant the entire optimization process is repeated again, taking into accont all the constraints and the estimation of the state vector.

4 Application to Linear Systems with Time Delay

Let consider the application of the proposed control design approach for linear systems with a time delay. The mathematical model of the controlled object is represented by the system of linear equations in the form of transfer functions:

$$y_i = \sum_{j=1}^{m} \frac{k_{ij}}{T_{ij}s+1}e^{-\tau_{ij}s}u_j + \sum_{q=1}^{n_w} \frac{\tilde{k}_{iq}}{\tilde{T}_{iq}s+1}e^{-\tilde{\tau}_{iq}s}f_q, \quad i = \overline{1,r}. \tag{16}$$

Here we use the notations: y_i, $i = \overline{1,r}$ are the output variables, u_j, $j = \overline{1,m}$ are the control input components, f_q, $q = \overline{1,n_w}$ are the external disturbances, k_{ij} and $\tilde{k}_{iq}$ are the gain coefficients, T_{ij} and $\tilde{T}_{iq}$ are the time constants, τ_{ij} and $\tilde{\tau}_{iq}$ are the transport delays. It is assumed that in model (16) output variables y_i, $i = \overline{1,r}$ and external disturbances f_q, $q = \overline{1,n_w}$ are measured, the external disturbances are limited by $|f_q(t)| < M$, $q = \overline{1,n_w}$ and are constant or slowly varying.

It is important to note that the linear model (16) is used here as a starting point for designing a control system, since it is often utilized in practice by engineers for an approximate description of dynamic processes, for example, in chemical industry for distillation columns. Such models allow to describe the process in the vicinity of some operating conditions and can be easily obtained by means of simple experiments, such as transient response. Such an approach to modeling is of particular importance in the case when the true nonlinear model of the considered processes is very complex and cannot be retrieved without expensive investigations.

Let introduce the notations

$$\mathbf{y} = (y_1, ..., y_r)^T, \quad \mathbf{u} = (u_1, ..., u_m)^T, \quad \mathbf{f} = (f_1, ..., f_{n_w})^T$$

and suppose that model (16) represents the dynamics of the system in the vicinity of the operating point $\mathbf{y}(0) = \mathbf{y}_s, \mathbf{u}(0) = \mathbf{u}_s$. So, the true values of the output and input vectors at time t are given by

$$\tilde{\mathbf{y}}(t) = \mathbf{y}_s + \mathbf{y}(t),$$
$$\tilde{\mathbf{u}}(t) = \mathbf{u}_s + \mathbf{u}(t). \tag{17}$$

As before in formulas (3) and (4), let us consider the constraints imposed on the input and output variables:

$$\tilde{\mathbf{y}}_{min} \le \tilde{\mathbf{y}}(t) \le \tilde{\mathbf{y}}_{max},$$
$$\tilde{\mathbf{u}}_{min} \le \tilde{\mathbf{u}}(t) \le \tilde{\mathbf{u}}_{max}, \tag{18}$$
$$\Delta\tilde{\mathbf{u}}_{min} \le \dot{\tilde{\mathbf{u}}}(t) \le \Delta\tilde{\mathbf{u}}_{max},$$

where $\tilde{\mathbf{y}}_{min}, \tilde{\mathbf{y}}_{max}, \tilde{\mathbf{u}}_{min}, \tilde{\mathbf{u}}_{max}, \Delta\tilde{\mathbf{u}}_{min}, \Delta\tilde{\mathbf{u}}_{max}$ are given vectors.

The control objective is to keep the output variables $\tilde{\mathbf{y}}$ within the given range, taking into account the limited control resources in accordance with the inequalities (18). Thus, let us consider the problem of synthesis of the control law, which ensures the achievement of this objective subject to the mathematical model (16), based on the MPC approach, which was discussed earlier.

To this end, let transform the mathematical model (16) to a discrete form. It can be noted that each term in the first sum of equation (16) can be represented by a scalar differential equation

$$\dot{x}_{ij} = -\frac{1}{T_{ij}}x_{ij} + \frac{k_{ij}}{T_{ij}}u_j(t - \tau_{ij}),$$
$$y_{ij} = x_{ij}. \tag{19}$$

A similar representation is valid for the terms of the second sum of model (16). Since the external disturbances are constant or slowly varying, each of them can be approximately described by the equation

$$\dot{f}_q = 0, \quad q = \overline{1, n_w}. \tag{20}$$

Let perform the discretization of the model (19) with the sampling interval T_s [12]. As a result, we get

$$x_{ij}[k + 1] = a_{ij}x_{ij}[k] + b_{ij}u_j[k - \tau_{ij}^d],$$
$$y_{ij}[k] = x_{ij}[k], \tag{21}$$

where $a_{ij} = e^{-T_s/T_{ij}}$, $b_{ij} = (1 - e^{-T_s/T_{ij}})k_{ij}$ are real numbers, $\tau_{ij}^d \ge 1$ is an integer number equal to the number of sampling intervals corresponding to the time delay τ_{ij}. A similar discrete-time model can be obtained for nonzero terms of the second sum of model (16).

Discretization of equation (20) leads to the equality

$$f_q[k + 1] = f_q[k], q = \overline{1, n_w}. \tag{22}$$

Gathering all equations (21) and (22) and extending the state vector with additional components representing the time delay, we obtain the following discrete linear time-invariant model

$$\mathbf{x}[k + 1] = \mathbf{A}\mathbf{x}[k] + \mathbf{B}\mathbf{u}[k],$$
$$\mathbf{y}[k] = \mathbf{C}\mathbf{x}[k]. \tag{23}$$

Here we use notations: $\mathbf{x} = \begin{pmatrix} \mathbf{x}_m & \mathbf{x}_d & \mathbf{x}_{de} \end{pmatrix}^T$ is a state vector, where $\mathbf{x}_m$ corresponds to equations (21) and (22), $\mathbf{x}_d$ and $\mathbf{x}_{de}$ are the vectors, representing the time delays for the control input and disturbances respectively.

Thus, the obtained discrete-time model (23) is equivalent to the original model (16). It is important to note that this model takes into account time delays by expanding the state vector.

Now, let us proceed with the control design based on the MPC approach that was discussed earlier. To this end, let form the predictive model. In accordance with (23) and (5), we get

$$\mathbf{x}[i+1] = \mathbf{Ax}[i] + \mathbf{Bu}[i], \ i = k+j, \ j = 0, 1, 2, ..., \ \mathbf{x}[k] = \tilde{\mathbf{x}}[k],$$
$$\mathbf{y}[i] = \mathbf{Cx}[i]. \tag{24}$$

In this particular case of the linear predictive model (24), the approach to control design proposed earlier can be described in more detail. So, it can be noted that there is a linear relationship between the control input $\overline{\mathbf{u}}$ on the prediction horizon P and the corresponding output of the predictive model $\overline{\mathbf{y}}$, determined by the equality

$$\overline{\mathbf{y}} = \mathbf{Lx}[k] + \mathbf{M}\overline{\mathbf{u}}. \tag{25}$$

Here the matrices $\mathbf{L}$ and $\mathbf{M}$ are given by

$$\mathbf{L} = \begin{pmatrix} \mathbf{CA} \\ \mathbf{CA}^2 \\ \vdots \\ \mathbf{CA}^P \end{pmatrix}, \ \mathbf{M} = \begin{pmatrix} \mathbf{CB} & 0 & \cdots & 0 \\ \mathbf{CAB} & \mathbf{CB} & \cdots & 0 \\ \cdots & \cdots & \ddots & \cdots \\ \mathbf{CA}^{P-1}\mathbf{B} & \mathbf{CA}^{P-2}\mathbf{B} & \cdots & \mathbf{CB} \end{pmatrix}.$$

Substituting (12) into (25), we get

$$\overline{\mathbf{y}} = \overline{\mathbf{y}}(\Delta\overline{\mathbf{u}}) = \mathbf{Lx}[k] + \tilde{\mathbf{M}}_0\mathbf{u}[k-1] + \tilde{\mathbf{M}}_1\Delta\overline{\mathbf{u}}, \tag{26}$$

where $\tilde{\mathbf{M}}_0 = \mathbf{MM}_0$, $\tilde{\mathbf{M}}_1 = \mathbf{MM}_1$ are auxiliary matrices. As a result, formula (26) allows us to calculate the predicted dynamic of the object $\overline{\mathbf{y}}$ for the given control actions $\Delta\overline{\mathbf{u}}$ on the horizon P.

Now let's form an admissible set on the prediction horizon. To this end, we represent inequalities (18) in discrete time:

$$\mathbf{y}_{min} \leq \mathbf{y}[k] \leq \mathbf{y}_{max}, \ \mathbf{u}_{min} \leq \mathbf{u}[k] \leq \mathbf{u}_{max}, \ \Delta\mathbf{u}_{min} \leq \Delta\mathbf{u}[k] \leq \Delta\mathbf{u}_{max}, \tag{27}$$

where, taking into account (17), the following auxiliary vectors are intoduced:

$$\mathbf{y}_{max} = \tilde{\mathbf{y}}_{max} - \mathbf{y}_s, \ \mathbf{y}_{min} = \tilde{\mathbf{y}}_{min} - \mathbf{y}_s,$$
$$\mathbf{u}_{max} = \tilde{\mathbf{u}}_{max} - \mathbf{u}_s, \ \mathbf{u}_{min} = \tilde{\mathbf{u}}_{min} - \mathbf{u}_s,$$
$$\Delta\mathbf{u}_{max} = T_s\Delta\tilde{\mathbf{u}}_{max}, \ \Delta\mathbf{u}_{min} = T_s\Delta\tilde{\mathbf{u}}_{min}.$$

Using (26) and (27), we can rewrite constraints (8) as follows

$$\overline{\mathbf{y}}_{min} - \varepsilon \leq \mathbf{L}\mathbf{x}[k] + \tilde{\mathbf{M}}_0\mathbf{u}[k-1] + \tilde{\mathbf{M}}_1\Delta\overline{\mathbf{u}} \leq \overline{\mathbf{y}}_{max} + \varepsilon, \varepsilon \geq \mathbf{0}. \qquad (28)$$

Based on representation (14) and inequalities (13), (27) and (28), we obtain a modified admissible set for the problem under consideration:

$$\Omega = \left\{ (\Delta\overline{\mathbf{u}}, \varepsilon) \in E^{(m+r)P} : \mathbf{A}\begin{pmatrix} \Delta\overline{\mathbf{u}} \\ \varepsilon \end{pmatrix} \leq \mathbf{b} \right\}. \qquad (29)$$

Here we use notations:

$$\mathbf{A} = \begin{pmatrix} \mathbf{M}_1 & \mathbf{0} \\ -\mathbf{M}_1 & \mathbf{0} \\ \mathbf{E} & \mathbf{0} \\ -\mathbf{E} & \mathbf{0} \\ \tilde{\mathbf{M}}_1 & -\mathbf{E} \\ -\tilde{\mathbf{M}}_1 & -\mathbf{E} \\ \mathbf{0} & -\mathbf{E} \end{pmatrix}, \ \mathbf{b} = \begin{pmatrix} \overline{\mathbf{u}}_{max} \\ -\overline{\mathbf{u}}_{min} \\ \Delta\overline{\mathbf{u}}_{max} \\ -\Delta\overline{\mathbf{u}}_{min} \\ \overline{\mathbf{y}}_{max} \\ -\overline{\mathbf{y}}_{min} \\ \mathbf{0} \end{pmatrix} + \begin{pmatrix} -\mathbf{M}_0 \\ \mathbf{M}_0 \\ \mathbf{0} \\ \mathbf{0} \\ -\tilde{\mathbf{M}}_0 \\ \tilde{\mathbf{M}}_0 \\ \mathbf{0} \end{pmatrix}\mathbf{u}[k-1] + \begin{pmatrix} \mathbf{0} \\ \mathbf{0} \\ \mathbf{0} \\ \mathbf{0} \\ -\mathbf{L} \\ \mathbf{L} \\ \mathbf{0} \end{pmatrix}\mathbf{x}[k].$$

It can be noted that the inequalities in (29) are linear, and the right part depends on the current state $\mathbf{x}[k]$ and control value $\mathbf{u}[k-1]$ on the previous step.

As a result, let us form the optimization problem of finding the best control strategy on the prediction horizon:

$$J_k = J_k\left(\Delta\overline{\mathbf{u}}, \varepsilon\right) = \left(\Delta\overline{\mathbf{u}} \ \varepsilon\right)^T \tilde{\mathbf{H}}\begin{pmatrix} \Delta\overline{\mathbf{u}} \\ \varepsilon \end{pmatrix} \rightarrow \min_{\left(\Delta\overline{\mathbf{u}} \ \varepsilon\right) \in \Omega \subseteq E^{(m+r)P}}, \qquad (30)$$

where Ω is an admissible set (29), $\tilde{\mathbf{H}} = \begin{pmatrix} \overline{\mathbf{Q}} & \mathbf{0} \\ \mathbf{0} & \rho\mathbf{R} \end{pmatrix}$ is a symmetric positive definite matrix. Optimization task (30) is a quadratic programming problem that can be efficiently solved numerically in real time.

In accordance with MPC strategy, the solution of the optimization problem (30) is applied to the control object only at the current step k, and at the next instant of time $k+1$ the entire process of the state estimation and optimization is repeated again.

5   Simulation Example

Let us consider the application of the proposed method to the oil refining process with rectification column as plant. To provide the required quality of the oil products there is no need to track the specific values, it is enough just to keep the controlled variables in the specific range. The mathematical model of the considered rectification column is the following:

$$y_1 = \frac{1}{0.5s+1}u_2 + \frac{1}{0.5s+1}u_3 + \frac{1}{0.5s+1}u_4 + \frac{1}{0.5s+1}u_5,$$

$$y_2 = \frac{600}{1.5s+1}e^{-0.2s}u_1 - \frac{0.05}{2s+1}e^{-0.2s}u_2,$$

$$y_3 = \frac{600}{1.5s+1}e^{-0.2s}u_1 - \frac{0.05}{2s+1}e^{-0.2s}u_3, \qquad (31)$$

$$y_4 = \frac{600}{1.5s+1}e^{-0.2s}u_1 - \frac{0.05}{2s+1}e^{-0.2s}u_4,$$

$$y_5 = \frac{600}{1.5s+1}e^{-0.2s}u_1 - \frac{0.05}{2s+1}e^{-0.2s}u_5.$$

Time and delay constants for the above model have minutes as units.

Initial values and constraints for output and control variables are given in Table 1 and Table 2. Let us note that all of the initial output variables are out of the constraints.

Table 1. Constraints for the output variables

Variable	$\mathbf{y}_s$	$\tilde{\mathbf{y}}_{min}$	$\tilde{\mathbf{y}}_{max}$
$\tilde{y}_1$	10069	9800	9900
$\tilde{y}_2$	841	830	835
$\tilde{y}_3$	841	830	835
$\tilde{y}_4$	841	830	835
$\tilde{y}_5$	841	830	835

Let us take the discretization step $T_s = 0.2$ min and the prediction horizon $P = 25$ which corresponds to 5 min. We will use the full vector of the slack variables (i.e. one slack variable for each output variable on each step of the prediction horizon). Dimension of the corresponding optimization problem is $N = 2 \cdot P \cdot n = 2 \cdot 25 \cdot 5 = 250$. Let us also take the following weight parameters for the optimization problem functional: $\rho = 100$, $\mathbf{Q} = \mathbf{E}_{5 \times 5}$, $\mathbf{R} = \mathbf{E}_{125 \times 125}$.

Table 2. Constraints for the control variables

Variable	$\mathbf{u}_s$	$\tilde{\mathbf{u}}_{min}$	$\tilde{\mathbf{u}}_{max}$	$\Delta\tilde{\mathbf{u}}_{min}$	$\Delta\tilde{\mathbf{u}}_{max}$
$\tilde{u}_1$	1.22	1.1	1.3	−0.01	0.01
$\tilde{u}_2$	2517	2400	2600	−50	50
$\tilde{u}_3$	2517	2400	2600	−50	50
$\tilde{u}_4$	2517	2400	2600	−50	50
$\tilde{u}_5$	2517	2400	2600	−50	50

The dynamics of the output variables for this case is shown on the Fig. 1 and the dynamics of the control variables is shown on the Fig. 2. It can be seen that the values of the output variables successfully move into the regions defined by the constraints and the control variables are kept in their own limits.

It is worth noting that the weight coefficient ρ defines the intensity of the control action – bigger values of ρ correspond to more intensive control. Experiments with the smaller values of ρ show that the output variables can be still put into the defined limits, but it takes more time.

Let us also consider the case when there are no delays in the prediction model (but there are delays in the plant still) with just the same parameters. Figure 3 and Fig. 4 demonstrate the dynamics in this case. Besides the overall difference in the dynamics we can mark out the important fact that in this case controlled variables y_3 and y_5 violate the constraints – y_3 reaches the value 829.619 and y_5 reaches the value 829.9075 with lower constraint of the value 830 for both. This is because the prediction model does not consider the delays and therefore not adequately represents the plant. As a result, the optimization problem gives a solution which leads to the violation of the constraints. The violation could have been even more significant if the delay value was bigger or if the value of the ρ was different.

In these experiments we used Matlab 2018 quadprog function for solving quadratic programming problem. It implements interior-point convex algorithm. Average time for solving the optimization problem at each control step was about 350–400 ms with AMD Ryzen 7 3700X CPU.

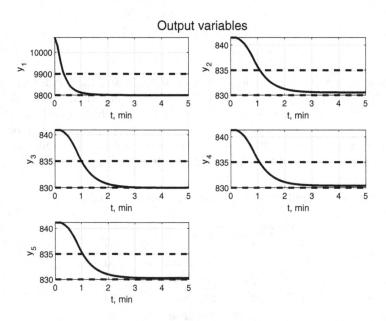

Fig. 1. Output variables dynamics.

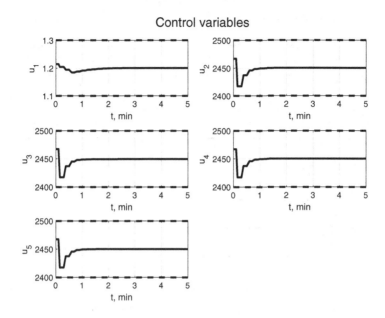

Fig. 2. Control variables dynamics.

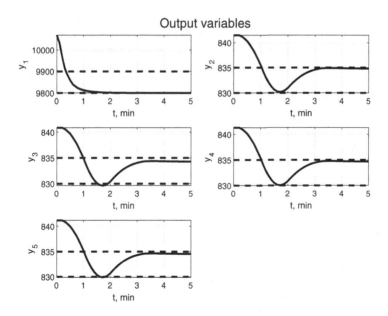

Fig. 3. Output variables dynamics with the prediction model without delays.

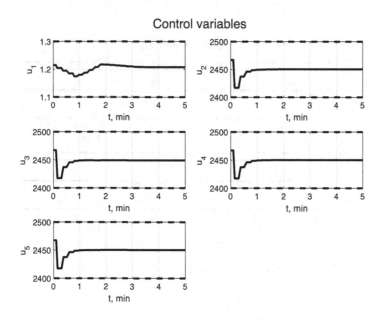

Fig. 4. Control variables dynamics with the prediction model without delay.

6   Conclusion

In this paper the control design approach to keep the output variables in a given range is proposed. This approach is based on the ideology of MPC and allows taking into account nonlinear dynamics, time delay and constraints imposed on input variables. The control action on the prediction horizon is chosen in such a way as to minimize the introduced functional, which includes a term responsible for the intensity of control and a penalty term for leaving the specified range. The particular case of a linear model with time delay is considered. It is shown that the optimization problem on the prediction horizon, which must be solved repeatedly at each instant of discrete time, in this case is a quadratic programming problem. The application of the approach to the oil refining process with a rectification column is demonstrated and the obtained simulation results are discussed.

References

1. Burdick, D.L., Leffler, W.L.: Petrochemicals in nontechnical language, 4th edn. PennWell Corporation, Oklahoma, USA (2010)
2. Lahiri, S.K.: Multivariable Predictive Control: Applications in Industry. John Wiley & Sons Publisher, Hoboken, NJ, USA (2017)
3. Kouvaritakis, B., Cannon, M.: Model Predictive Control: Classical, Robust and Stochastic. Spinger, Cham (2016)

4. Camacho, E.F., Bordons, C.: Model Predictive Control, 2nd edn. Springer-Verlag, London (2007)
5. Faulwasser, T., Müller, M.A., Worthmann, K. (eds.): Recent Advances in Model Predictive Control. LNCIS, vol. 485. Springer, Cham (2021). https://doi.org/10.1007/978-3-030-63281-6
6. Sotnikova, M.: Plasma stabilization based on model predictive control. Int. J. Mod. Phys. A **24**(5), 999–1008 (2009)
7. Sotnikova, M.: Ship dynamics control using predictive models. IFAC Proc. Volumes (IFAC-PapersOnline) **45**(27), 250–255 (2012)
8. Corriou, J.-P.: Process Control. Springer, Cham (2018). https://doi.org/10.1007/978-3-319-61143-3
9. Sotnikova, M.V.: Digital control design based on predictive models to keep the controlled variables in a given range. Vestnik Sankt-Peterburgskogo Universiteta, Prikladnaya Matematika, Informatika, Protsessy Upravleniya **15**(3), 397–409 (2019)
10. Sotnikova M.V., Sevostyanov R.A.: Digital control of output variables in a given range considering delay. Vestnik of Saint Petersburg University, Applied Mathematics, Computer Science, Control Processes 17(4), 449–463 (2021)
11. Nocedal, J., Wright, S.J.: Numerical Optimization, 2nd edn. Springer-Verlag, New York (2006)
12. Donzellini, G., Oneto, L., Ponta, D., Anguita, D.: Introduction to Digital Systems Design. Springer, Cham (2019). https://doi.org/10.1007/978-3-319-92804-3

An Optimization Approach to the Robust Control Law Design for a Magnetic Levitation System

Anastasia Tomilova(✉) ⓘD

Saint-Peterburg State University, Saint-Peterburg, Russia
nt4815@gmail.com

Abstract. This paper is devoted to the problem of robust control law design taking into account constraints imposed on the performance of the closed-loop nominal system. The magnetic levitation system is chosen as the control object, since its mathematical model is not accurate due to significant difficulties in describing the magnetic field, especially near the surface of the magnet. The proposed approach to control design is aimed at providing the best robust properties of the closed-loop system while maintaining acceptable performance indices of the nominal model.

This aim is achieved by solving the optimization problem of expanding the frequency boundaries of robust stability on the admissible set, including regulators that provide the eigenvalues of the closed-loop system for nominal model in the given region at left half-plane. The specific feature of the developed controller is its multi-purpose structure that allows to achieve desired system performance in different regimes. It is shown that the solution of the stated optimization problem is reduced to unconstrained nonlinear programming. The obtained results is compared with linear quadratic regulator. The simulation examples are presented in MATLAB/Simulink environment and discussed.

Keywords: Robust stability · Control system · Magnetic levitation plant · Optimization · Frequency boundary · Multi-purpose structure

1 Introduction

Robustness analysis of control system is one of the significant question of control theory. This question is paid attention when modeling dynamics of moving objects. A lot of scientific research, articles and books are being conducted and devoted to this topic, for example [1–5]. Usually a mathematical model defines only an approximate dynamics of a real control object, and parameters values of a real object often differ from parameters values of a mathematical model. Such differences can influence on the object dynamic. Therefore the question arises about the preservation of certain properties of the control system. If small variation of model parameters don't lead to fundamental changes in dynamic properties of the closed-loop system then this control system can be believed as operational.

© The Author(s), under exclusive license to Springer Nature Switzerland AG 2022
Y. Kochetov et al. (Eds.): MOTOR 2022, CCIS 1661, pp. 286–299, 2022.
https://doi.org/10.1007/978-3-031-16224-4_20

The design of a mathematical model is a very complex process in practice. This is because in the design process a number of difficulties arise, such as insufficient information about the control object, the desire to simplify the mathematical formalization, the presence of various unaccounted factors and so on. All this affects the operation of the control system and leads to the need to analyze robust properties and synthesize a robust control law.

There are two main classes of uncertainties in a mathematical model: parametric and unstructured. In the first case, approaches based on linear matrix inequalities [6,7], Kharitonov's theorem on the stability of a family of polynomials [8] and others [9–12] are used for the analysis and synthesis of control systems. In the second case, frequency methods are usually applied [13,14].

The main contribution of this paper is the development of the approach to optimizing the robust properties of the control law. Its aim is to provide the best robust properties of the closed-loop system while maintaining acceptable performance indices of the nominal model. This aim is achieved by solving the optimization problem of expanding the frequency boundaries of robust stability on the admissible set, including regulators that provide the eigenvalues of the closed-loop system for nominal model in the given region at left half-plane. The specific feature of the developed controller is its multi-purpose structure [4,15, 16] that allows to achieve desired system performance in different regimes. It is shown that the solution of the stated optimization problem is reduced to unconstrained nonlinear programming.

In this paper, the magnetic levitation system is chosen as the control object [3,17], since its mathematical model is not accurate due to significant difficulties in describing the magnetic field, especially near the surface of the magnet. The obtained results are illustrated by the problem of ball vertical position stabilization. The simulation examples are presented in MATLAB/Simulink environment and compared with linear quadratic regulator [18].

2   Problem Formulation

In this section mathematical model of magnetic levitation system and robust control law for this system are considered. And also the problem formulation of robust control law synthesis is carried out.

2.1   Mathematical Model of the Magnetic Levitation System

Let consider the mathematical model of the magnetic levitation system [3,17]. The general scheme of the magnetic levitation system is shown on Fig. 1. The operation of this system is as follows: a steel ball is in the air under the action of two forces: gravitational force F_g directed vertically downwards and electromagnetic field force F_m created by an electromagnet.

On the scheme of the magnetic levitation system the following designations are accepted: I – current of the electromagnet, V – voltage applied to electromagnet, R – resistance of the electromagnet loop, L – inductance of the electromagnet loop, x_b – ball position.

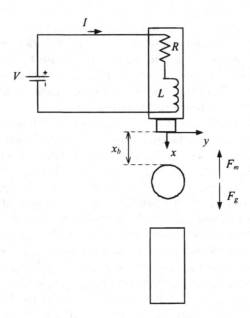

Fig. 1. Scheme of the magnetic levitation system

The control objective is the stabilization of the ball position in the particular point between the electromagnet and the pedestal by means of the controlled voltage which is applied to the electromagnet.

The origin of the coordinate system Oxy is displaced on the electromagnetic surface so that axe Ox is directed downward. Control input is a voltage applied to electromagnet.

Let introduce variables: $x_1 = x_b$, $x_2 = \dot{x}_b$, $x_3 = I$, $u = V$ and write the system of nonlinear differential equations describing the mathematical model of the magnetic levitation system [3,17]. Let compose these equations based on the laws of electrical circuits and Newton's second law, while taking into account the expressions for the electromagnetic field force. As a result, we get:

$$\dot{x}_1 = x_2,$$
$$\dot{x}_2 = g - \frac{1}{2} \cdot \frac{K_m x_3^2}{M x_1^2},$$
$$\dot{x}_3 = -\frac{R}{L} x_3 + \frac{1}{L} u. \tag{1}$$

Here M – mass of the ball, K_m – magnetic field constant, g – gravitational constant. In the presented system the ball position x_b is measured with using the optical sensor built into the pedestal. Let write the measurement equation for system (1):

$$y = x_1. \tag{2}$$

Equations (1), (2) only approximately reflect the dynamics of the system, since there are significant difficulties in the formalized description of the electromagnetic field, taking into account additional unaccounted for effects on its part. In this regard, let consider the system in deviations for system (1) and compose the linear approximation equations:

$$\dot{\bar{x}}_1 = \bar{x}_2,$$
$$\dot{\bar{x}}_2 = \frac{2g}{x_{b0}}\bar{x}_1 - \frac{2g}{I_0}\bar{x}_3,$$
$$\dot{\bar{x}}_3 = -\frac{R}{L}\bar{x}_3 + \frac{1}{L}\bar{u},$$
$$\bar{y} = \bar{x}_1.$$

(3)

The variables $\bar{x}_1 = x_1 - x_{b0}$, $\bar{x}_2 = x_2$, $\bar{x}_3 = x_3 - I_0$, $\bar{u} = u - u_0$ describe the system in deviations from the equilibrium position $(x_{b0}, 0, I_0)$, where x_{b0} is the position of the ball corresponding to the voltage $u = u_0$, the value of the current I_0 is found from the relationship: $I_0 = \sqrt{2gM/K_m}x_{b0}$.

Let write the system (3) in matrix form

$$\dot{\bar{x}} = A\bar{x} + B\bar{u},$$
$$\bar{y} = C\bar{x}.$$

(4)

Here we use the notations:

$$A = \begin{pmatrix} 0 & 1 & 0 \\ a_{21} & 0 & a_{23} \\ 0 & 0 & a_{33} \end{pmatrix}, \quad B = \begin{pmatrix} 0 \\ 0 \\ b \end{pmatrix}, \quad C = \begin{pmatrix} 1 & 0 & 0 \end{pmatrix},$$

where $a_{21} = \dfrac{2g}{x_{b0}}$, $a_{23} = -\dfrac{2g}{I_0}$, $a_{33} = -\dfrac{R}{L}$, $b = \dfrac{1}{L}$ are model coefficients.

2.2 Structure of the Dynamic Regulator

In this paper a multi-purpose approach to the control law synthesize is used [4, 15, 16]. This makes it possible to provide the desired quality of a closed-loop system functioning in a various modes. The main indicators of the quality considered in this paper are the transient time and oscillability index.

Let construct a dynamic regulator with a multi-purpose structure [4, 15, 16] represented by the formulas

$$\dot{z} = Az + B\bar{u} + G(\bar{y} - Cz),$$
$$\bar{u} = Kz,$$

(5)

where z is the state vector of the asymptotic observer, which allows to recover information about the unknown components of the state vector $\bar{x}$.

For the nonlinear system (1), the control action u takes the following form

$$u = u_c(x_{b0}) + Kz,$$

where $u_c = u_c(x_{b0})$ is a constant term that compensates for the force of gravity in the equilibrium position x_{b0}.

Configurable elements of the control law (5) with multi-purpose structure are

- components of the vector $\mathbf{K}$ of the basic control law $\bar{u} = \mathbf{K}\bar{x}$, which determine the dynamics of the proper motion of the control system;
- components of the vector $\mathbf{G}$ of the asymptotic observer, which ensures the Hurwitz of the matrix $\mathbf{A} - \mathbf{GC}$.

The search for configurable elements is carried out in such a way as to achieve the desired quality of functioning of the closed-loop control system (3), (5) in various modes, and at the same time ensure the best robust properties of the control system.

2.3   Problem Formulation of the Robust Control Law Synthesis

We will assume that the components of the matrix $\mathbf{A}$ and vector $\mathbf{B}$ of the mathematical model (4) are determined inaccurately and vary within the specified limits, namely $\pm 15\%$ of the nominal values, that is

$$a_{21} \in [\underline{a}_{21}, \bar{a}_{21}], \ a_{23} \in [\underline{a}_{23}, \bar{a}_{23}], \ a_{33} \in [\underline{a}_{33}, \bar{a}_{33}], \ b \in [\underline{b}, \bar{b}]. \tag{6}$$

Here $\underline{a}_{21}, \underline{a}_{23}, \underline{a}_{33}, \underline{b}, \bar{a}_{21}, \bar{a}_{23}, \bar{a}_{33}, \bar{b}$ are the lower and upper boundaries of the segments, determined by the following formulas:

$$\underline{a}_{21} = a_{21}^0 - 0.15a_{21}^0, \ \bar{a}_{21} = a_{21}^0 + 0.15a_{21}^0,$$

$$\underline{a}_{23} = a_{23}^0 - 0.15a_{23}^0, \ \bar{a}_{23} = a_{23}^0 + 0.15a_{23}^0,$$

$$\underline{a}_{33} = a_{33}^0 - 0.15a_{33}^0, \ \bar{a}_{33} = a_{33}^0 + 0.15a_{33}^0,$$

$$\underline{b} = b_0 - 0.15b_0, \ \bar{b} = b_0 + 0.15b_0,$$

where $a_{21}^0, a_{23}^0, a_{33}^0, b_0$ are nominal values of the mathematical model (4).

The problem is to develop a multi-purpose control law (5) with the best robust properties for a magnetic levitation system (4) subject to the constraints imposed on the quality of the transient processes for the nominal model, namely the transient time and oscillability index. The considered optimization problem is to expand the frequency boundaries of robust stability and at the same time keep the quality of transient processes within acceptable limits, taking into account the uncertainty (6).

3   An Optimization Approach to the Robust Control Law Synthesis

3.1   Robust Stability Analysis of the Control System

Let apply the Laplace transform and represent the mathematical model (4) and the control law (5) in the tf-form. As a result, we get

$$\bar{y} = P_n(s)\bar{u}, \tag{7}$$

$$\bar{u} = K(s)\bar{y}. \tag{8}$$

Here $P_n(s)$ is the nominal transfer function of the object and $K(s)$ is the transfer function of the controller, which are determined by the equalities

$$P_n(s) = \mathbf{C}(s\mathbf{E} - \mathbf{A})^{-1}\mathbf{B},$$

$$K(s) = (1 - \mathbf{K}(s\mathbf{E} - \mathbf{A} + \mathbf{GC})^{-1}\mathbf{B})^{-1}\mathbf{K}(s\mathbf{E} - \mathbf{A} + \mathbf{GC})^{-1}\mathbf{G}.$$

We will assume that the transfer function $K(s)$ of the dynamic controller (5) does not change during operation, and the transfer function $P_n(s)$ of the control object (4) has uncertainty. Therefore, the controller (8) actually closes not an object with a model (7), but another object with a model:

$$\bar{y} = P(s)\bar{u},$$

the transfer function $P(s)$ of which differs from the nominal ones.

For the robust control law (8), we can construct the boundaries of robust stability [1,4] represented by the formulas:

$$A_{up}(\omega) = \left(1 + \frac{1}{|T(j\omega)|}\right)|P_n(j\omega)|,$$
$$A_{lo}(\omega) = \left(1 - \frac{1}{|T(j\omega)|}\right)|P_n(j\omega)|. \tag{9}$$

Here $A_{up}(\omega)$ and $A_{lo}(\omega)$ are the upper and lower bounds, $T(j\omega)$ is an auxiliary function given by

$$T(j\omega) = K(j\omega)[1 + P_n(j\omega)K(j\omega)]^{-1}P_n(j\omega).$$

Now let build the frequency boundaries within which the frequency response of the control object (4) varies when the model coefficients change in the given ranges (6). To this end, let introduce into consideration the following functions

$$M_{up}(\omega) = \max_{\mathbf{a}\in\Omega} M(\omega, \mathbf{a}),$$
$$M_{lo}(\omega) = \min_{\mathbf{a}\in\Omega} M(\omega, \mathbf{a}), \tag{10}$$

where $M(\omega, \mathbf{a}) = |P(j\omega, \mathbf{a})|$ is the frequency response, $\mathbf{a}$ is an auxiliary vector consisting of variable elements of the object model (4), Ω is an admissible set defined by expressions (6), $M_{up}(\omega)$ and $M_{lo}(\omega)$ are upper and lower bounds respectively.

Thus, in order to guarantee the preservation of the robust stability for the closed-loop system (4), (5), it is necessary and sufficient that the boundaries (9) of the robust stability for the controller (5) include the frequency range (10).

3.2   Synthesis of Robust Control Law

Let consider the problem of optimizing the robust control law (5) for the magnetic levitation system (4) under constraints on the quality of the transient processes, namely, on the transient time and oscillability index. To simplify the formalization of constraints, let assume that they are set by introducing the special region of the roots location on the complex plane.

It can be noticed that the transfer function $K(s)$ depends on the choice of configurable elements of the multi-purpose control law (5). Let combine these elements into a vector of parameters $\mathbf{h} = \{\mathbf{K}, \mathbf{G}\}$, then $K(s, \mathbf{h})$ is the transfer function of the regulator.

In order to optimize the robust properties of the control system, we will maximize the width of the frequency range of robust stability, which, according to [1], is a function of frequency and is equal to

$$W(\omega, \mathbf{h}) = \frac{1}{|T(j\omega, \mathbf{h})|}.$$

Maximizing the function $W(\omega, \mathbf{h})$ with respect to the vector $\mathbf{h}$ is equivalent to minimizing $|T(j\omega, \mathbf{h})|$. In this regard, we will consider functional:

$$J(\mathbf{h}) = \int_{\omega_1}^{\omega_2} |T(j\omega, \mathbf{h})| d\omega, \tag{11}$$

where $\omega_1 \geq 0$ and $\omega_2 \geq 0$ are the boundaries of operating range. Functional (11) determines the integral characteristic of the width of frequency range over the entire frequency interval $[\omega_1, \omega_2]$.

Thus, let consider the problem of modal parametric optimization, which can be represented as

$$J = J(\mathbf{h}) \longrightarrow \min_{\mathbf{h} \in \Omega_H}. \tag{12}$$

Here the set Ω_H of admissible parameters is such that all the roots of the characteristic polynomial of a closed-loop system (4), (5) lie in a given region of the complex plane. This region is shown in Fig. 2 and is set as follows:

$$C_\Delta = \{s = x \pm jy \in \mathbf{C}^1 : x \leq -\bar{\alpha}, 0 \leq y \leq (-x - \bar{\alpha}) \tan \bar{\beta}\}, \tag{13}$$

where $x = Re(s)$, $y = Im(s)$, s is an arbitrary point on the complex plane, $\bar{\alpha} > 0$ and $0 \leq \bar{\beta} < \pi/2$ are given real numbers.

Let formulate the statement that allows us to parametrize the region C_Δ using vectors of a real Euclidean space. In the book [1] a similar statement is given for a simpler definition of the region C_Δ.

Theorem 1. *For any vector $\gamma \in E^{n_d}$ all the roots of the polynomial*

$$\Delta^*(s, \gamma) = \begin{cases} \widetilde{\Delta}^*(s, \gamma), & \text{if } n_d \text{ is even;} \\ (s + a_{d+1}(\gamma, \bar{\alpha})) \widetilde{\Delta}^*(s, \gamma), & \text{if } n_d \text{ is odd,} \end{cases} \tag{14}$$

belong to the region C_Δ and, vice versa, if all the roots of some polynomial $\Delta(s)$ belong to this region then we can specify such a vector $\gamma \in E^{n_d}$ that the identity $\Delta(s) \equiv \Delta^*(s, \gamma)$ holds. The polynomial $\widetilde{\Delta}^*(s, \gamma)$ is given by the formula

$$\widetilde{\Delta}^*(s, \gamma) = \prod(s^2 + a_i^1(\gamma, \bar{\alpha})s + a_i^0(\gamma, \bar{\alpha})), \quad d = [n_d/2],$$

$$a_i^1(\gamma, \bar{\alpha}) = 2\bar{\alpha} + 2\gamma_{i1}^2,$$

$$a_i^0(\gamma, \bar{\alpha}) = \bar{\alpha}^2 + 2\gamma_{i1}^2\bar{\alpha} + f(\gamma_{i2})\frac{\gamma_{i1}^4}{\cos^2(\bar{\beta})}, \quad i = \overline{1, d}, \tag{15}$$

$$a_{d+1}(\gamma, \bar{\alpha}) = \gamma_{d0}^2 + \bar{\alpha},$$

$$\gamma = \{\gamma_{11}, \gamma_{12}, \gamma_{21}, \gamma_{22}, \ldots, \gamma_{d1}, \gamma_{d2}, \gamma_{d0}\}.$$

The function $f(\cdot) : (-\infty, +\infty) \longrightarrow (0,1)$ satisfies the condition for the existence of an inverse function in the entire region of the definition.

Thus, for any choice of vector $\gamma \in E^{n_d}$ the roots of the polynomial $\Delta^*(s, \gamma)$ are located inside the region C_Δ and, vice versa, each fixed distribution of roots in the region C_Δ corresponds to a single vector $\gamma \in E^{n_d}$ [1]. That is, there is a one-to-one correspondence between the location of the roots of the characteristic polynomial of the closed-loop system (4), (5) inside the region C_Δ and the values of the vector $\gamma \in E^{n_d}$.

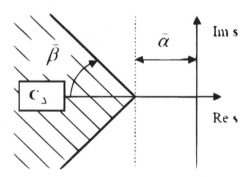

Fig. 2. The region C_Δ

Let use Theorem 1 to develop a computational method for solving the parametric synthesis problem (12). Let define an arbitrary vector $\gamma \in E^{n_d}$ and form corresponding auxiliary polynomial $\Delta^*(s, \gamma)$ using the formulas (14), (15).

We will search configurable elements of the regulator (5), united in the vector **h**, so that the following identity is ensured

$$\Delta(s, \mathbf{h}) \equiv \Delta^*(s, \gamma), \tag{16}$$

where $\Delta(s, \mathbf{h})$ is a closed-loop system characteristic polynomial of degree n_d. Equating the coefficients at the same powers of s in (16), let obtain a system of nonlinear equations

$$\mathbf{Q}(\mathbf{h}) = \chi(\gamma). \tag{17}$$

In the general case, the system (17) has a non-unique solution [1]. Then the vector $\mathbf{h}$ can be represented as a set of two vectors $\mathbf{h} = \{\bar{\mathbf{h}}, \mathbf{h_c}\}$, where $\mathbf{h_c}$ is a free component, $\bar{\mathbf{h}}$ is a vector uniquely determined by the solution of system (17) for a given vector $\mathbf{h_c}$. Thus, the solution of the system of equations (17) with respect to the unknown components of the vector $\mathbf{h}$ can be written as

$$\mathbf{h} = \mathbf{h}^* = \{\bar{\mathbf{h}}^*(\mathbf{h_c}, \gamma), \mathbf{h_c}\} = \mathbf{h}^*(\gamma, \mathbf{h_c}) = \mathbf{h}^*(\epsilon),$$

where $\epsilon = \{\gamma, \mathbf{h_c}\} \in E^\lambda$ is a vector of independent parameters of dimension λ:

$$\lambda = dim(\epsilon) = dim(\gamma) + dim(\mathbf{h_c}) = n_d + n_c.$$

Then the functional (11) can be represented as follows

$$J = J(\mathbf{h}) = J^*(\mathbf{h}^*(\epsilon)) = J^*(\epsilon). \tag{18}$$

In the book [1] it is shown that the problem (12) is equivalent to the problem of an unconditional extremum

$$J^* = J^*(\epsilon) \longrightarrow \inf_{\epsilon \in E^\lambda}. \tag{19}$$

As a result, we can go from minimizing the functional (11) with respect to the variable $\mathbf{h}$ on the set Ω_H to unconditional minimization with respect to the variable ϵ. That is, the optimization problem (12) is reduced to the unconditional extremum problem using the parameterization of the region C_Δ.

Thus, the following algorithm can be applied to solve the optimization problem (12).

1. Set an arbitrary point $\gamma \in E^{n_d}$ and form a polynomial $\Delta^*(s, \gamma)$ using formulas (14), (15).

2. In accordance with the identity $\Delta(s, \mathbf{h}) \equiv \Delta^*(s, \gamma)$, form a system of equations

$$\mathbf{Q}(\mathbf{h}) = \chi(\gamma), \tag{20}$$

which is always compatible and, if its solution is not unique, assign an arbitrary vector of free variables $\mathbf{h_c} \in E^{n_c}$.

3. For a given vector $\epsilon = \{\gamma, \mathbf{h_c}\} \in E^\lambda$, solve system (20) and obtain the point $\mathbf{h}^*(\epsilon)$ as its solution.

4. Form the equation of the closed-loop system (4), (5) and calculate the value of the minimized functional $J^*(\epsilon)$ (18) on the corresponding motion.

5. Using any numerical method of an unconditional extremum set a new point ϵ and minimize the function $J^*(\epsilon)$, repeating steps 3, 4.

6. After finding the point $\epsilon_0 = \arg\min J^*(\epsilon)$ determine the vector $\mathbf{h_0} = \mathbf{h}^*(\epsilon_0)$ which is accepted as a solution to problem (12).

7. If the value $J_0^* = \inf J^*(\epsilon)$ is not reached at the end point ϵ_0 then use the same numerical method to form any minimizing sequence $\{\bar{\epsilon}_i\}$ that ensures convergence to the number J_0^*. Since for any ϵ we have $J^*(\epsilon) \geq 0$ such a sequence is sure to exist.

Thus, solving the problem of unconstrained minimization (19) with respect to the variable ϵ, let find the values of configurable elements $\mathbf{h}$ of the control law with a multi-purpose structure (5).

As a result, the dynamic regulator (5) with the configurable elements obtained by solving the optimization problem (12) provides the best robust properties of the closed-loop system (4), (5).

4   Simulation Examples

In this example, we use the following parameters of the Quanser MAGLEV device:

$$L = 0.4125\,N, \ R = 11\,Om, \ K_m = 6.5308e - 005\,N \cdot m^2/A^2, \ M = 0.068\,kg.$$

Let us introduce the region C_Δ (13) with the following parameters: $\bar{\alpha} = 70$, $\bar{\beta} = 30°$ and design the dynamic controller with the multi-purpose structure (5), whose configurable elements are obtained by solving the optimization problem (12). On Fig. 3 the robust stability boundaries (9) of the control law (5) and the frequency range (10) within which the frequency response of the control object varies, are shown. Comparing these curves, we can conclude that the resulting controller (5) guarantees the preservation of robust stability when the model parameters change within the given limits (6). This is because the frequency range (10) is located inside the robust stability boundaries (9).

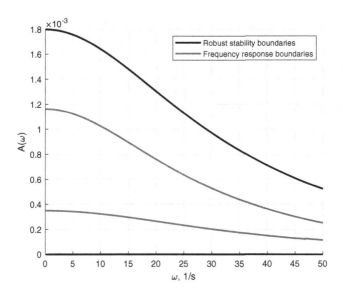

Fig. 3. Comparison of frequency ranges

Suppose that the control objective is to stabilize the ball position at the point $x_{b0} = 0.006$. Figure 4 shows the change in the distance x_b from the electromagnet to the ball for a closed-loop system with the control law (5). It can be seen that the desired objective is achieved in about 0.2 s.

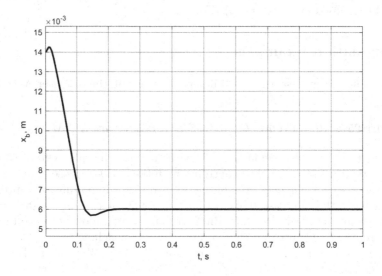

Fig. 4. Ball displacement.

For a comparison, let consider the synthesis of the control law using a linear quadratic regulator (LQR) [18]. On Fig. 5 the boundaries of robust stability for LQR regulator and the frequency range (10) are presented. And on Fig. 6 the corresponding transient process of changing the ball position x_b is shown. These figures demonstrate the correct functioning and robust stability of the closed-loop system with LQR regulator.

Comparing the obtained results, we can conclude that the transient process proceeds more slowly for control law (5) than when using the linear quadratic regulator. But at the same time, it provides the wider frequency range of robust stability (Fig. 7). Thus, the system closed by the designed robust control law (5) has better robust properties than the system closed by LQR regulator.

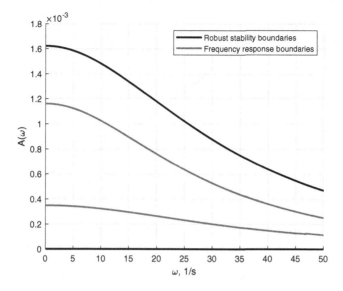

Fig. 5. Comparison of frequency ranges for linear quadratic regulator.

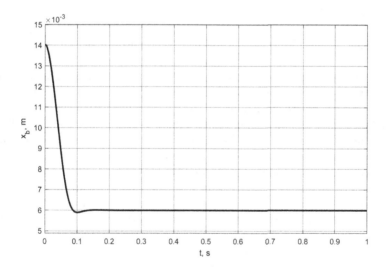

Fig. 6. Ball displacement for linear quadratic regulator.

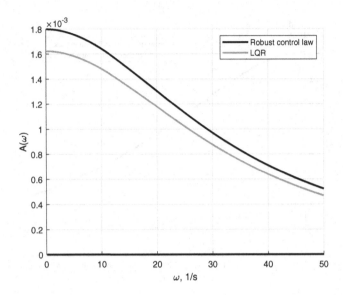

Fig. 7. Comparison of robust stability boundaries.

5 Conclusion

In this paper, an optimization approach to the synthesis of the robust control law is proposed. The main objective of this approach is to improve the robust properties of the closed-loop system and at the same time maintain the acceptable quality of the nominal system performance. This is achieved by introducing the special region, where eigenvalues of the closed-loop system should be placed, and the functional, characterizing the frequency range of the robust stability. It is shown that the corresponding optimal control problem can be stated in the form of modal parametric optimization and then reduced to the search for an unconditional extremum. As a result, we provide the computational algorithm, which allows to design feedback control with the mentioned properties.

The simulation examples for the magnetic levitation system are considered. Two different controllers are designed: optimal robust control law and the linear quadratic regulator. The obtained results are presented and compared. It can be seen from the examples that the proposed approach allows to build the control law with the best robust properties.

References

1. Veremey, E.I.: Lineynye sistemy s obratnoi svyaz'yu. Izdatel'stvo "Lan", Saint Peterburg (2013)
2. Polyak, B.T., Shcherbakov, L.S.: Robastnaya ustojchivost' i upravlenie. Nauka, Moscow (2002)

3. Sotnikova, M.V.: Sintez robastnogo cifrovogo regulyatora dlya sistemy magnitnoj levitacii. Sovremennye informacionnye tehnologii i IT-obrazovanie **8**, 1033–1040 (2012)
4. Sotnikova, M.V., Tomilova, A.S.: Algoritmy analiza robastnyh svojstv mnogocelevyh zakonov upravleniya podvizhnymi ob'ektami. Sovremennye informacionnye tehnologii i IT-obrazovanie **14**(2), 374–381 (2018)
5. Bhattacharyya, S.P., Datta, A., Keel, L.H.: Linear control theory: Structure, Robustness, and Optimization, 1st edn. CRC Press, Boca Raton (2009)
6. Boyd, S., Ghaoui, E., Feron, E., Balakrishnan, V.: Linear Matrix Inequalities in Systems and Control Theory. Society for Industrial and Applied Mathematics, Philadelphia (1994)
7. Garcia, P., Ampountolas, K.: Robust disturbance rejection by the attractive ellipsoid method - part I: continuous-time systems. IFAC-PapersOnLine **51**(32), 34–39 (2018)
8. Kharitonov, V.L.: The asymptotic stability of the equilibrium position of a family of systems of differential equations. Diferentsial 'nye Uravneniya **14**, 2086–2088 (1978)
9. Matušu, R., Prokop, R.: Graphical analysis of robust stability for systems with parametric uncertainty: an overview. Trans. Inst. Measur. Cont. **33**(2), 274–290 (2011)
10. Bhattacharyya, S.P.: Robust control under parametric uncertainty: an overview and recent results. Annu. Rev. Control. **44**, 45–77 (2017)
11. Iwasaki, T.: Robust performance analysis for systems with structured uncertainty. Int. J. Robust Nonlinear Control **6**(2), 85–99 (1996)
12. Zhou, K., Khargonekar, P.P., Stoustrup, J., Niemann, H.H.: Robust performance of systems with structured uncertainties in state space. Automatica **31**(2), 249–255 (1995)
13. Doyle, J.C.: Analysis of feedback systems with structured uncertainties. IEE Proc. Pt. D Control Theor. Appl. **129**(6), 242–250 (1982)
14. Doyle, J., Francis, B., Tannenbaum, A.: Feedback Control Theory. MacMillan, New York (1992)
15. Veremey, E.I., Sotnikova, M.V.: Mnogocelevaya struktura zakonov upravleniya morskimi podvizhnymi ob'ektami. In: XII Vserossijskoe soveshchanie po problemam upravleniya, pp. 3289–3300. IPU RAS, Moscow (2014)
16. Veremei, E.I., Korchanov, V.M.: Multiobjective stabilization of a certain class of dynamic systems. Autom. Remote Control **49**(9), 1210–1219 (1989)
17. Sotnikova, M.V., Veremey, E.I.: Synthesis of astatic MPC-regulator for magnetic levitation plant. WSEAS Trans. Syst. Control **12**, 355–361 (2017)
18. Kwakernaak, H., Sivan, R.: Linear Optimal Control Systems. Wiley, New York (1972)

Optimal Strategies of Consumers with Energy Storages in Electricity Market

Alexander Vasin⬤ and Olesya Grigoryeva$^{(\boxtimes)}$⬤

Lomonosov Moscow State University, Moscow 119991, Russia
olesyagrigez@gmail.com

Abstract. In the present work, we consider a problem of energy storages optimal control by consumers buying energy at the wholesale markets or under fixed tariff rates. Since the maximal prices are usually several times less than the minimal prices, the efficient control of the storage may provide a valuable profit. We determine the optimal consumption schedule and the optimal storage control with account of hour specific consumption as well as a moving load, which can be redistributed throughout the day. The model takes into account random factors, including the energy production by renewable energy sources and the demand depending on the weather. First, we study the case where the storage control is based on the reliable forecast of the random factors for the planning interval, and then examine the corresponding stochastic optimization problem, where future values of random factors are characterized by a probability distribution. We determine the optimal strategy for the markets where a consumer can sell the excess energy back to the market at the same price, and also consider the problem where such sale is impossible.

Keywords: Energy storages · Optimal control · Lagrange theorem · Stochastic optimization.

1 Introduction

The problem of energy storages optimal employment is of interest for a wide set of consumers buying energy at the wholesale markets or under fixed tariff rates. In the both cases, the nighttime prices are usually several times less than the daytime prices, so accumulation of the cheap energy and its consumption at the pike periods may provide a valuable profit. Storages also facilitate the efficient use of renewable energy sources (RESs) by consumers, since the volume of power that a RES supplies is a random variable depending on weather conditions.

Optimal algorithms for energy storage systems in microgrid applications are widely discussed in the literature. According to the recent survey [1], about 7000 papers on this subject were published from 2010 till 2020. Consider some works related to our study.

Paper [2] investigates the benefits of demand resources in buildings for optimal energy trading in day-ahead and real-time energy markets. The building flexible demand resources considered are electric vehicles and batteries. The paper examines the combined optimization of electric vehicles and batteries charging and discharging with the objective of maximizing the total profit of the building microgrid including RES, the energy storage system, electric vehicles charging/discharging station, and several houses. The model takes into account flexible and nonflexible loads for each house. The microgrid has to submit its hourly bids to the day-ahead market a few hours ahead of the real-time power transfer. The bid can be to sell power to the market or to buy power from the market. Any imbalance between the real-time power transfer and the day-ahead submitted bid is penalized. The paper does not formulate a corresponding stochastic optimization problem, but proposes the heuristic two-stage optimization method. At the first stage, the bidding strategy is determined proceeding from the forecast of the day-ahead market prices. At the second stage, the other components of the microgrid strategy are optimized according to the real-time market prices and other random factors important for the profit maximization. The paper formulates the optimization problem as a mixed integer linear program which can be solved by the CPLEX solver. Simulation experiments based on the real data demonstrate that the optimal coordination of energy storage batteries and electric vehicles with renewable generation can improve significantly the profit of the microgrid in the energy market. However, the practical implementation of this strategy is unclear since the future values of the random factors stay unknown at each time of the planning interval, and the setting of the problem does not take it into account.

Paper [3] considers the multi-energy microgrid with wind-solar power generation and electricity, heat, and gas loads. It proposes an energy storage optimization method of multi-energy coupling demand response. The multi-objective optimization model of multiple energy storage capacity planning is examined for minimizing economic cost and carbon emission. Adaptive dynamic weighting factors are used to adapt the strategy to the flexibility of planning scenarios. Paper [4] reviews many types of existing energy storage systems, discusses their characteristics and tendencies of their development. Based on their architectures, capacities and operation characteristics, potential application fields are identified. Finally, research fields that are related to energy storage systems are studied with their impacts on the future of power systems.

In [5], inelastic demand from consumers is considered, which includes the hourly components of the required volume, as well as the shiftable load, which can be redistributed during the day, taking into account the cost of transferring from the most favorable time at less convenient. Using the theory of contracts, the authors study the problem of optimizing the operation of the energy system by introducing tariffs that encourage consumers to shift the shiftable load at off-peak times. In [6], the author considers the problem of minimizing the cost of electricity consumption by the end user, who may have RES, energy storage and electric vehicles at his disposal. A mixed-integer linear programming framework is implemented to solve this problem. Paper [7] describes a decentralized method of intelligent management of microgrid, which may include photovoltaic arrays, controllable loads, energy storage and plug-in electric vehicles. All components of the microgrid interact by simply exchanging messages with their neighbors, as a result of which the method rapidly converges to an optimal solution for the entire microgrid.

Papers [8–10] consider several particular problems of storage systems control for different types of consumers and their purposes.

Our previous research [11,12] examined a problem of the social welfare maximization by means of large energy storages located at several nodes of the wholesale electricity market. The model takes into account random factors, including the energy production by renewable energy sources and the demand depending on the weather. We determined the optimal control of the energy storages.

The present study examines a problem of optimal consumption schedule and energy storage control for a local consumer who does not make any impact on the prices and aims to maximize his profit. First, we determine the optimal consumption schedule with account of hour specific consumption as well as a shiftable load, which can be redistributed throughout the day, taking into account the cost of transferring from the most favorable time at less convenient. Section 3 examines a similar problem for a consumer with an energy storage and determines the optimal storage control. In Sect. 4, the model takes into account random factors, including the energy production by RESs and the demand depending on the weather. We study the case where the storage control is based on the reliable forecast of the random factors for the planning interval (in particular, for the day ahead), as well as the corresponding stochastic optimization problem, where future values of random factors are characterized by a probability distribution. In contrast to [2], our main focus is on the markets where a consumer any time can sell the excess energy back to the market at the same price without preliminary bidding to the day-ahead market. We also consider the problem where such sail is impossible (Sect. 5).

Our most important new results relate to the model with account of random factors, including the amounts of free energy obtained by the consumer from the environment and from the RES. For the markets where the consumer any time can sell the excess energy back to the market at the same price, we provide an effective method for calculating the optimal consumption and storage control strategy. For the case where the energy resale is impossible, we find the optimal consumer strategy for two particular settings of the problem without random factors. These results are also new. However, in contrast to the previous case, these solutions do not correspond to the optimal strategy of the consumer for the stochastic optimization problem. It is more sophisticated and requires a different approach to solve it in this case.

2   Consumer Model and Optimal Consumption Strategy

Since the demand for electricity varies significantly during the day, we consider consumption depending on time $t = \overline{1,T}$, where t is a period of the time with approximately constant needs, T is the length of the planning interval. In particular, an hour of the day may be denoted by t, then $T = 24$. A typical consumer has needs associated with a specific time t (for example, space heating and lighting), as well as several types of needs, which can be satisfied at different times (cooking, washing and so on). In order to simplify formulas, we consider below just one such shiftable load. Then the utility function of a consumer may be described as follows. His consumption schedule is characterized by vectors $\overrightarrow{v} = (v_0^t, v_1^t, t = \overline{1,T})$, where the volume $v_0^t \geq 0$ relates to the consumption associated with the needs for a given hour t. Functions $u_0^t(v_0^t, \psi^t), t = \overline{1,T}$, show the utility of such consumption depending on the random factor ψ^t characterizing the weather conditions and other random events affecting the need for electricity. Volume v_1^t determines the energy consumption associated with the shiftable load in period t. The maximal power of the corresponding device limits this volume:

$$0 \leq v_1^t \leq V_{M1}, \ t = \overline{1,T}. \tag{1}$$

The utility of consumption for this purpose depends on the total amount of energy allocated to this target and takes into account the costs e^t of distributing this consumption over different times t:

$$u_1\left(\sum_{t=1}^{T} v_1^t\right) - \sum_{t=1}^{T} v_1^t e_1^t.$$

Thus, the total utility of the consumer is as follows:

$$UT(\overrightarrow{v}_0, \overrightarrow{v}_1, \overrightarrow{\psi}) = \sum_{t=1}^{T} u_0^t(v_0^t, \psi^t) + \left(u_1\left(\sum_{t=1}^{T} v_1^t\right) - \sum_{t=1}^{T} v_1^t e_1^t\right).$$

In this section, we consider the model without random factors. Then a consumer strategy determines consumption vector $\overrightarrow{v}$ depending on price vector $\overrightarrow{\pi} = (\pi^t, t = \overline{1,T})$, where π^t is the energy price at time t. The optimal strategy maximizes the total utility with account of the energy costs:

$$(\overrightarrow{v}_0^*, \overrightarrow{v}_1^*) \rightarrow \max(U(\overrightarrow{v}_0, \overrightarrow{v}_1) - \sum_{t=1}^{T} \pi^t(v_0^t + v_1^t)), \tag{2}$$

where $U(\overrightarrow{v}_0, \overrightarrow{v}_1) = UT(\overrightarrow{v}_0, \overrightarrow{v}_1, 0)$.

In the whole paper below, utility functions u_0^t, u_1 meet standard assumptions for micro-economic models (see [13]): they are monotonously increasing and concave by consumption volumes.

If the utility functions are also differentiable, $\overrightarrow{v}_0^* > 0$, $V_1^* = \sum v_1^* > 0$, and constraint (1) is never binding then the first order conditions for problem (2) are:

$$\forall t \quad u_0^{t'}(v_0^{t*}) = \pi^t, (v_1^{t*} > 0) \Rightarrow \pi^t + e^t = \arg\min_{\tau=\overline{1,T}}(\pi^\tau + e_1^\tau) = u_1'(V_1^*), \tag{3}$$

If vector v_1^* obtained from (3) does not meet constraint (1) for some t then let us order the time by increasing of $\pi^t + e^t$: let $\pi^{t_1} + e_1^{t_1} \leq \dots \leq \pi^{t_T} + e_1^{t_T}$, and $\pi^{t_0} + e_1^{t_0} = 0$. Then there exists $k \in Z$ such that

$$kV_{M1} \geq v_1^*(\pi^{t_k} + e_1^{t_k}) \geq (k-1)V_{M1}, \tag{4}$$
$$v_1^{t_l*} = V_{M1} \text{ under } l = \overline{1, k-1},$$
$$u_1'(V_1^*) = \pi^{t_k} + e_1^{t_k}, \text{ or } V_1 = (k-1)V_{M1},$$
$$\pi^{t_k} + e_1^{t_k} \geq u_1'(V_1^*) \geq \pi^{t_{k-1}} + e_1^{t_{k-1}}.$$

Theorem 1. *Problem (2) is a convex programming problem. The optimal consumption strategy $(\overrightarrow{v}_0^*, \overrightarrow{v}_1^*)$ proceeds from the first order conditions (3,4).*

Proof. A standard approach is to derive the result from the Lagrange theorem (see the next proof). Consider the economic explanation. Assume from the contrary that, for some t, $v_0^t < v_0^{t*}$ specified by condition (3). Then the consumer can buy more energy and use it for the time specific consumption. Since the marginal utility exceeds the price, the payoff function would increase. If $v_0^t > v_0^{t*}$ then the marginal utility is less than the price, and the inverse operation would increase the total payoff. In order to minimize the cost of energy volume V_1 for the shiftable load, the consumer should buy it at times $t_1, ..., t_{k-1}$ in volumes V_{M1}, and the rest — at time t_k. If the marginal utility $u_1'(V_1) < \pi^{t_k} + e^{t_k}$ and $v_1^{t_k} > 0$ then it is possible to increase the total payoff by reduction of $v_1^{t_k}$ and selling the energy. The inverse operation increases the total payoff if $u_1'(V_1) > \pi^{t_k} + e^{t_k}$ and $v_1^{t_k} < V_{M1}$.

3   The Problem for Consumer with Energy Storage

Consider a consumer who can use an energy storage. The characteristics of the storage are the capacity, the rates and the efficiency coefficients of charging and discharging. Following [13,14], describe them as follows. Denote as E^{min} and E^{max} respectively, the minimum and the maximum allowable charge levels of the storage, V_{ch} and V_{dis} — maximum rates of charging and discharging, η_{ch} and η_{dis} — as charging and discharging efficiency coefficients, respectively. Denote as v_{Bat}^t the amount of energy the storage charges or discharges during period t, a positive value corresponds to charging; v_{Bat}^0 shows the initial charge. A storage control strategy is specified by vector $\overrightarrow{v}_{Bat} = (v_{Bat}^t, t = \overline{1, T})$. Feasible controls satisfy the following constraints: $\forall t$

$$0 \leq -v_{Bat}^t \leq V^{dis} \text{ for discharging rates}; \tag{5}$$

$$0 \leq v_{Bat}^t \leq V^{ch} \text{ for charging rates}; \tag{6}$$

$$E^{min} \leq \sum_{k=0}^{t} v_{Bat}^{k} \leq E^{max} \quad \forall t = \overline{0,T}; \tag{7}$$

and the condition of the storage energy balance in the planning interval:

$$\sum_{t=1}^{T} v_{Bat}^{t} = 0. \tag{8}$$

Denote $\overrightarrow{\eta} = (\eta^{t}(v_{Bat}^{t}), t = \overline{1,T})$, where

$$\eta^{t}(v_{Bat}^{t}) = \begin{cases} \eta_{ch}, & \text{if } v_{Bat}^{t} > 0 \text{ (battery is charging)}; \\ \frac{1}{\eta_{dis}}, & \text{if } v_{Bat}^{t} < 0 \text{ (battery is discharging)}; \\ 0, & \text{if } v_{Bat}^{t} = 0 \text{ (there is no energy exchange with the battery)}. \end{cases}$$

The setting of the optimal consumption problem changes as follows:

$$(\overrightarrow{v}_0, \overrightarrow{v}_1, \overrightarrow{v}_{Bat}) \rightarrow \max[U(\overrightarrow{v}_0, \overrightarrow{v}_1) - \sum_{t=1}^{T} \pi^{t}(v_0^{t} + v_1^{t} + v_{Bat}^{t} \cdot \eta(v_{bat}^{t}))]. \tag{9}$$

Thus, consumer strategy $\overrightarrow{v}$ includes also vector $\overrightarrow{v}_{Bat}$, meeting constraints (5)–(8), and the energy costs are determined with account of the storage impact. Below we examine two cases: 1) the case where a consumer can sell the excess energy back to the market at the same price; 2) the problem where such sail is impossible. The latter condition implies the next constraint on the consumer strategy:

$$v_0^{t} + v_1^{t} + v_{bat}^{t}/\eta_{dis} \geq 0, t = \overline{1,T}. \tag{10}$$

Below we use the following notations:

$$U1(\overrightarrow{v}) = U(\overrightarrow{v}_0, \overrightarrow{v}_1) - \sum_{t=1}^{T} \pi^{t}(v_0^{t} + v_1^{t} + v_{Bat}^{t} \cdot \eta(v_{Bat}^{t}));$$

$$g_1^{t}(v_{Bat}^{t}) = \frac{v_{Bat}^{t}}{\eta_{dis}} + V_{dis}^{max}, \quad t = \overline{1,T};$$

$$g_2^{t}(v_{Bat}^{t}) = V_{ch}^{max} - \eta_{ch}v_{Bat}^{t}, \quad t = \overline{1,T};$$

$$g_3^{t}(\overrightarrow{v}_{Bat}) = \sum_{k=0}^{t} v_{Bat}^{k} - E^{min}, \quad t = \overline{1,T};$$

$$g_4^{t}(\overrightarrow{v}_{Bat}) = E^{max} - \sum_{k=0}^{t} v_{Bat}^{k}, \quad t = \overline{1,T};$$

$$g_5^{t}(\overrightarrow{v}_{Bat}) = v_0^{t} + v_1^{t} + v_{bat}^{t}/\eta_{dis}, \quad t = \overline{1,T};$$

$$g_6(\overrightarrow{v}_{Bat}) = \sum_{t=1}^{T} v_{Bat}^{t}.$$

Inequalities $g_k^t(\overrightarrow{v}_{Bat}) \geq 0, t = \overline{1,T}, k = 1,5$, correspond to constraints (5–7, 10) imposed on the storage control, and equality $g_6(\overrightarrow{v}_{Bat}) = 0$ — to constraint (8). Consider the Lagrange function for problem (9) under the given constraints:

$$L(\overrightarrow{v}_0, \overrightarrow{v}_1, \overrightarrow{v}_{Bat}, \overrightarrow{\lambda}) = U1(\overrightarrow{v}) + \sum_{t=1}^{T}(\lambda_1^t g_1^t(v_{Bat}^t) + \lambda_2^t g_2^t(v_{Bat}^t) +$$

$$+\lambda_3^t g_3(\overrightarrow{v}_{Bat}) + \lambda_4^t g_4(\overrightarrow{v}_{bat}) + \lambda_5^t g_5(\overrightarrow{v}_{Bat})) + \lambda_6 g_6(\overrightarrow{v}_{Bat}).$$

Theorem 2. *Problem (9) is a convex programming problem. If utility functions $u_0^t(v), u_1(v)$ are continuously differentiable and $v_0^{t*} > 0, V_1^* > 0$ then there exist Lagrange coefficients $\lambda_k^t \geq 0$ and $\lambda_6 \in \mathbb{R}$ such that the optimal consumption strategy $\overrightarrow{v}^* = (\overrightarrow{v}_0^*, \overrightarrow{v}_1^*, \overrightarrow{v}_{Bat}^*)$ meets the following system:*

$$\begin{cases} \dfrac{\partial L}{\partial v_0^t}(v^*) = 0, \quad t = \overline{1,T}; \\[2mm] \dfrac{\partial L}{\partial v_1^t}(v^*) = 0, \quad if \ v_1^{t*} \in (0, V_{M1}), t = \overline{1,T}; \\[2mm] \dfrac{\partial L}{\partial v_1^t}(v^*) \geq 0, \quad if \ v_1^{t*} = V_{M1}, t = \overline{1,T}; \\[2mm] \dfrac{\partial L}{\partial v_1^t}(v^*) \leq 0, \quad if \ v_1^{t*} = 0, t = \overline{1,T}; \\[2mm] \dfrac{\partial L}{\partial v_{Bat}^t}(v^*) = 0, \quad t = \overline{0,T}. \end{cases} \quad (11)$$

Proof. Proceeding from constraints (5–7, 10), the set of strategies for problem (9) is convex and closed. The mentioned properties of the utility functions imply that payoff function $U1$ is concave on this set. Thus, problem (9) is a convex programming problem, and, according to the known Lagrange theorem (see [15]), the optimal strategy $\overrightarrow{v}^*$ meets the first order conditions (11).

In this section below we examine problem (9) without constraint (10) and assuming that $V_{ch} = V_{dis} = V_M$, $E_{min} = 0$, $E_{max} = E_M$, $\eta_{ch} = \eta_{dis}$ for formal simplicity. Thus, any time t each consumer can buy and sell the energy at the same price π^t.

Theorem 3. *Under the given conditions, the optimal consumption strategy $(\overrightarrow{v}_0^*, \overrightarrow{v}_1^*)$ for problem (9) does not depend on the storage parameters and coincides with the strategy specified by Theorem 1.*

Proof. The first order conditions for vectors $\overrightarrow{v}_0^*, \overrightarrow{v}_1^*$ in system (11) are:

$$u_0^{t'}(v_0^t) - \pi^t + \lambda_5^t = 0 \ \ \forall t = \overline{1,T}; \quad u'(\sum_{t=1}^{T} v^t) - e^t - \pi^t + \lambda_5^t = 0 \ \ \forall t = \overline{1,T}.$$

Since constraint (10) is not binding, $\lambda_5^t = 0 \; \forall t = \overline{1,T}$. So the optimal vectors $\overrightarrow{v}_0^*, \overrightarrow{v}_1^*$ for problem (9) meet the same conditions as the optimal strategy for problem (2).

Consider a particular case of problem (9) — the search for the optimal strategy of the energy resale using the storage:

$$\overrightarrow{v}_{Bat}^* \to \max_{\overrightarrow{v}_{Bat}} \left(- \sum_{t=1}^{T} \pi^t v_{Bat}^t \eta(v_{Bat}^t)\right) \tag{12}$$

Theorem 4. *The optimal consumer strategy for problem (9) includes the optimal consumption strategy $(\overrightarrow{v}_0^*, \overrightarrow{v}_1^*)$ for problem (2) and the optimal storage control $\overrightarrow{v}_{Bat}^*$ for problem (12).*

Proof. Under the given constraints on $\overrightarrow{v}$ for problem (9), on $(\overrightarrow{v}_0, \overrightarrow{v}_1)$ for problem (2), and on $\overrightarrow{v}_{Bat}$ for problem (12),

$$\max_{\overrightarrow{v}}[U(\overrightarrow{v}_0, \overrightarrow{v}_1) - \sum_{t=1}^{T} \pi^t(v_0^t + v_1^t + v_{Bat}^t \cdot \eta(v_{bat}^t))] \leq \max_{(\overrightarrow{v}_0, \overrightarrow{v}_1)} (U(\overrightarrow{v}_0, \overrightarrow{v}_1) - \tag{13}$$

$$- \sum_{l=1}^{T} \pi^t(v_0^t + v_1^t)) + \max_{\overrightarrow{v}_{Bat}} \left(- \sum_{t-1}^{T} \pi^t v_{Bat}^t \eta(v_{Bat}^t)\right)$$

On the other hand, strategy $(\overrightarrow{v}_0^*, \overrightarrow{v}_1^*, \overrightarrow{v}_{Bat}^*)$ is feasible for problem (9) without constraint (10). Hence, the left-hand side equals the right-hand side in (13).

Note 1. In the practical implementation of the optimal strategy, for any time where $v_{Bat}^{t*} < 0$, the energy from the storage first substitutes the energy purchase at the current price, and then the consumer sells (if $v_0^{t*} + v_1^{t*} + v_{bat}^{t*}/\eta_{dis} < 0$) or buys (if $v_0^{t*} + v_1^{t*} + v_{bat}^{t*}/\eta_{dis} > 0$) the rest.

Consider solutions of problem (12) for some particular cases. First, assume that constraint (7) on the storage volume is never binding. Let $\{\tau_1 \leq \tau_2... \leq \tau_T\}$ denote the ordering of time periods in ascending order of the prices (that is, $\pi^{\tau_1} \leq \pi^{\tau_2} \leq ... \leq \pi^{\tau_T}$), and $\{\overline{\tau}_1 \leq \overline{\tau}_2... \leq \overline{\tau}_T\}$ denote the ordering of time periods in descending order of the prices, l denote the maximal number such that $\pi^{\tau_l} \leq \pi^{\overline{\tau}_l}/\eta^2$.

Proposition 1. *The optimal strategy for problem (12) in this case is: $v_{Bat}^{t*} = V_M, \; t = \overline{\tau_1, \tau_l}; \; v_{Bat}^{t*} = -V_M, \; t = \overline{\overline{\tau}_1, \overline{\tau}_l}, \; v_{Bat}^{t*} = 0$ for any other t.*

Next, consider the opposite case where constraints (5–6) on the charging rate are never binding, while constraint (7) may be binding. We define significant local extremes $t_1 < \overline{t}_1 < t_2 < \overline{t}_2 < ... < t_k < \overline{t}_k$, such that $t_1, ..., t_k$ are local minima and $\overline{t}_1, ..., \overline{t}_k$ are local maxima meeting the following conditions:

$$\forall t \in (t_l, \overline{t}_l) \ \pi_0^{\overline{t}_l} \geq \pi_0^t \geq \pi_0^{t_l}, \ \forall t' \in (t, \overline{t}_l) \ \eta^2 \pi_0^{t'} > \pi_0^t,$$

$$\forall t \in (\overline{t}_l, t_{l+1}) \ \pi_0^{\overline{t}_l} \geq \pi_0^t \geq \pi_0^{t_{l+1}}, \ \forall t' \in (t, t_{l+1}) \ \pi_0^{t'} < \eta^2 \pi_0^t, \tag{14}$$

$$\forall l \ \pi_0^{\overline{t}_l} > \eta^2 \pi_0^{t_l}, \pi_0^{\overline{t}_l} > \eta^2 \pi_0^{t_{l+1}}, \ under \ \ k+1 := 0.$$

Proposition 2. *The optimal strategy for problem (12) in this case is:* $v_{Bat}^{t*} = E, \ t = \overline{t_1, t_k}; \ v_{Bat}^{t*} = -E, \ t = \overline{\overline{t}_1, \overline{t}_k}, \ v_{Bat}^{t*} = 0$ *for any other* t.

So, according to this strategy, the consumer buys the maximal amount of energy at each significant local minimum and sells it at the next significant local maximum.

4   Model with Random Factors

In this section, we take into account random factors important for determination of the optimal consumption strategy. We characterize them by vector $\psi = (\psi^t, t = \overline{1, T})$, where each component ψ^t represents the total amount of free energy obtained by the consumer from the environment in the form of light and heat, and from the RES, if he owns anyone. Below we assume that, for any t, this value does not exceed the optimal consumption volume specified by Theorem 1:

$$0 \leq \psi^t \leq v_0^{t*} + v_1^{t*}, t = \overline{1, T}. \tag{15}$$

First, we examine the case where values $\psi^t, t = 1, T$ are known for the entire planning interval. Then, for any consumption strategy $(\overrightarrow{v}_0, \overrightarrow{v}_1)$, the necessary amount of energy for time t is reduced by the amount ψ^t, and the optimal consumption strategy is a solution of the following problem similar to (2):

$$\max_{\overrightarrow{v}}(U(\overrightarrow{v}_0, \overrightarrow{v}_1) - \sum_{t=1}^{T} \pi^t \cdot \max(0, v_0^t + v_1^t - \psi^t)). \tag{16}$$

Theorem 5. *For any vector $\overrightarrow{\psi}$ meeting constraints (15), solution $(\overrightarrow{v}_0^*, \overrightarrow{v}_1^*)$ of problem (2) is a solution of problem (16).*

In practice future values $\psi^\tau, \tau > t$, are typically unknown to the consumer, and only the probability distribution is available for these values. Then, for every time t, components $(v_0^t, v_1^t, v_{Bat}^t)$ are determined depending on this distribution and past values $\psi^\tau, \tau \leq t$. Each consumer strategy is given by vector function $(v^t(\overrightarrow{\psi}^t), t = \overline{1, T})$, where $\overrightarrow{\psi} = (\psi^1, ..., \psi^t)$. The optimal consumer strategy is a solution to the following stochastic optimization problem:

$$(\overrightarrow{v}, \overrightarrow{v}_{Bat})^*(\overrightarrow{\psi}) \to \max_{(\overrightarrow{v}, \overrightarrow{v}_{Bat})(\overrightarrow{\psi})} \mathbb{E}\{u(\overrightarrow{v}(\overrightarrow{\psi})) - \sum_{t=1}^{T} \pi^t(v_0^t(\overrightarrow{\psi}) + v_1^t(\overrightarrow{\psi}) + \tag{17}$$

$$+v_{Bat}^t(\overrightarrow{\psi})\eta(v_{Bat}^t(\overrightarrow{\psi})) - \psi^t)\}$$

under constraints (1, 5–7), where $\mathbb{E}(...)$ is the mathematical expectation of the total payoff under the given distribution of $\overrightarrow{\psi}$. In general, the problem is rather hard, its solving is possible only by time-consuming methods such as dynamic programming. However, Theorem 5 implies the following important result.

Theorem 6. *If any possible vector $\overrightarrow{\psi}$ meets constraints (15), then any solution $(\overrightarrow{v}^*)$ of problem (9) determines the solution of problem (17): $\forall \overrightarrow{\psi}, t = \overline{1,T}$ $(v_0^{t*}, v_1^{t*}, v_{Bat}^t)(\overrightarrow{\psi}^t) = (v_0^{t*}, v_1^{t*}, v_{Bat}^{t*}).$*

5 Optimal Consumer Strategy When the Energy Resale is Impossible

In this section, we study problem (9) with account of constraint (10). First, consider the case where constraints (1) on the the energy consumption associated with the shiftable load (5–6) on the charging rates, and (7) on the storage volume are never binding. In order to find the optimal consumer strategy, determine $t^* = \arg\min_t \pi^t$, $t^{**} = \arg\min_t e^t$, $\tilde{t} = \arg\min_t (\pi^t + e^t)$. Let v_{Bat0}^{t*} and v_{Bat1}^{t*} denote the optimal discharges of the storage for the time specific consumption and for the consumption associated with the shiftable load.

Proposition 3. *The optimal consumer strategy $\overrightarrow{v}^*$ for this case meets the following conditions: for every t, if $\pi^t/\pi^{t^*} > \eta^2$ then $u_0'(v_0^{t*}) = \eta^2\pi^{t^*}$, $v_{Bat0}^{t*} = \eta v_0^{t*}$, otherwise $u_0'(v_0^{t*}) = \pi^t$, $v_{Bat0}^{t*} - 0$, the consumer buys the energy at the current price;*

$$\text{if } (\pi^{t^*} + e^{t^{**}})\eta^2 < \pi^{\tilde{t}} + e^{\tilde{t}} \text{ then } u_1'(v_1^{t^{***}}) = (\pi^{t^*} + e^{t^{**}})\eta^2, \; v_{Bat1}^{t^{***}} = \eta v_1^{t^{***}},$$

otherwise $u_1'(v_1^{\tilde{t}*}) = \pi^{\tilde{t}} + e^{\tilde{t}}$, $v_{Bat1}^{t*} = 0$ for any other t, $v_{Bat1}^{t*} = 0$ for every t;

$$v_{Bat}^{t^*} = v_{Bat1}^{t^{***}} + \sum_{t=1}^{T} v_{Bat0}^{t*}; \; v_{Bat}^{t*} = -v_{Bat0}^{t*} - v_{Bat1}^{t*} \text{ for any } t \neq t^*.$$

Proof. The proof follows from the first order conditions (11) for this case. The proposition shows that the energy storage charges only at time t^* when the price is minimal. In contrast to the case where the energy resale is possible, for every t such that $\pi^t/\pi^{t*} > \eta^2$, the optimal consumption volume v_0^* provides the marginal utility $\eta^2\pi^{t*}$ instead of π^t. Note that if such t exists, a similar problem for that case has no formal solution: the energy resale provides the unlimited profit.

If we take into account random factors then, for a given vector $\overrightarrow{\psi}$ meeting constraint (15), the solution proceeds from similar conditions: the optimal consumption strategy $(\overrightarrow{v}_0^*, \overrightarrow{v}_1^*)$ stays the same, while the optimal energy storage charge $v_{Bat}^{t*}(\overrightarrow{\psi})$ reduces in $\sum_{t=1}^{T} \psi^t$. For the case where future values $\psi^\tau, \tau > t$, are unknown to the consumer, we cannot obtain any result similar to Theorem 6: the solution of the corresponding stochastic optimization problem essentially depends on the probability distribution for $\overrightarrow{\psi}$.

The rest of this section examines the case where constraints (1) on the energy consumption associated with the shiftable load, and constraints (5–6) on the charging rates may be binding, while constraint (7) on the storage volume is never binding. The algorithm uses the ordering $\{\tau^1 \leq \tau^2 ... \leq \tau^T\}$ of time periods in ascending order of the prices, and the time ordering by increasing of $\pi^t + e^{t_1}$:
$\pi^{t_1} + e^{t_1} \leq ... \leq \pi^{t_T} + e^{t_T}$, employed above. Let τ_c denote the current optimal time for charging the storage, q_0^t, q_1^t, q_2^t — volumes discharged at time t to increase v_0^t, v_1^t and substitute the energy purchase at price π^t at this time, respectively, $mu_i^t, i = 1, 2, 3$ — the current marginal utility of energy expenses for each of these variants with account of the constraints: $\forall t \ v_1^t \leq V_{M1}, \ q_0^t + q_1^t + q_2^t \leq V_M,$ $q_2^t \leq V_{M2}^t := \eta(v_0^{t*} + v_1^{t*})$.

We take solution $(\overrightarrow{v}_0^*, \overrightarrow{v}_1^*)$ of problem (2) in order to determine the initial point of the search:

Stage 0. Set $\overrightarrow{v}_0 := \overrightarrow{v}_0^*$, $\overrightarrow{v}_1 := \overrightarrow{v}_1^*$, $\overrightarrow{v}_{Bat} := 0$, $\tau_c := \tau_1$, $q_i^t := 0 \ \forall t, i$. Note that, for this point, $\forall t \ mu_0^t = mu_2^t = \pi^t$, $mu_1^{t_l} = 0$ for $l = \overline{1, k-1}$, $mu_1^{t_l} = u_1'(v_1^*) - e^{t_l}$ for $l = \overline{k, T}$.

Stage 1. If $\eta^2 \pi^{\tau_c} \geq \max\limits_t \max\limits_{i=1,2,3} mu_i^t(\overrightarrow{v})$ then the current vector $\overrightarrow{v}$ is a solution of the problem.

Stage 2. Otherwise continuously increase $v_{Bat}^{\tau_c}$ and distribute it among $(t, i) \in A(\overrightarrow{v}) := \text{Arg} \max\limits_{t,i} mu_i^t(\overrightarrow{v})$: $dv_{Bat}^{\tau_c} = \sum\limits_{(t,i)\in A(\overrightarrow{v})} dq_i^t$, keeping these marginal utilities equal. In particular, if $(t, 2) \in A(\overrightarrow{v})$ for some t then distribute $dv_{Bat}^{\tau_c}$ among such $(t, 2)$ since the marginal utilities stay the same in this case. Change vector $\overrightarrow{v}$ according to this process, in particular, set $v_{Bat}^t := -(q_0^t + q_1^t + q_2^t)$ if this value is negative, $v_i^t := v_i^{t*} + q_i^t/\eta, \ i = 0, 1$. Interrupt the process as soon as one of the following events happens:

a) $v_{Bat}^{\tau_c}$ reaches V_M. Then set $c := c + 1$ and go to Stage 1.
b) q_2^t reaches V_{M2}^t for every $(t, 2) \in A(\overrightarrow{v})$ Then set $mu_2^t := 0$ for every $(t, 2) \in A(\overrightarrow{v})$ and go to Stage 1.
c) v_1^t reaches V_{M1} for some $(t, 1) \in A(\overrightarrow{v})$. Then set $mu_1^t := 0$ and go to Stage 1.
d) v_{Bat}^t reaches $-V_M$ for some t such that $(t, i) \in A(\overrightarrow{v})$ for some $i \in \{1, 2, 3\}$. Then set $mu_i^t := 0$ for every $i \in \{1, 2, 3\}$ and go to Stage 1.
e) $\max\limits_t \max\limits_{i=1,2,3} mu_i^t(\overrightarrow{v})$ reaches $\eta^2 \pi^{\tau_c}$. Then the current vector $\overrightarrow{v}$ is a solution of the problem.

After less than $3T$ iterations of Stages 1,2, the method finds the solution.

6   Conclusion

In the present work, we examine a problem of the energy storage optimal control by a consumer who buys energy at fixed prices for all time periods in the planning interval. The model considers the consumption associated with the needs of each period, as well as the shiftable load that may be redistributed throughout this interval. We take into account the maximum storage capacity, the maximum rates and the efficiency coefficients of charging and discharging for the storage. For the case where any time a consumer can sell the excess energy back to the market at the same price, we prove that the optimal consumption does not depend on the parameters of the storage, and the storage optimal control does not depend on the consumer utility functions. The both problems relate to the convex programming, and we provide the first order conditions for computation of the optimal strategies. We also consider the model with account of random factors important for determination of the optimal consumer strategy, including the amounts of free energy obtained by the consumer from the environment in the form of light and heat, and from the RES. We examine the corresponding stochastic optimization problem and determine the optimal strategy proceeding from the solutions of the two problems for the previous model. This result reveals the qualitative properties of consumer optimal strategies for this case, and permits to examine the impact of the storage systems wide spread on the market equilibrium. This is an important task for future researchs.

We also consider the case where the energy resale is impossible, and find the optimal consumer strategy for two particular settings of the problem without random factors. The method permits rather obvious generalization for the model with external factors that are known for any time till the end of the planning interval. However, our results show that, in contrast to the previous case, these solutions do not correspond to the optimal strategy of the consumer without the energy storage, and the stochastic optimization problem requires a different approach to solve it in this case.

References

1. Reza, M.S., et al.: Optimal algorithms for energy storage systems in microgrid applications: an analytical evaluation towards future directions. IEEE Access **10**, 10105–10123 (2022). https://doi.org/10.1109/ACCESS.2022.3144930
2. Eseye, A.T., Lehtonen, M., Tukia, T., Uimonen, S., Millar, R.J.: Optimal energy trading for renewable energy integrated building microgrids containing electric vehicles and energy storage batteries. IEEE Access **7**, 106092–106101 (2019). https://doi.org/10.1109/ACCESS.2019.2932461
3. Shen, Y., Hu, W., Liu, M., Yang, F., Kong, X.: Energy storage optimization method for microgrid considering multi-energy coupling demand response. J. Energy Storage **45**, 103521 (2022). https://doi.org/10.1016/j.est.2021.103521
4. Nadeem, F., Suhail Hussain, S.M., Kumar Tiwari, P., Kumar Goswami, A., Selim Ustun, T.: Comparative review of energy storage systems, their roles, and impacts on future power systems. IEEE Access **7**, 4555–4585 (2018). https://doi.org/10.1109/ACCESS.2018.2888497
5. Aizenberg, N., Stashkevich, T., Voropai, N.: Forming rate options for various types of consumers in the retail electricity market by solving the adverse selection problem. Int. J. Public Adm. **2**(5), 99–110 (2019)
6. Narimani, M.R.: Demand side management for homes in smart grids. In: North American Power Symposium (NAPS), pp. 1–6 (2019). https://doi.org/10.1109/NAPS46351.2019.9000233

7. Wang, T., O'Neill, D., Kamath, H.: Dynamic control and optimization of distributed energy resources in a Microgrid. IEEE Trans. Smart Grid **6**(6), 2884–2894 (2015). https://doi.org/10.1109/TSG.2015.2430286

8. Dolatabadi, A., Jadidbonab, M., Mohammadi-ivatloo, B.: Shortterm scheduling strategy for wind-based energy hub: a hybrid stochastic/IGDT approach. IEEE Trans. Sustain. Energy **10**(1), 438–448 (2019). https://doi.org/10.1109/TSTE. 2017.2788086

9. Das, M., Singh, A.K., Biswas, A.: Techno-economic optimization of an off-grid hybrid renewable energy system using metaheuristic optimization approaches-case of a radio transmitter station in India. Energy Convers. Manage **185**, 339–352 (2019). https://doi.org/10.1016/j.enconman.2019.01.107

10. Yan, Q., Zhang, B., Kezunovic, M., Fellow, L.: Optimized operational cost reduction for an EV charging station integrated with battery energy storage and PV generation. IEEE Trans. Smart Grid **10**(2), 2096–2106 (2019). https://doi.org/10. 1109/TSG.2017.2788440

11. Vasin, A.A., Grigoryeva, O.M.: On optimizing electricity markets performance. Optim. Appli. **12422**, 272–286 (2020)

12. Vasin, A.A., Grigoryeva, O.M., Shendyapin, A.S.: New tools and problems in regulating the electricity market. J. Comput. Syst. Sci. Int. **60**(3), 422–434 (2021)

13. Denzau A.T.: Microeconomic Analysis: Markets and Dynamics. Richard d Irwin. 1 Jan 1992

14. Gellings, C.W.: The concept of demand-side management for electric utilities. Proc. IEEE **73**(10), 1468–1570 (1985). https://doi.org/10.1109/PROC.1985.13318

15. Walsh, G.R.: Saddle-point Property of Lagrangian Function. Methods of Optimization, pp. 39–44, John Wiley & Sons, New York (1975)

Operational Research Applications

Variational Autoencoders for Precoding Matrices with High Spectral Efficiency

Evgeny Bobrov[1,2(✉)] ⓘ, Alexander Markov[5], Sviatoslav Panchenko[5], and Dmitry Vetrov[3,4]

[1] M. V. Lomonosov Moscow State University, Moscow, Russia
eugenbobrov@ya.ru
[2] Moscow Research Center, Huawei Technologies, Moscow, Russia
[3] National Research University Higher School of Economics, Moscow, Russia
[4] Artificial Intelligence Research Institute, Moscow, Russia
[5] Moscow Institute of Physics and Technology, Moscow, Russia

Abstract. Neural networks are used for channel decoding, channel detection, channel evaluation, and resource management in multi-input and multi-output (MIMO) wireless communication systems. In this paper, we consider the problem of finding precoding matrices with high spectral efficiency (SE) using variational autoencoder (VAE). We propose a computationally efficient algorithm for sampling precoding matrices with minimal loss of quality compared to the optimal precoding. In addition to VAE, we use the conditional variational autoencoder (CVAE) to build a unified generative model. Both of these methods are able to reconstruct the distribution of precoding matrices of high SE by sampling latent variables. This distribution obtained using VAE and CVAE methods is described in the literature for the first time.

Keywords: MIMO · Precoding · Optimization · DL · VAE · SE · SINR

1 Introduction

Massive multiple-input multiple-output (MIMO) is one of the core technologies of the fifth-generation (5G) wireless systems as it promises significant improvements in spectral efficiency (SE) [27], compared to the traditional, small-scale MIMO. Massive MIMO scales up the number of antennas in a conventional system by orders of magnitude [23]. In multi-user (MU) massive MIMO, the infrastructure base stations (BSs) are equipped with hundreds of antenna elements that simultaneously serve tens of user equipments (UEs) in the same frequency band.

A key advantage of the massive number of antennas is that the SE quality is significantly improved by employing simple linear signal processing techniques [1,2,5,14]. In multiple-input multiple-output (MIMO) systems with a large number of antennas, precoding is an important part of downlink signal processing,

Y. Kochetov et al. (Eds.): MOTOR 2022, CCIS 1661, pp. 315–326, 2022.
https://doi.org/10.1007/978-3-031-16224-4_22

since this procedure can focus the transmission signal energy on smaller areas and allows for greater spectral efficiency with less transmitted power [3, 6, 18].

Various linear precodings allow to direct the maximum amount of energy to the user as Maximum Ratio Transmission (MRT) or completely get rid of the inter-user interference as Zero-Forcing (ZF) [20]. There are also different non-linear precoding techniques such as Dirty Paper Coding (DPC), Vector Perturbation (VP), and L-BFGS SE Optimization, which achieve better SE quality than linear methods but have higher implementation complexity [24]. The authors of L-BFGS SE Optimization reduce the task of finding precoding matrices with high spectral efficiency to an unconstrained optimization problem using a special differential projection method and solve it by the Quasi-Newton iterative procedure to achieve gains in capacity [34]. We take this work as a basis and accelerate the solution of the iterative procedure using machine learning methods.

The application of machine learning in the field of wireless communication has been widely studied [8, 29, 32, 33] and deep neural network technology has recently become a promising tool in the field of wireless physical layer technology [30]. In [31], deep reinforcement learning is applied to a new decentralized resource allocation mechanism for communication between vehicles. In [4], deep learning tools are used to automatically configure 1-bit precoders in complex MU precoding algorithms. The [26] offers a deep energy autoencoder for incoherent MU-MIMO systems with multiple carriers, achieving higher reliability and flexibility. [10] has developed a new collaborative hybrid processing platform based on deep learning for end-to-end optimization. A beam selection method based on a deep neural network is proposed by [22]. In [12], the choice of a delayed repeater based on deep learning for secure cognitive relay networks with buffer management is investigated.

Among the machine learning methods [29, 30], variational autoencoder (VAE) is a generative model that is widely applied in unsupervised and supervised learning [32]. VAE is used in many areas, including image processing, [25], text data processing [17], and recommendation systems [15]. A method using VAE for blind channel equalization has also been proposed in [9], and this study demonstrates the benefits of VAE for solving the precoding (beamforming) problem in the wireless MIMO system. A method using VAE in the field of data transmission is proposed in [16].

This work studies the potential of variational autoencoder (VAE) and conditional variational autoencoder (CVAE) in wireless MIMO systems as deep learning precoding methods. Both VAE and CVAE methods can be used to learn the distribution of precoding matrices with high spectral efficiency. Models of VAE and CVAE are based on the encoder-decoder structure and also utilize latent variables sampled from the standard normal distribution in order to assist decoding and sampling. The framework of variational optimization is constructed. Based on the theory of Bayesian inference, the parameters of the VAE model are obtained through the maximization of Variational Lower Bound (VLB). The proposed VAE/CVAE approach is compared with the complex and highly efficient optimization of L-BFGS [7, 34]. In the case of a single antenna, which is a special case of MIMO, high values of average SE similar to that of L-BFGS can be achieved using precoding matrices sampled from VAE/CVAE with less execution time than the iterative optimization L-BFGS method. We measure the quality of the models using Signal-to-Interference-and-Noise-Ratio (SINR) [28] and spectral efficiency (SE) [27].

The remainder of this paper is organized as follows. In Sect. 2 we describe the variational autoencoder model. In Sect. 3 we describe the system model of the studied Massive MIMO network. In Sect. 4 we set up the problem that is being solved. In Sect. 5 we investigate the distribution of precoding matrices corresponding to the same spectral efficiency values. Here we also consider the problem of finding a good precoding matrix using a particular deep learning model — so-called conditional variational autoencoder. Section 6 contains numerical results and Sect. 7 contains the conclusion.

2   Variational Autoencoder

Given a set of independent and identically distributed samples from true data distribution $W_i \sim p_d(W)$, $i = 1, \ldots, N$, variational autoencoder (VAE) is used to build a probabilistic model $p_\theta(W)$ of the true data distribution $p_d(W)$.

VAE is a generative model that uses latent variables for the distribution reconstruction and sampling. Using VAE, it is possible to sample new objects from the model distribution $p_\theta(W)$ in two steps:

1. Sample $Z \sim p(Z)$.
2. Sample $W \sim p_\theta(W|Z)$.

Here Z is a latent variable from a fixed prior distribution $p(Z)$. The parameters of the distribution $p_\theta(W|Z)$ are obtained using a neural network with weights θ and Z as an input, namely $p_\theta(W|Z) = \mathcal{N}(W|\mu_\theta(Z), \sigma_\theta^2(Z)I)$. In this case the output of the network is a pair of mean $\mu_\theta(Z)$ and dispersion $\sigma_\theta^2(Z)$ of the reconstructed distribution. This network is called a generator, or a decoder.

To fit the model to data we maximize marginal log-likelihood $\log p_\theta(W)$ of the train set. However, $\log p_\theta(W)$ cannot be optimized straightforwardly, because there is an integral in high-dimensional space inside the logarithm which cannot be computed analytically or numerically estimated with enough accuracy in a reasonable amount of time [13]. So, in order to perform optimization, we instead maximize the variational lower bound (VLB) on log-likelihood:

$$\log p_\theta(W) \geqslant \mathbb{E}_{Z \sim q_\phi(Z|W)} \log p_\theta(W|Z) - KL(q_\phi(Z|W)\|p(Z)) = L(W; \phi, \theta) \to \max_{\phi, \theta}. \quad (1)$$

Here $KL(q\|p)$ denotes the Kullback-Leibler divergence which qualifies the distance of two distributions q and p [13]. For two normal distributions, KL-divergence has a closed-form expression. Also $q_\phi(Z|W)$ is called a proposal, recognition or variational distribution. It is usually defined as a Gaussian with parameters from a neural network with weights ϕ which takes W as an input:

$q_\phi(Z|W) = \mathcal{N}(Z|\mu_\phi(W), \sigma_\phi^2(W)I)$. Again, in this case, the output of the network is a pair of mean $\mu_\phi(W)$ and dispersion $\sigma_\phi^2(W)$ of the proposal distribution. This network is called a proposal network, or an encoder.

It is also possible to make variational autoencoder dependent on some additional input H. Prior distribution over Z is now conditioned on H, i.e. $p_\psi(Z|H) = \mathcal{N}(Z|\mu_\psi(H), \sigma_\psi^2(H)I)$, where once again mean $\mu_\psi(H)$ and distribution $\sigma_\psi^2(H)$ are the output of the neural network with parameters ψ, which takes H as an input. This network is called a prior network and this modification of VAE is called conditional variational autoencoder (CVAE). Note that CVAE uses three neural networks while VAE uses only two.

The sampling process of CVAE is as follows:

1. Sample $Z \sim p_\psi(Z|H)$.
2. Sample $W \sim p_\theta(W|Z, H)$.

Here Z is a latent variable from the prior distribution $p_\psi(Z|H)$, but this distribution is no longer fixed and now depends on the condition H and parameters ψ. Proposal and generative networks also have H as an input in CVAE.

As before, to fit the model to the data we maximize marginal conditional log-likelihood $\log p_{\psi,\theta}(W|H)$ of the train set:

$$\log p_{\psi,\theta}(W|H) \geqslant \mathbb{E}_{Z \sim q_\phi(Z|W,H)} \log p_\theta(W|Z, H) - KL(q_\phi(Z|W, H)||p_\psi(Z|H)) =$$
$$= L(W; \phi, \psi, \theta) \to \max_{\phi, \psi, \theta}. \quad (2)$$

3   Optimization Function

Consider a MIMO setting with a single antenna for each user, where the total number of users is q, and $k = 1 \ldots q$ is a user index. The setting is characterized by the channel matrix H; $h_k \in \mathbb{C}^n$ is k-th row of $H \in \mathbb{C}^{q \times n}$, and n is the number of base station antennas. Given a channel matrix H we need to find precoding matrix W; $w^k \in \mathbb{C}^n$ is k-th column of $W \in \mathbb{C}^{n \times q}$.

The quality of transmission is determined by $\mathrm{SINR}_k(W)$ — signal-to-interference-and-noise ratio (SINR), the ratio of useful signal to interference signal plus noise:

$$\mathrm{SINR}_k(W) = \frac{|h_k \cdot w^k|^2}{\sum_{l \neq k} |h_k \cdot w^l|^2 + \sigma^2}, \quad (3)$$

where σ^2 is the noise power of the system.

The criterion for estimating the quality of precoding is spectral efficiency (SE) $f(W)$ [27]. We should account for a constraint on the signal power emitted from i-th antenna $\sum_k |w_i^k|^2$. The optimization problem for finding precoding $W \in \mathbb{C}^{n \times q}$:

$$f(W) = \sum_k \log_2 \left(1 + \mathrm{SINR}_k(W)\right) \to \max_W \quad (4)$$

$$\max_{1 \leq i \leq n} \sum_l |w_i^l|^2 \leq p$$

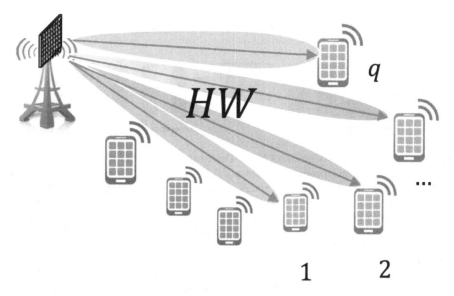

Fig. 1. Multi-User precoding allows to transmit different information to different users simultaneously. Using the matrix W we can configure the amplitude and phase of the beams presented on the picture. The problem is to approximate the precoding matrix W using machine learning as fast as possible, given the SE target function (4) assuming that the data set W is constructed in advance.

An approximate solution of (4) can be obtained using L-BFGS method [7].

One can notice that if W is the solution matrix for the above optimization problem then for any diagonal matrix Q which elements lie on the unit circle in the complex plane, the matrix $W \cdot Q$ produces the same spectral efficiency and satisfies per-antenna power constraints. To study the distribution of precoding matrices with high SE we decided to use a variational autoencoder.

4 Problem Setup and Research Questions

To sum up, in this task our goal is to train a variational autoencoder neural network, conditioned on channel matrix H, that generates precoding matrices for MIMO problems with channel matrix H. Our research questions are as follows:

- Can we produce new high-SE precoding matrices for H by training variational autoencoder on a set of precoding matrices W_j, corresponding to a fixed channel matrix H?
- Can we produce high-SE precoding matrices using variational autoencoder in the case of arbitrary channel matrix H, i.e. by conditioning VAE on H?

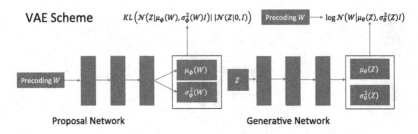

Fig. 2. The neural network model of VAE. **Training**. Encoder (proposal network) takes matrix W as input and produces vector $\mu_\phi(W)$ of shape d_{model} and a single number $\sigma_\phi^2(W)$. A latent code Z is sampled from $\mathcal{N}(Z|\mu_\phi(W), \sigma_\phi^2(W)I)$ using reparametrization $Z = \mu_\phi(W) + \varepsilon \cdot \sigma_\phi(W)$, where ε is standard gaussian noise. Decoder (generative network) takes previously sampled latent code Z and produces vector $\mu_\theta(Z)$ of shape d_{data} and a single number $\sigma_\theta^2(Z)$ which are the parameters of a normal distribution $\mathcal{N}(W|\mu_\theta(Z), \sigma_\theta^2(Z)I)$. Finally, VAE VLB (1) is computed and optimizer step is made. **Sampling**. Using VAE model we sample latent code Z from standard complex gaussian noise $\mathcal{N}(Z|0, I)$. We then use this code to obtain decoder output $p_\theta(Z|W)$ and treat it as a sampled precoding matrix W. Remember that VAE model can be trained and produce samples of W only for one specific channel matrix H. To produce precoding matrix W for the arbitrary channel H a CVAE model should be used (see Fig. 3).

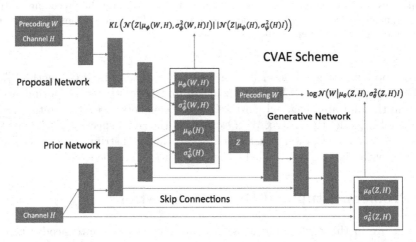

Fig. 3. The neural network model of CVAE. **Training**. Encoder (proposal network) takes matrices W and H as input and produces vector $\mu_\phi(W, H)$ of shape d_{model} and a single number $\sigma_\phi^2(W, H)$. Prior network takes matrix H as input and produces vector $\mu_\psi(W, H)$ of shape d_{model} and a single number $\sigma_\psi^2(W, H)$. A latent code Z is sampled from $\mathcal{N}(Z|\mu_\phi(W, H), \sigma_\phi^2(W, H)I)$, using reparametrization. Decoder (generative network) takes previously sampled latent code Z, matrix H and produces vector $\mu_\theta(Z, H)$ of shape d_{data} and a single number $\sigma_\theta^2(Z, H)$ which are the parameters of a normal distribution $\mathcal{N}(W|\mu_\theta(Z, H), \sigma_\theta^2(Z, H)I)$. Finally, CVAE VLB (2) is computed and optimizer step is made. Also skip connections solve the degradation problem in neural network training [11]. **Sampling**. Given matrix H we sample latent code Z from prior distribution $p_\psi(Z|H)$ parametrized with prior network. We then use this code to obtain decoder output $p_\theta(W|Z, H)$ and treat it as a sampled precoding W for channel H.

5 Approach

In this section, we describe the training procedure for (conditional) variational autoencoder.

We begin with the unconditional one. Variational autoencoder consists of two networks: encoder $q_\phi(Z|W)$ and decoder $p_\theta(W|Z)$. During training, for each matrix W:

1. Encoder takes matrix W as input and produces vector $\mu_\phi(W)$ of shape d_{model} and a single number $\sigma^2_\phi(W)$.
2. A latent code Z is sampled from $\mathcal{N}(Z|\mu_\phi(W), \sigma^2_\phi(W)I)$ using reparametrization $Z = \mu_\phi(W) + \varepsilon \cdot \sigma_\phi(W)$, where ε is standard gaussian noise.
3. Decoder takes previously sampled latent code Z and produces vector $\mu_\theta(Z)$ of shape d_{data} and a single number $\sigma^2_\theta(Z)$ which are the parameters of a normal distribution $\mathcal{N}(W|\mu_\theta(Z), \sigma^2_\theta(Z)I)$.
4. Compute VLB (1) and make optimizer step.

We fix d_{model} — the dimension of latent space — to be equal to 64 and prior distribution $p(Z)$ to be standard gaussian distribution in d_{model}-dimensional space. We also represent complex matrix W of shape 16×4 as a real vector of shape $d_{data} = 16 \times 4 \times 2 = 128$ of real and imaginary parts of W.

To generate possible matrices W after training, we simply sample Z from prior distribution $p(Z)$ and then pass it through the decoder network $p_\theta(W|Z)$, treating $\mu_\theta(Z)$ as a sampled matrix W.

Having trained unconditional autoencoder, it is almost straightforward to obtain a conditional one. In addition to encoder and decoder networks $p_\theta(W|Z, H)$ and $q_\phi(Z|W, H)$, CVAE also contains a prior network $p_\psi(Z|H)$. As previously, d_{data} is equal to $128 = 16 \times 4 \times 2$ and $d_{model} = 64$. To make a training step, we do the following:

1. Encoder (proposal network) takes matrices W and H (represented as two real-valued vectors of shape 128 each) as input and produces vector $\mu_\phi(W, H)$ of shape d_{model} and a single number $\sigma^2_\phi(W, H)$.
2. Prior network takes matrix H as input and produces vector $\mu_\psi(H)$ of shape d_{model} and a single number $\sigma^2_\psi(H)$.
3. A latent code Z is sampled from $\mathcal{N}(Z|\mu_\phi(W, H), \sigma^2_\phi(W, H)I)$ using reparametrization.
4. Decoder (generative network) takes previously sampled latent code Z, matrix H (represented as real-valued vector of shape 128) and produces vector $\mu_\theta(Z, H)$ of shape d_{data} and a single number $\sigma^2_\theta(Z, H)$ which are the parameters of a normal distribution $\mathcal{N}(W|\mu_\theta(Z, H), \sigma^2_\theta(Z, H)I)$.
5. Computing VLB for CVAE (2) and making optimizer step.

Usage of CVAE for sampling matrices W, however, is slightly more difficult. Given matrix H we sample latent code Z from prior distribution $p_\psi(Z|H)$ parametrized with prior network. We then use this code to obtain decoder output $\mu_\theta(W|Z, H)$ and treat it as a sampled precoding matrix W for channel H. The neural network model of CVAE is presented in Fig. 3.

6   Experimental Results

In this section, we present the results of training VAE and CVAE to produce precoding matrices with high spectral efficiency for a given channel matrix H. First of all, we report the results of training VAE to fit a distribution of precoding matrices W for a fixed channel H. We consider two approaches: VAE and a CVAE with the fixed matrix H as a condition.

To obtain target precoding matrices W for a given H, we optimize spectral efficiency using the L-BFGS optimization approach [7], starting from randomly initialized matrix W, until convergence. The L-BFGS method is implemented with the PyTorch [21] framework.

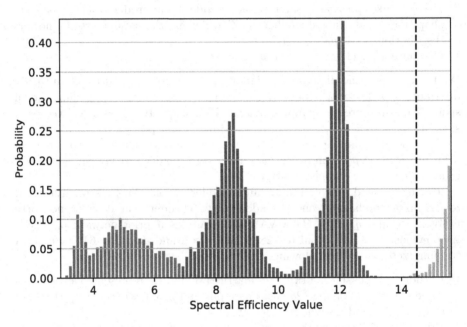

Fig. 4. Distribution of spectral efficiency for precoding matrices W_i obtained with random initialization of L-BFGS method for a fixed channel matrix H. Orange corresponds to samples with high efficiency (14 SE and more), blue and green — samples with average efficiency (from 7 to 14 SE), red — samples with low efficiency (less than 7 SE). **Training.** Samples with SE above threshold $t = 14.5$ (dashed line) were selected for training VAE/CVAE neural networks. Using such training data and following the experiment shown in Sect. 6.1 and Fig. 5, it is possible to generate highly efficient precoding matrices W. (Color figure online)

6.1   Generating High SE Precoding Matrices W_i for the Fixed H

For a given channel matrix $H \in \mathbb{C}^{q \times n}$, we find the precoding $W \in \mathbb{C}^{n \times q}$ using L-BFGS approach [7] with a random initialization. This method returns precoding

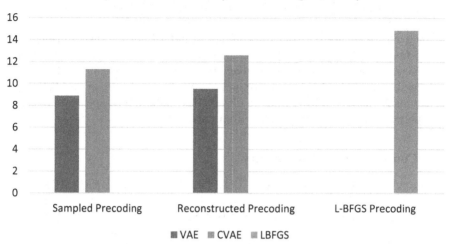

Fig. 5. Spectral efficiency precoding quality produced by the algorithms.

matrices with high spectral efficiency using the iterative optimization approach and measures the SE quality using the formula of SE (4). We repeat this from different starting points to get different matrices W_j for a fixed H, filtering out precoding matrices with SE lower than a certain threshold t in the process (see Fig. 4), obtaining a set $\{W_j\}_{j=1}^N$ of size N.

We measured the quality of generated precoding matrices $W \in \mathbb{C}^{n \times q}$ simply by evaluating spectral efficiency (4) $f(W)$ with a given channel $H \in \mathbb{C}^{q \times n}$. For a given channel matrix H and a set of $N = 5000$ precoding matrices W_i, the average SE value for W_i and H is 14.85 (Fig. 5 L-BFGS Precoding). We fix matrix dimensions to be equal to $q = 4, n = 16$, and the value of filtering threshold $t = 14.5$.

We managed to obtain the mean SE value of 8.87 with VAE and 11.31 with CVAE for a fixed H matrix for sampled precoding matrices W_i (Fig. 5 Sampled Precoding). This method samples from the distribution of precoding matrices with high SE (4) using the deep learning approach of VAE/CVAE. We also measured mean SE in reconstructed objects (precodings W_i passed through encoder and decoder) and obtained 9.53 and 12.61 for VAE and CVAE-fixed-H respectively (Fig. 5 Reconstructed Precoding).

6.2 Generating Matrices $W_{i,j}$ Using H_i as Condition for CVAE

To study how MIMO solutions change with respect to H, we perturb H with random matrices Δ_i, obtaining K different channel matrices $H_i = H + \Delta_i$, where each Δ_i satisfies $||\Delta_i||_F^2 = \delta$ and repeat the exact same process of generating a set of N precoding matrices for each H_i. Overall, we have a set of pairs

$\{H_i, W_{i,j}\}_{i,j=1}^{K+1,N}$ of channel matrices and corresponding precodings with high spectral efficiency, where $H \in \mathbb{C}^{q \times n}$ and $W \in \mathbb{C}^{n \times q}$. We fix hyper-parameters to be equal to $K = 15, q = 4, n = 16, N = 500, \delta = 5$. For this task, channel matrices H are generated from the complex standard normal distribution.

We train CVAE on a set $\{H_i, W_{i,j}\}$ with $H_i = H + \Delta_i$ and $H_0 = H$ for convenience. However, we do not manage to get high-quality samples from such a model, having mean SE for reconstructed objects equal to 6.49 and 6.45 if we use one of H_i as a condition and we observe the same fall in spectral efficiency for a perturbation of H which differ from one of Δ_i.

6.3   Computational Complexity

The proposed VAE/CVAE solution works much faster than the L-BFGS reference method. We estimate the inference of the proposed two-layers neural network as one iteration of the L-BFGS procedure. Which includes two matrix-vector operations inside the layers and one matrix-vector multiplication of SE (4). The complexity of the L-BFGS iteration is estimated as three SE (5) estimations [27]. Since in practice the number of iterations reaches up to a hundred [34], the inference of the proposed neural network is faster up to a hundred times. Both VAE and CVAE methods require a training procedure in the background, but we mainly focus on the inference time in a real system. The learning time of the CVAE method is higher than that of the VAE method [19]. If we denote the complexity of the VAE/CVAE inference as y, the relative complexity of the L-BFGS method will be My, where M is the number of L-BFGS iterations.

7   Conclusion

In this paper, we propose a deep learning algorithm based on the theory of variational autoencoders. This algorithm is designed to generate precoding matrices with high spectral efficiency for massive MIMO systems. With its help, we successfully sampled precoding matrices using both versions of the autoencoder: VAE and CVAE. The first one must be trained separately for different channel matrices H, and the second one can be conditioned on the matrix of the channel H and thus can be used for different H as a unified model. In the case of a fixed channel matrix H, both approaches were able to generate precoding matrices with high spectral efficiency, but CVAE surpasses VAE in terms of quality and versatility. However, both approaches can be used to study the distribution of precoding matrices with high spectral efficiency.

Acknowledgements. The authors are grateful to Mr. D. Kropotov and Prof. O. Senko for discussions.

References

1. Abdallah, A., Mansour, M.M., Chehab, A., Jalloul, L.M.: MMSE detection for 1-bit quantized massive MIMO with imperfect channel estimation. In: 2018 IEEE 19th International Workshop on Signal Processing Advances in Wireless Communications (SPAWC), pp. 1–5. IEEE (2018)
2. Albreem, M.A., Juntti, M., Shahabuddin, S.: Massive MIMO detection techniques: a survey. IEEE Commun. Surv. Tutorials **21**(4), 3109–3132 (2019)
3. Andrews, J.G., et al.: What will 5G be? IEEE J. Sel. Areas Commun. **32**(6), 1065–1082 (2014)
4. Balatsoukas-Stimming, A., Castañeda, O., Jacobsson, S., Durisi, G., Studer, C.: Neural-network optimized 1-bit precoding for massive MU-MIMO. In: 2019 IEEE 20th International Workshop on Signal Processing Advances in Wireless Communications (SPAWC), pp. 1–5. IEEE (2019)
5. Björnson, E., Larsson, E.G., Marzetta, T.L.: Massive MIMO: ten myths and one critical question. IEEE Commun. Mag. **54**(2), 114–123 (2016)
6. Bobrov, E., Kropotov, D., Lu, H., Zaev, D.: Massive mimo adaptive modulation and coding using online deep learning algorithm. IEEE Commun. Lett. **26**(4), 818–822 (2022). https://doi.org/10.1109/LCOMM.2021.3132947
7. Bobrov, E., Kropotov, D., Troshin, S., Zaev, D.: Study on Precoding Optimization Algorithms in Massive MIMO System with Multi-Antenna Users (2021)
8. Bobrov, E., et al.: Machine learning methods for spectral efficiency prediction in massive mimo systems (2021)
9. Caciularu, A., Burshtein, D.: Blind channel equalization using variational autoencoders. In: 2018 IEEE International Conference on Communications Workshops (ICC Workshops), pp. 1–6. IEEE (2018)
10. Dong, P., Zhang, H., Li, G.Y.: Framework on deep learning-based joint hybrid processing for mmWave massive MIMO systems. IEEE Access **8**, 106023–106035 (2020)
11. He, K., Zhang, X., Ren, S., Sun, J.: Deep residual learning for image recognition. In: Proceedings of the IEEE Conference on Computer Vision and Pattern Recognition, pp. 770–778 (2016)
12. Huang, C., Chen, G., Gong, Y., Xu, P.: Deep reinforcement learning based relay selection in delay-constrained secure buffer-aided CRNs. In: GLOBECOM 2020–2020 IEEE Global Communications Conference, pp. 1–6. IEEE (2020)
13. Kingma, D.P., Welling, M.: Auto-encoding variational bayes. arXiv preprint arXiv:1312.6114 (2013)
14. Larsson, E.G., Edfors, O., Tufvesson, F., Marzetta, T.L.: Massive MIMO for next generation wireless systems. IEEE Commun. Mag. **52**(2), 186–195 (2014)
15. Liang, D., Krishnan, R.G., Hoffman, M.D., Jebara, T.: Variational autoencoders for collaborative filtering. In: Proceedings of the 2018 World Wide Web Conference, pp. 689–698 (2018)
16. Lopez-Martin, M., Carro, B., Sanchez-Esguevillas, A.: Variational data generative model for intrusion detection. Knowl. Inf. Syst. **60**(1), 569–590 (2018). https://doi.org/10.1007/s10115-018-1306-7
17. Miao, Y., Yu, L., Blunsom, P.: Neural variational inference for text processing. In: International Conference on Machine Learning, pp. 1727–1736. PMLR (2016)
18. Ngo, H.Q., Larsson, E.G., Marzetta, T.L.: Energy and spectral efficiency of very large multiuser MIMO systems. IEEE Trans. Commun. **61**(4), 1436–1449 (2013)

19. Pagnoni, A., Liu, K., Li, S.: Conditional variational autoencoder for neural machine translation. arXiv preprint arXiv:1812.04405 (2018)
20. Parfait, T., Kuang, Y., Jerry, K.: Performance analysis and comparison of ZF and MRT based downlink massive mimo systems. In: 2014 Sixth International Conference on Ubiquitous and Future Networks (ICUFN), pp. 383–388. IEEE (2014)
21. Paszke, A., et al.: Automatic differentiation in pytorch (2017)
22. Rezaie, S., Manchón, C.N., De Carvalho, E.: Location-and orientation-aided millimeter wave beam selection using deep learning. In: ICC 2020–2020 IEEE International Conference on Communications (ICC), pp. 1–6. IEEE (2020)
23. Rusek, F., et al.: Scaling up MIMO: Opportunities and challenges with very large arrays. IEEE Signal Process. Mag. **30**(1), 40–60 (2012)
24. Tran, L.N., Juntti, M., Bengtsson, M., Ottersten, B.: Beamformer designs for MISO broadcast channels with zero-forcing dirty paper coding. IEEE Trans. Wireless Commun. **12**(3), 1173–1185 (2013)
25. Turhan, C.G., Bilge, H.S.: Variational autoencoded compositional pattern generative adversarial network for handwritten super resolution image generation. In: 2018 3rd International Conference on Computer Science and Engineering (UBMK), pp. 564–568. IEEE (2018)
26. Van Luong, T., Ko, Y., Vien, N.A., Matthaiou, M., Ngo, H.Q.: Deep energy autoencoder for noncoherent multicarrier MU-SIMO systems. IEEE Trans. Wireless Commun. **19**(6), 3952–3962 (2020)
27. Verdú, S.: Spectral efficiency in the wideband regime. IEEE Trans. Inf. Theory **48**(6), 1319–1343 (2002)
28. Wang, B., Chang, Y., Yang, D.: On the SINR in massive MIMO networks with MMSE receivers. IEEE Commun. Lett. **18**(11), 1979–1982 (2014)
29. Xia, W., Zheng, G., Zhu, Y., Zhang, J., Wang, J., Petropulu, A.P.: A deep learning framework for optimization of MISO downlink beamforming. IEEE Trans. Commun. **68**(3), 1866–1880 (2019)
30. Ye, H., Li, G.Y., Juang, B.H.: Power of deep learning for channel estimation and signal detection in OFDM systems. IEEE Wireless Commun. Lett. **7**(1), 114–117 (2017)
31. Ye, H., Li, G.Y., Juang, B.H.F.: Deep reinforcement learning based resource allocation for V2V communications. IEEE Trans. Veh. Technol. **68**(4), 3163–3173 (2019)
32. Zhao, T., Li, F.: Variational-autoencoder signal detection for MIMO-OFDM-IM. Digital Signal Process. **118**, 103230 (2021)
33. Zhao, T., Li, F., Tian, P.: A deep-learning method for device activity detection in mMTC under imperfect CSI based on variational-autoencoder. IEEE Trans. Veh. Technol. **69**(7), 7981–7986 (2020)
34. Zhu, C., Byrd, R.H., Lu, P., Nocedal, J.: Algorithm 778: L-BFGS-B: Fortran subroutines for large-scale bound-constrained optimization. ACM Trans. Mathem. Softw. (TOMS) **23**(4), 550–560 (1997)

Spectrum Allocation in Optical Networks: DSatur Coloring and Upper Bounds

Mikhail Krechetov$^{(\boxtimes)}$, Mikhail Kharitonov, and Dmitri Shmelkin

Huawei Technologies, Russian Research Institute,
Moscow Research Center, Moscow, Russia
`Mikhail.Krechetov1@huawei.com`

Abstract. Routing and Spectrum Allocation (RSA) is one of the central problems in modern optical networks. In this setting, also called Flexible Grid, we want to find paths and allocate non-overlapping frequency slots for as many demands as possible. As opposed to the Routing and Wavelength Assignment (RWA) problem, where all frequency slots are of the same width, demands in Flexible Grids may require slots of different sizes, which makes the original NP-hard problem even more challenging. In this paper, we consider Spectrum Allocation (when routing is known), develop an efficient greedy algorithm based on "degree of saturation"-coloring, and by means of Constraint Programming, we show that the greedy algorithm is almost optimal even for hard real-world instances.

Keywords: Elastic optical networks · Spectrum Allocation · Graph coloring · Constraint Programming

1 Introduction

Due to the continuous growth of internet traffic together with emerging high-rate services, such as high-resolution video and cloud computing, the telecom industry needs cost-effective and scalable network planning solutions. Flex-grid optical networks have been introduced in the last few years to enhance the spectrum efficiency and thus the network capacity, In Flex-grids, the frequency spectrum of an optical link is divided into narrow frequency slots of 12.5 GHz width, which allows for more efficient bandwidth allocation.

The central problem in Flex-Grids is called Routing and Spectrum Assignment/Allocation (RSA) and concerns assigning contiguous and non-overlapping frequency slots to network demand. The main goal of RSA is to maximize the number of routed demands while optimizing the spectrum used. Usually, the routing in Flex-Grids and related problems are solved in real-time and have severe runtime requirements of around a few seconds. The main motivation behind this research paper is to introduce a novel and fast greedy algorithm for (re)allocation of Spectrum Resources.

Supported by Huawei Technologies.

In papers [1–3] and [4] authors suggest different Integer Linear Programming (ILP) formulations and ILP-based heuristics the static version of RSA when all demands are known in advance. See also [5–7] and references therein for more recent algorithms and heuristics including genetic algorithms and machine learning. These methods stay out of the scope of this paper since they inevitably require significantly larger running times while our main goal is to (re)optimize the network resources on the fly.

In this paper, we develop a provably efficient algorithm for Spectrum Allocation (when Routing is known in advance). By provably efficient algorithm we mean that the results of our heuristic have a few percent of optimality gap with respect to the best upper bound we know. This paper is organized as follows. In Sect. 2 we define k-Interval Coloring problem (see [8] and references therein), Spectrum Allocation (SA) problem and mathematical programming formulation for SA. We describe our greedy algorithm in Sect. 3 and provide an upper bound procedure in Sect. 4. We illustrate the performance of our algorithms and input data in Sect. 5.

2   Background

First, we give basic definitions that we use throughout the paper:

Definition 1 (Interval Coloring). *Let $G = (V, E)$ be an undirected graph with vertex set V and edge set E and let $\omega_i \geq 1$ be a positive integer parameter (width) at every vertex $i \in V$. A k-interval coloring of G is a function $I_G(i)$ that assigns an interval $I_G(i) \subseteq \{1, ..., k\}$ of w_i consecutive integers (called colors) to each vertex $i \in V$. A coloring function I is "legal" if $I_G(i) \cap I_G(j) = \emptyset$ for $(i, j) \in E$.*

Definition 2 (Path Graph). *Let $G = (V, E)$ be an undirected graph with vertex set V and edge set E and let P be a set of simple paths in that graph. We introduce a new undirected graph $PG = (P, PE)$. The vertex set P of PG is in one-to-one correspondence with paths from the set P and two vertices $i \in P, j \in P$ are connected by an edge in PE if and only if $i \cap j \neq \emptyset$ (as subsets of E).*

2.1   Spectrum Allocation

In this section, we describe (some of) the engineering requirements for Spectrum Allocation in elastic optical networks (flexible grids). In this paper we consider a mathematical model for a highly complex engineering problem, so we inevitably omit engineering constraints that are not relevant to the algorithms we develop.

Definition 3 (Wavelengths). *In this paper, we assume that every edge of a network is divided into $k = 320$ wavelengths of size $12.5\,GHz$ resulting in 4 THz of total frequency at every edge. We need to allocate the set of consecutive wavelengths $\omega_d = \{\omega_d^1, ..., \omega_d^{w_d}\}$ of size w_d for every demand d. Note that demands can not share wavelengths, thus we require interval coloring constraints for sets ω_d (also called the spectrum of a demand d).*

Definition 4 (Spectrum Allocation). *As an input, we are given an optical cable (network topology) graph $G = (V, E)$ and the set of demands D. Every demand $d \in D$ is a pair (p, ω) of a path $p \subset E$ and an integer bandwidth parameter $\omega \geq 1$. We are also given a parameter k that equals the total bandwidth of edges in E. We need to allocate wavelengths for as many demands as possible.*

Proposition 1. *Given an instance of Spectrum Allocation problem with a graph G and a set of demands D, let's consider the set of paths $P = \{p | (p, \omega) \in D\}$ and the corresponding path graph PG (see Definition 2). Then the problem of Spectrum Allocation is equivalent to the problem of finding the largest k-interval colorable subgraph of PG.*

2.2 Mathematical Programming

Now we are ready to provide a mathematical programming formulation of the Spectrum Allocation problem. For every demand $d \in D$ we introduce a binary variable $r_d \in \{0, 1\}$. If r_d equals one, the demand d will be assigned to some wavelengths (routed), otherwise, the demand d is ignored. We also introduce integer variables f_d and l_d for every demand; f_d stands for the first occupied wavelength and l_d is for the last occupied wavelength.

The first constraint in Fig. 1 enforces the bound on the total number of wavelengths in the network. The second constraint ensures that every demand should occupy exactly ω_d consecutive wavelengths. Finally, the last two constraints guarantee that demands can not share wavelengths; this is ensured by big-M constraints - only one of the inequalities is relaxed depending on the value of decision variable δ. These constraints are enforced if both demands are 'routed' and share at least one edge $e \in E$.

$$\max \sum_{d \in D} r_d$$
$$\text{s.t. } 0 \leq f_d \leq l_d < k \quad \text{for } d \in D$$
$$l_d - f_d = \omega_d \quad \text{for } d \in D$$
$$\delta(d_1, d_2) \in \{0, 1\} \quad \text{for } d_1, d_2 \in D$$
$$l_{d_1} + 1 \leq f_{d_2} + M \cdot \delta(d_1, d_2) \quad \text{for } d_1, d_2 \in D, \ (p_{d_1}, p_{d_2}) \in PE, \ r_{d_1} = r_{d_2} = 1$$
$$l_{d_2} + 1 \leq f_{d_1} + M \cdot (1 - \delta(d_1, d_2)) \quad \text{for } d_1, d_2 \in D, \ (p_{d_1}, p_{d_2}) \in PE, \ r_{d_1} = r_{d_2} = 1$$

Fig. 1. Mathematical programming formulation for spectrum allocation problem.

The mathematical programming formulation can be approached by any modern MIP solver, for example, Gurobi [9]. The alternative approach is to use Constraint Programming solvers such as [10], in this case, the last two constraints can be substituted by a single non-overlapping constraint (Fig. 2).

$$\max \sum_{d \in D} r_d$$

$$\text{s.t. } 0 \le f_d \le l_d < k \quad \text{for } d \in D$$

$$l_d - f_d = w_d \quad \text{for } d \in D$$

$$[f_{d_1}, l_{d_1}] \cap [f_{d_2}, l_{d_2}] = \emptyset \quad \text{for } d_1, d_2 \in D, \ (p_{d_1}, p_{d_2}) \in PE, \ r_{d_1} = r_{d_2} = 1$$

Fig. 2. Constraint programming formulation for spectrum allocation problem.

Unfortunately, both "exact" formulations are not efficiently solvable by both state-of-the-art MIP and CP solvers even for medium-size networks with up to one hundred nodes. That is why we develop an efficient greedy algorithm in Sect. 3. We also use CP solver to solve a much easier optimization program that delivers an upper bound on the number of routed demands at the optimum, see Sect. 4.

3   Main Algorithm

In this section, we develop a (greedy) heuristic algorithm that provides a lower bound on the maximal number of routed services for the Spectrum Allocation problem. The main idea is that finding the optimal spectrum allocation is equivalent to finding the correct allocation order, which is summarized in the following trivial lemma:

Lemma 1. *For every Spectrum Allocation instance* (G, D, k) *and its optimal solution* I, *there exists a permutation* **Per** *(D) of the demands, such that demands from* **Per** *(D) are allocated one by one with the lowest available wavelengths will deliver the optimal solution* I.

That is why we approach the Spectrum Allocation problem with the heuristic that tries to guess a good allocation order, see Sect. 3.1. Finally, we strengthen that heuristic by reversing the wavelength direction, see Sect. 3.2.

3.1   DSatur Coloring

The basic greedy approach to the Spectrum Allocation problem consists of the following steps:

1. Define a comparator function over the set of demands.
2. At every step select the minimal demand w.r.t. comparator.
3. Allocate the first available frequency slot to that demand.
4. Return to step 2.

Definition 5 (Saturation). *The number of available wavelengths for a demand* $d = (p_d, w_d)$ *is called Saturation and is denoted* **satur**(d). *Let's denote by* D_r *the set of demands that are already has assigned wavelengths (routed). Following*

Fig. 1, we denote by f_d and l_d the first and the last occupied wavelength of a demand:

$$satur(d) = \left| \bigcap_{e \in p_d} \{\{0, ..., k-1\} \setminus \bigcup_{\{d_r \in D_r | e \in p_{d_r}\}} [f_{d_r}, l_{d_r}]\} \right| \tag{1}$$

By $deg_P(d)$ we denote the degree of d in the path graph P ($deg_P(d)$ equals the number of other demands that share at least one edge of the network topology with d).

Definition 6 (Comparator). *For two demands $d_1, d_2 \in D$ we say that $d_1 \leq d_2$ if and only if*

$$(-satur(d_1), deg_P(d_1), id(d_1)) \leq (-satur(d_2), deg_P(d_2), id(d_2)) \tag{2}$$

3-tuples above are compared in lexicographical order. First, we select demands with the largest saturation value. If saturations are equal we select demand with the lowest number of conflicts and so on. Finally, we compare by indices so that the order produced by the comparator is total.

3.2 Bi-Loop

Now, we introduce a Bi-loop heuristic that is used to improve the results of DSatur coloring even further:

For a given number of iterations repeat:

1. $D_0 = \{d \in D, \ r(d) = 0\}$.
2. Using DSatur coloring allocate as many demands from D_0 as possible.
3. $D_1 = \{d \in D, \ r(d) = 1\}$.
4. Clear all allocated slots.
5. Using reversed DSatur coloring allocate as many demands from D_1 as possible.

The goal is to re-allocate frequencies for routed demands making them more compact and possibly freeing some space for new demands. We do it by simply reversing the comparator in DSatur coloring. For the experimental results, we refer to Sect. 5.

4   Upper Bounds

We consider a network topology graph $G = (V, E)$, a set of paths P and the corresponding path graph PG (see Definition 2). We consider an instance of Interval Coloring problem (see Definition 1) over PG and relax it to the Integer Linear Problem (ILP) below. For every path $p \in P$ we introduce a binary variable $r_p \in \{0, 1\}$ which indicates whether we select the path p or ignore it:

$$\max \sum_{p \in P} \omega_p \cdot r_p, \ s.t. \sum_{p: \ e \in p} \omega_p \cdot r_p \leq k \ for \ e \in E \tag{3}$$

In the ILP above we maximize the total traffic in the network under bandwidth constraints. Here we denote the total bandwidth of an edge by k; in all our experiments (and due to engineering requirements) $k = 320$. This ILP is much easier for ad-hoc solvers and provides an efficiently computable upper bound on traffic value.

4.1   Clique-Based Upper-Bound

Apparently, we may construct a better upper bound on traffic value by implementing tighter bandwidth constraints.

Lemma 2 (Clique Width). *A k-coloring I_G is "legal" (see Definition 1) if and only if intervals $\bigcap_{i \in C} I_G(i) = \emptyset$ are disjoint for every maximal clique $C \subseteq G$*

Proof. Trivial in both directions since every edge belongs to at least one maximal clique.

First, we compute the set of maximal cliques MC in PG and utilize the lemma above:

$$\max \sum_{p \in P} \omega_p \cdot r_p, \; s.t. \sum_{p \in P(C)} \omega_p \cdot r_p \leq k \; for \; C \in MC \tag{4}$$

We find the list of all maximal cliques using NetworkX [12]; see [13–15] for further details. The algorithm is worst-case exponential, however for the test cases below we did not observe such behavior. For larger cases, it is possible to consider a trade-off between the tightness of the upper bound and the number of maximal clique constraints.

5   Experiments

In this section, we describe the test cases and experimental results for both lower and upper bound methods. The data and algorithms may be found at our GitHub: https://github.com/mkrechetov/SpectrumAllocation.

5.1   Synthetic Instances

We test our manually prepared realistic dense instances with many demands so that cases are difficult for heuristic algorithms as well as solvers. These cases were developed by Mikhail Kharitonov for a student contest[1] and we call them "Voronovo" instances 49–64. We illustrate all test networks in Fig. 3; note that such graph structure with many cycles is typical for so-called core optical networks.

---

[1] https://ssopt.org/2019.

Case	Demands	Nodes	Links	Max demands per edge	Avrg demands per edge
49	258	75	110	46	17
50	454	27	36	50	34
51	574	97	130	50	18
52	871	179	228	49	21
53	846	43	60	54	36
54	763	80	104	51	26
55	579	44	58	50	34
56	298	43	48	48	22
57	364	57	96	50	20
58	239	19	25	52	23
59	223	15	19	49	26
60	199	18	27	45	26
61	309	18	41	49	19
62	347	75	111	47	22
63	343	18	25	49	27
64	343	18	25	56	28

5.2 Lower Bounds

In the table below we illustrate both heuristics from Sect. 3. For most cases, Bi-Loop heuristics is unable to "pack" more demands; however, for some cases, the improvement is significant while being computationally cheap.

Case	Demands	DSatur (Demands)	Bi-loop (Demands)	Case	Demands	DSatur (Demands)	Bi-loop (Demands)
49	258	182	**183**	57	364	**253**	253
50	454	362	**363**	58	239	**194**	194
51	574	**435**	435	59	223	**174**	174
52	871	**660**	**660**	60	199	150	**153**
53	846	**632**	**632**	61	309	**233**	233
54	763	606	**619**	62	347	**247**	247
55	579	**445**	445	63	343	**263**	263
56	298	236	**240**	64	343	**255**	255

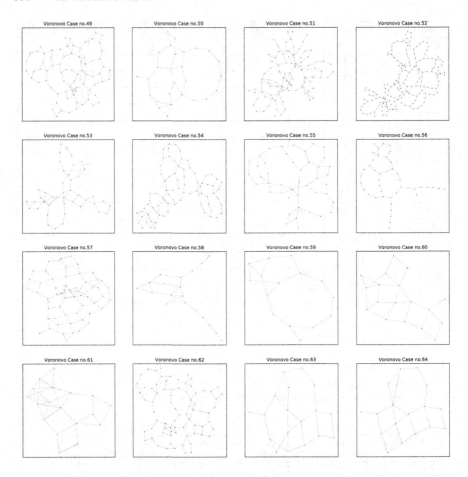

Fig. 3. Topologies of all test cases.

5.3   Upper Bounds

In this subsection, we give the details about maximal cliques and upper bounds. By "Edge Bound" we denote the result (optimal value) of the first relaxation from Sect. 4. In the "Clique Bound" column we list the objective values of the second relaxation from Sect. 4.1; as we may see, the results of this relaxation are tighter than the previous one. In the next columns, we show the number of maximal cliques, the size of the largest maximal clique, and the runtime of the max-clique algorithm from NetworkX.

Case	Edge Bound	Clique Bound	Maximal Cliques	Maximum Clique Size	Time
49	200	192	438	56	0.31s
50	377	373	57	56	0.10s
51	455	450	196	50	0.27
52	690	681	590	54	0.66s
53	646	646	154	54	0.37s
54	640	635	128	52	0.15s
55	461	459	367	63	0.45s
56	241	241	17	48	0.03s
57	272	266	802	55	0.22s
58	200	200	61	52	0.04s
59	175	175	12	50	0.03s
60	161	158	112	45	0.05s
61	243	243	64	49	0.07s
62	273	259	195	54	0.19s
63	266	265	23	49	0.04s
64	262	259	28	56	0.05s

5.4 Optimality Gap

Finally, we show the optimality gap provided by our lower and upper bound algorithms. The below demonstrates that the Bi-Loop algorithm is fast and provably efficient in solving difficult Spectrum Allocation instances.

Case	Demands (Total)	Bi-loop (Objective, Time)	Upper Bound (Objective, Time)	Optimality Gap
49	258	183, 0.9s	192, 1.4s	4.91%
50	454	363, 2.1s	373, 1.4s	2.75%
51	574	435, 3.3s	450, 2.0s	3.44%
52	871	660, 7.1s	681, 4.8s	3.18%
53	846	632, 6.1s	646, 5.9s	2.21%
54	763	619, 6.0s	635, 4.7s	4.78%
55	579	445, 4.7s	459, 3.0s	3.14%
56	298	236, 1.3s	241, 0.5s	2.11%
57	364	253, 2.2s	266, 0.8s	5.13%
58	239	194, 0.7s	200, 0.3s	3.09%
59	223	174, 0.5s	175, 0.5s	0.57%
60	199	150, 0.8s	158, 0.3s	5.33%
61	309	233, 1.4s	243, 0.7s	3.86%
62	347	247, 1.2s	259, 0.8s	4.85%
63	343	263, 1.8s	265, 0.5s	0.76%
64	343	255, 1.8s	259, 0.7s	1.56%

5.5   Real-world Telecommunication Networks

We also test our approach on real-world data from the Survivable fixed telecommunication Network Design library (SNDlib) [11]. We consider only networks with undirected demand and link models, single-path routing, no fix-charge cost, no hops, and no survivability requirements. Among all such instances, we select "Atlanta", "France", "Germany50", "India35", "NewYork", and "Norway" scenarios. We keep all the links and demands, and the only difference with SND problem is that we consider preinstalled link capacity with no possibility to add extra capacity. We precompute paths for all demands by the same algorithm as it was done for Synthetic Instances (the routing algorithm is beyond the scope of this paper).

Case	Demands	Nodes	Links	Max demands/edge	Avrg demands/edge
Atlanta	210	15	22	54	27
France	300	25	45	54	10
Germany50	662	50	88	60	30
India35	595	35	80	43	24
NewYork	240	16	49	22	9
Norway	702	27	51	109	47

For real-world test cases we observe the same situation as for Synthetic instances: if there are enough network resources to route a significant part of the demands, i.e. the problem is not over-constrained (what is usually the case in Flex-Grids), our greedy algorithm shows a small optimality gap with respect to clique-based upper bounds.

Case	Demands (Total)	Bi-loop (Objective)	Upper Bound (Objective)	Optimality Gap
Atlanta	210	182	193	6.04%
France	300	277	286	3.24%
Germany50	662	623	655	5.13%
India35	595	583	587	0.68%
NewYork	240	162	172	6.17%
Norway	702	633	645	1.89%

6   Conclusions

In this paper, we have developed lower and upper bounds for the Spectrum Allocation problem in optical networks (Flexible Grids). In this setting, we are given network topology, a set of demands, a path for every demand (optical channel), and bandwidth requirements and the goal is to correctly allocate wavelengths for as many demands as possible. We developed a fast greedy algorithm based on DSatur coloring and verified the results by computing tight upper bounds using a Constraint Programming solver. We tested our approach on difficult realistic network planning scenarios and showed only a few percent of optimality gap for all input networks.

Author Contributions Statement

M.Kh. developed greedy algorithms and prepared the input data, M.Kr. introduced constraint programming upper bounds and prepared the text of the paper. D.Sh. analyzed the results. All authors contributed to the writing of the manuscript.

References

1. Christodoulopoulos, K., Tomkos, I., Varvarigos, E.: Elastic bandwidth allocation in flexible OFDM-based optical networks. IEEE J. Lightwave Technol. **29**, 354–1366 (2011)
2. Klinkowski, M., Walkowiak, K.: Routing and spectrum assignment in spectrum sliced elastic optical path network. IEEE Commun. Lett. **15**(8), 884–886 (2011)
3. Wang, Y., Cao, X., Pan, Y.: A study of the routing and spectrum allocation in spectrum-sliced elastic optical path networks. In: INFOCOM, 2011 Proceedings IEEE, pp. 1503–1511. IEEE (2011)
4. Velasco, L., Klinkowski, M., Ruiz, M., Comellas, J.: Modeling the routing and spectrum allocation problem for flex grid optical networks. Photon. Netw. Commun. **24**, 1–10 (2012). https://doi.org/10.1007/s11107-012-0378-7
5. Chen, X., Zhong, Y., Jukan, A.: Multipath routing in elastic optical networks with distance-adaptive modulation formats. In: IEEE ICC 2013-Optical Networks and Systems, pp. 3915–3920. IEEE (2013)
6. Chatterjee, B.C., Sarma, N., Oki, E.: Routing and spectrum allocation in elastic optical networks: a tutorial. IEEE Commun. Surv. Tutor. **17**(3), 1776–1800 (2015)
7. Salani, M., Rottondi, C., Tornatore, M.: Routing and spectrum assignment integrating machine-learning-based QoT estimation in elastic optical networks. In: IEEE INFOCOM 2019 - IEEE Conference on Computer Communications, pp. 1738–1746. IEEE, Paris, France (2019)
8. Bouchard, M., Čangalović, M., Hertz, A.: About equivalent interval colorings of weighted graphs. Discr. Appl. Math. **157**(17), 3615–3624 (2009)
9. Gurobi Optimization, LLC: Gurobi Optimizer Reference Manual (2022)
10. Perron, L., Furnon, V.: OR-Tools 7.2, Google (2019)
11. SNDlib. http://sndlib.zib.de/
12. Hagberg, A., Schult, D., Swart, P.: Exploring network structure, dynamics, and function using NetworkX. In: Proceedings of the 7th Python in Science Conference (SciPy 2008), pp. 11–15, Pasadena, CA, USA (2008)
13. Bron, C., Kerbosch, J.: Algorithm 457: finding all cliques of an undirected graph. Commun. ACM **16**(9), 575–577 (1973)
14. Tomita, E., Tanaka, A., Takahashi, H.: The worst-case time complexity for generating all maximal cliques and computational experiments. Theoret. Comput. Sci. **363**(1), 28-42 (2006). (Computing and Combinatorics, 10th Annual International Conference on Computing and Combinatorics (COCOON 2004). Elsevier (2006))
15. Cazals, F., Karande, C.: A note on the problem of reporting maximal cliques. Theoret. Comput. Sci. **407**(1–3), 564–568 (2008)

Comparison of Reinforcement Learning Based Control Algorithms for One Autonomous Driving Problem

Stepan Kabanov$^{(\boxtimes)}$ ⓘ, German Mitiai$^{(\boxtimes)}$ ⓘ, Haitao Wu$^{(\boxtimes)}$ ⓘ, and Ovanes Petrosian$^{(\boxtimes)}$ ⓘ

Saint Petersburg State University, Petersburg, Russia
phoenixgenera@gmail.com, germansck@gmail.com, a89173627785@gmail.com,
petrosian.ovanes@yandex.ru

Abstract. Autonomous driving systems include modules of several levels. Thanks to deep learning architectures at the moment technologies in most of the levels have high accuracy. It is important to notice that currently in autonomous driving systems for many tasks classical methods of supervised learning are no longer applicable. In this paper we are interested in a specific problem, that is to control a car to move along a given reference trajectory using reinforcement learning algorithms. In control theory, this problem is called an optimal control problem for moving along the reference trajectory. Airsim environment is used to simulate a moving car for a fixed period of time without obstacles. The purpose of our research is to determine the best reinforcement learning algorithm for a formulated problem among state-of-the-art algorithms such as DDPG, PPO, SAC, DQN and others. As a result of the conducted training and testing, it was revealed that the best algorithm for this problem is A2C.

Keywords: Autonomous driving · Reinforcement learning · Control algorithms · Optimal control of moving along the reference trajectory

1 Introduction

Except helping a human with driving process, autonomous vehicle can be a solution for problem such as a risk caused by the driver. Autonomous vehicles improve traffic safety significantly, increase fuel efficiency, and help reconnect the elderly and disabled to our society [28]. The earliest autonomous vehicles can be dated back to [8, 20, 26]. As an important milestone, The Defense Advanced Research Project Agency (DARPA) Grand Challenges has led and accelerate development of autonomous vehicles [27].

There are several approaches for autonomous driving problems:

The article was carried out under the auspices of a grant from the President of the Russian Federation for state support of young Russian scientists - candidates of science, project number MK-4674.2021.1.1.

Reinforcement Learning. RL is one of the machine learning directions where the agent or system learns the optimal actions by interacting with the environment [25]. The use of reinforcement learning in autonomous driving is discussed in details here [31]. Other research papers dedicated to the tracking problem [5,15,21].

Imitation Learning. The goal of imitation learning is to learn a controller that imitates the behavior of an expert. Applications for autonomous driving are presented here [6,7,20]. The paper [6] shows that imitation learning can achieve high performance in realistic simulations of urban driving scenarios.

MPC and Numerical Optimal Control. In order to approach the autonomous vehicle movement problem, it is possible to consider the model predictive control (MPC) approach, which is being developed within the framework of numerical optimal control, [9,14,22,29]. In the MPC approach, the current control action is achieved by solving a finite-horizon open-loop optimal control problem at each sampling instant. For linear systems there exists a solution in explicit form [2,11].

Optimal Tracking Control Problem. There exists the approach from the classical control theory. An optimal tracking control problem has been considered in the papers [1,3,4,12]. But the approaches of control theory are not applicable for complex environments that cannot be modeled as a system of difference or differential controls. Therefore, these solutions are not suitable for our problem.

Other related results are presented in the paper [13]. Here the continuous updating approach is applied to a special and practical case of an inverse optimal control problem for determining the driver's behavior while driving along a reference trajectory. Previously, the results of an optimal feedback control approach based on the paradigm of continuous updating were obtained in [19].

There also are some research papers dedicated to a similar direction. In [38] reinforcement learning techniques applied to the problem of autonomous driving. The control of the vehicle is based on the joint work by PID and PP controllers, the agent in turn weighs the contribution of each controller, trying to achieve the best control. In the paper [36] the authors apply a reinforcement learning approach to driving a real full-size car. Training is performed by using a synthetic data from a simulator, which makes it fast and relatively inexpensive. In [34], the authors show how to achieve safer and comfortable driving in a multi-agent environment. In [35], the authors proposed Deep Q-network for deciding when to perform a maneuver based on safety considerations, and Deep Q-Learning framework for completing a longitudinal maneuver.

In this paper we consider the problem of driving a car to move along a given reference trajectory using reinforcement learning algorithms in simulation environment close to reality. The purpose of our research is to determine the best reinforcement learning algorithm for a formulated problem among state-of-the-art algorithms such as DDPG, PPO, SAC, DQN and others. The simulation environment is Microsoft Airsim, it is built on Unreal Engine that offers physically and visually realistic simulations [24]. Several roads are created on which training and subsequent comparison of algorithms are presented.

2   Reinforcement Learning Algorithms

In this paper, we apply Reinforcement Learning algorithms to solve an autonomous driving problem, then compare the performances for each algorithm. RL refers to learning how to take actions in different states of an environment to maximize cumulative future reward, where the reward is a form of feedback associated with a task and received in response to an action. A reinforcement learning process is usually modeled as Markov Decision Processes (MDPs). MDP process is defined using a tuple $< S, A, P, r >$. S is the state space of the environment, where a state $s \in S$ represents a situation of the environment. A is the set of actions that can be taken by an agent. $r : S \times A \rightarrow \mathbb{R}$ is the numerical reward obtained as a function of state and action. The goal of RL is to maximize future discounted return. The value of taking action a_t in s_t is calculated as the expected cumulative reward obtained over all possible trajectories, as follows:

$$Q^\pi (s_t, a_t) = r (s_t, a_t) + \mathbb{E}_{a \sim \pi(s)} \left[\sum_{i=1}^{\infty} \gamma^i r (s_{t+i}, a_{t+i}) \mid s_t, a_t \right] \tag{1}$$

Here, $Q^\pi (s_t, a_t)$ is called the Q-value of the state-action pair while an agent follows a policy π. Below is a brief description of these algorithms.

Actor to Critic is a synchronous, deterministic variant of Asynchronous Advantage Actor Critic (A3C). It uses multiple workers to avoid the use of a replay buffer [17] (Algorithm 1).

Algorithm 1. A2C

Initialize thread step counter $t \leftarrow 0$
Initialize target network weights $\theta^- \leftarrow \theta$
Initialize network gradients $d\theta \leftarrow 0$
Get initial state s
repeat
    Take action a with ε-greedy policy based on $Q(s, a, \theta)$
    Receive new state s' and reward r

$$y = \begin{cases} r & \text{for terminal } s' \\ r + \gamma \max_{a'} Q(s', a', \theta^-) & \text{for non-terminal } s' \end{cases}$$

    Accumulate gradients wrt θ: $d\theta + \frac{\partial (y - Q(s, a, \theta))^2}{\partial \theta}$
    $s = s'$
    $T \leftarrow T + 1$ and $t \leftarrow t + 1$
    if T mod $I_{target} == 0$ **then**
        Update the target network $\theta^- \leftarrow \theta$
    end if
    if t mod $I_{AsyncUpdate} == 0$ **then**
        Update the target network $\theta^- \leftarrow \theta$
    end if
    if t mod $I_{AsyncUpdate} == 0$ or s is terminal **then**
        Perform asynchronous update of θ using $d\theta$
        Clear gradients $d\theta \leftarrow 0$
    end if
until $T > T_{max}$

Soft Actor Critic (SAC) [10] is an off policy actor-critic deep RL algorithm based on the maximum entropy reinforcement learning framework. SAC is proposed to alleviate hyper-parameter tuning caused by high sample complexity and convergence property (Algorithm 2).

Algorithm 2. SAC

Input: initial policy parameters θ. Q-function parameters φ_1, φ_2, empty replay buffer $\mathcal{D}$

Set target parameters equal to main parameters $\varphi_{\text{targ},1} \leftarrow \varphi_1$, $\varphi_{\text{targ},2} \leftarrow \varphi_2$

repeat

  Observe state s and select action $a \sim \pi_\theta(\cdot|s)$

  Execute a in the environment

  Observe next state s', reward r, and done signal d to indicate whether s' is terminal

  Store (s, a, r, s', d) in replay buffer $\mathcal{D}$

  In s' is terminal, reset environment state.

  if it's time to update **then**

    for j in range (however many updates) **do**

      Randomly sample a batch of transitions, $B = \{(s, a, r, s', d)\}$ form $\mathcal{D}$

      Compute targets for the Q functions:

$$y(r, s', d) = r + \gamma(1 - d)\left(\min_{i=1,2} Q_{\varphi_{\text{targ},i}}(s', \tilde{a}') - \alpha \log \pi_\theta(\tilde{a}'|s')\right), \quad \tilde{a}' \sim \pi_\theta(\cdot|s')$$

Update Q-functions by one step of gradient descent using

$$\nabla_{\varphi_i} \frac{1}{|B|} \sum_{(s,a,r,s',d)\in B} (Q_{\varphi_i}(s,a) - y(r, s', d))^2 \quad \text{for } i = 1, 2$$

Update policy by one step of gradient ascent using

$$\nabla_\theta \frac{1}{|B|} \sum_{s\in B} \left(\min_{i=1,2} Q_{\varphi_i}(s, \tilde{a}_\theta(s)) - \alpha \log \pi_\theta(\tilde{a}_\theta(s)|s)\right),$$

      where $\tilde{a}_\theta(s)$ is a sample from $\pi_\theta(\cdot|s)$ which is differentiable wrt θ via the reparametrization trick.

      Update target networks with

$$\varphi_{\text{targ},i} \leftarrow \rho\varphi_{\text{targ},i} + (1 - \rho)\varphi_i \quad \text{for} i = 1, 2$$

    end for

  end if

until convergence

Proximal Policy Optimization (PPO) [23] is policy gradient method for reinforcement learning which alternate between sampling data through interaction with the environment, and optimizing a surrogate objective function using stochastic gradient ascent. The variant of PPO suggested by OpenAI is applied in this research [23] (Allgorithm 3)

Algorithm 3. PPO-Clip

Input: initial policy parameters θ_0, initial value function parameters φ_0
for $k = 0, 1, 2, \ldots$ **do**

Collect set of trajectories $\mathcal{D}_k = \{\tau_i\}$ by running policy $\pi_k = \pi(\theta_k)$ in the environment

Compute rewards-to-go $\hat{R}_t$

Compute advantage estimates, $\hat{A}_t$ (using any method of advantage estimation) based on the current value function V_{φ_k}.

Update the policy by maximizing the PPO-Clip objective:

$$\theta_{k+1} = \arg\max_{\theta} \frac{1}{|\mathcal{D}_k|T} \sum_{\tau \in \mathcal{D}_k} \sum_{t=0}^{T} \min\left(\frac{\pi_\theta(a_t|s_t)}{\pi_{\theta_k}(a_t|s_t)} A^{\pi_{\theta_k}}(s_t, a_t), g(\varepsilon, A^{\pi_{\theta_k}}(s_t, a_t)) \right),$$

typically via stochastic gradient ascent with Adam
Fit value function by regression on mean-squared error:

$$\varphi_{k+1} = \arg\min_{\varphi} \frac{1}{|\mathcal{D}_k|T} \sum_{\tau \in \mathcal{D}_k} \sum_{t=0}^{T} \left(V_\varphi(s_t) - \hat{R}_t \right)^2,$$

typically via some gradient descent algorithm.
end for

Deep Q-learning (DQN) [18] was proposed by Mnih el al., which was able to achieve superhuman level performance on different Atari games by combining Q-learning and deep neural network. In autonomous driving domain, DQN is a popular solution for lane change decision making problem since its ability to learn complex control policies in high-dimensional tasks (Algorithm 4).

Algorithm 4. DQN

Initialize replay memory $\mathcal{D}$ to capacity N
Initialize action-value function Q with random weights
for episode= $\overline{1, M}$ **do**

Initialise sequence $s_1 = \{x_1\}$ and preprocessed sequenced $\varphi_1 = \varphi(s_1)$
for $t = \overline{1, T}$ **do**

with probability ε select a random action a_t
otherwise select $a_t = \max_a Q^*(\varphi(s_t), a, \theta)$
Execute action a_t in emulator and observe reward r_t and image x_{t+1}
Set $s_{t+1} = s_t, a_t, x_{t+1}$ and preprocess $\varphi_{t+1} = \varphi(s_{t+1})$
Store transition $(\varphi_t, a_t, r_t, \varphi_{t+1})$ in $\mathcal{D}$
Sample random minibatch of transitions $(\varphi_j, a_j, r_j, \varphi_{j+1})$ from $\mathcal{D}$

$$y_j = \begin{cases} r_j & \text{for terminal } \varphi_{j+1} \\ r_j + \gamma \max_{a'} Q(\varphi_{j+1}, a', \theta) & \text{for non-terminal } \varphi_{j+1} \end{cases}$$

Perform a gradient descent step on $(y_j - Q(\varphi_j, a_j, \theta))^2$ according to differentiated the loss function
end for
end for

Deep Deterministic Policy Gradient (DDPG) [16] is an algorithm which concurrently learns a Q-function and a policy. It uses off-policy data and the Bellman equation to learn the Q-function, and uses the Q-function to learn the policy (Algorithm 5).

Algorithm 5. Deep Deterministic Policy Gradient

Input: initial policy parameters θ, Q-function parameters φ, empty replay buffer $\mathcal{D}$
Set target parameters equal to main parameters $\theta_{\text{targ}} \leftarrow \theta$, $\varphi_{\text{targ}} \leftarrow \varphi$
repeat
 Observe state s and select action $a = \text{clip}(\mu_\theta(s) + \varepsilon, a_{\text{Low}}, a_{\text{High}})$, where $\varepsilon \sim \mathcal{N}$
 Execute a in the environment
 Observe next state s', reward r, and done signal d to indicate whether s' is terminal
 Store (s, a, r, s', d) in replay buffer $\mathcal{D}$
 If s' is terminal, reset environment state.
 if it's time to update **then**
  for however many updates **do**
   Randomly sample a batch of transitions, $B = \{(s, a, r, s', d)\}$ from $\mathcal{D}$
   Compute targets

$$y(r, s', d) = r + \gamma(1 - d)Q_{\varphi_{\text{targ}}}(s', \mu_{\text{targ}}(s'))$$

   Update Q-function by one step of gradient descent using

$$\nabla_\varphi \frac{1}{|B|} \sum_{(s,a,r,s',d)\in B} (Q_\varphi(s, a) - y(r, s', d))^2$$

   Update policy by one step of gradient ascent using

$$\nabla_\varphi \frac{1}{|B|} \sum_{s\in B} Q_\varphi(s, \mu_\theta(s))$$

   Update target networks with

$$\varphi_{\text{targ}} \leftarrow \rho\varphi_{\text{targ}} + (1 - \rho)\varphi$$

$$\theta_{\text{targ}} \leftarrow \rho\theta_{\text{targ}} + (1 - \rho)\theta$$

  end for
 end if
until convergence

---

3 Simulation Environment

The Microsoft Airsim simulator [24] based on the Unreal Engine 4 is chosen as the environment for driving simulation. The environment provides a required level of realism of motion physics, which allows for reliable research and experiments in the field of autonomous control of ground vehicles. The choice fell on this simulator due to the low system requirements compared to competing analogues and at the same time with the equivalent operational functionality. A screenshot from the environment can be seen on Fig. 1.

Fig. 1. Screenshot from AirSim

The agent's reward directly depends on how accurately the car follows a given route. More the car deviates from the desired trajectory, the smaller the reward. When a certain threshold value of the deviation of the car from the route is reached, we finish the training iteration and restart the simulation. At each iteration, the training takes place on one randomly selected road. Examples of roads are presented on Fig. 2. The PID controller keeps the speed of the car around 21 km/h.

4   State-Action-Reward Model

According to the reinforcement learning paradigm the RL agent iteratively learns the optimal policy by interacting with the environment. The state-action-space model is vital to understand of what are the control parameters and how the learning process is organized. Figure 3 below represents the basic idea and elements involved in a reinforcement learning model.

- **State:** current situation of the agent. The agent's observations are set as follows. These are the 6(N) points closest to the car of the desired trajectory, measured relative to the coordinate system of the car. The coordinate system is oriented so that the X-axis is directed along the car body, and the Y-axis is perpendicular to it and directed along the right side of the car. After the transformation, we sort these points in ascending order of the X coordinate in the coordinate system of the car.

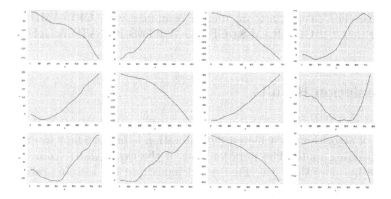

Fig. 2. Examples of roads

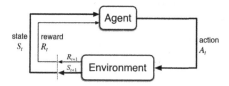

Fig. 3. An idea of reinforcement learning

– **Action:** something an agent does. The agent directly sets the angle of rotation of the wheels of the car within $[-0.5, 0.5]$. For the DQN algorithm, unlike the other algorithms used, it was necessary to specify a discrete state space. The discrete state space consists of 5 actions corresponding to the conditional units of wheel rotation: $\{-0.5, -0.2, 0, 0.2, 0.5\}$.
– **Reward:** feedback from the environment. The reward is awarded to the agent depending on the distance to the nearest point of the road. Since the road is given by a polyline, we can consider the distance to the nearest point of the road as the distance to the nearest link of the polyline by the formula

$$dist = \frac{\|(x_0 - x_1) \times (x_0 - x_2)\|}{\|x_2 - x_1\|},$$

where x_0 is the position of the car, and x_2 and x_1 are points from the nearest link of the polyline. Knowing the distance to the route, it is possible to calculate the agent's load at a specific time using the following formula

$$R(d) = e^{\frac{1}{3}d} - \frac{1}{2},$$

where d is the distance from the car to the nearest link of the polyline.

In addition to Reinforcement Learning algorithms for wheel rotation control, we also use a PID controller to maintain a constant speed of the car. Thus, we free our agent from the need to control the speed of movement.

Configure: For calculations we used a computer with CPU-AMD Ryzen 5 3550H CPU @ 2.10 GHz; RAM-8.0 GB; OS-Windows 10; NVIDIA GeForce GTX 1650.

5   Simulation Results

We have trained and tested the algorithms outlined in the section above. Convergence of algorithms is achieved. One iteration is one attempt to drive along a given training route. The iteration ends if the car deviates from the specified route by more than 3 m. Figures 5 demonstrate the learning rate and the convergency of the algorithms.

As the main learning metrics used in the evaluation of algorithms, we chose the reward for passing the test route and the average deviation of the car from the test route. It is clearly shown how the agents drove the test road on Fig. 6. The test results are presented on Fig. 4

Comparison of algorithms			
Algorithm	Max dist	Mean dist	Time
DQN	1.774	0.3910	154000
A2C	1.1620	0.3019	138000
SAC	1.74556	0.3081	158000
DDPG	1.3229	0.33847	148000
PPO	1.2844	0.2745	82000

Fig. 4. Test results

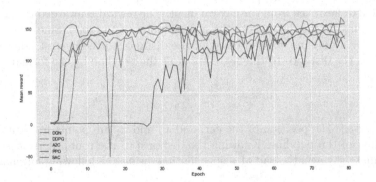

Fig. 5. Learning curves

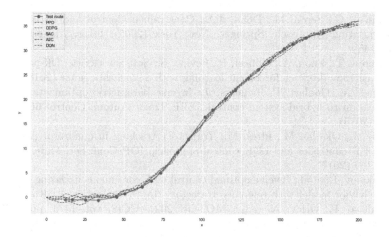

Fig. 6. Road test plot

6   Conclusion

The best results for optimal moving control along the reference trajectory were shown by PPO and AC2. PPO showed a very fast convergence rate, the best average distance. A2C has been converging longer than PPO, but at the same time, it has the best performance in max distance.

References

1. Aghasadeghi, N., Bretl, T.: Inverse optimal control for differentially flat systems with application to locomotion modeling. In: 2014 IEEE International Conference on Robotics and Automation (ICRA), pp. 6018–6025. IEEE (2014)
2. Bemporad, A., Morari, M., Dua, V., Pistikopoulos, E.: The explicit linear quadratic regulator for constrained systems. Automatica **38**(1), 3–20 (2002)
3. Bohner, M., Wintz, N.: The linear quadratic tracker on time scales. Int. J. Dyn. Syst. Differ. Equ. **3**(4), 423–447 (2011)
4. Botan, C., Ostafi, F., Onea, A.: A solution to the optimal tracking problem for linear systems. Matrix. **1**, 1–5 (2001)
5. Buhrle, E., Kopf, F., Inga, J., Hohmann, S.: Adaptive optimal trajectory tracking control of continuous-time systems. In: 19th European Control Conference, June 29–July 2, 2021, Rotterdam, Virtual Conference (2021)
6. Chen, J., Yuan, B., Tomizuka, M.: Deep imitation learning for autonomous driving in generic urban scenarios with enhanced safety. In: 2019 IEEE/RSJ International Conference on Intelligent Robots and Systems (IROS), pp. 2884–2890 (2019)
7. Codevilla, F., Muller, M., Lopez, A., Koltun, V., Dosovitskiy, A.: End-to-end driving via conditional imitation learning (2018)
8. Dickmanns, E., Zapp, A.: Autonomous high speed road vehicle guidance by computer vision1. In: IFAC Proceedings Volumes, vol. 20(5, Part 4), pp. 221–226 (1987). 10th Triennial IFAC Congress on Automatic Control - 1987 Volume IV, Munich, Germany, 27–31 July https://www.sciencedirect.com/science/article/pii/S1474667017553203, https://doi.org/10.1016/S1474-6670(17)55320-3

9. Goodwin, G., Seron, M., Dona, J.A.: Constrained Control and Estimation: An Optimisation Approach. Springer, New York (2005). https://doi.org/10.1007/b138145

10. Haarnoja, T., Zhou, A., Abbeel, P., Levine, S.: Soft actor-critic: Off-policy maximum entropy deep reinforcement learning with a stochastic actor (2018)

11. Hempel, A., Goulart, P., Lygeros, J.: Inverse parametric optimization with an application to hybrid system control. IEEE Trans. Autom. Control 60(4), 1064–1069 (2015)

12. Islam, M., Okasha, M., Idres, M.: Trajectory tracking in quadrotor platform by using PD controller and LQR control approach. IOP Conf. Ser. Mater. Sci. Eng. 260, 1–9 (2017)

13. Kuchkarov, I., et al.: Inverse optimal control with continuous updating for a steering behavior model with reference trajectory. In: Strekalovsky, A., Kochetov, Y., Gruzdeva, T., Orlov, A. (eds.) MOTOR 2021. CCIS, vol. 1476, pp. 387–402. Springer, Cham (2021). https://doi.org/10.1007/978-3-030-86433-0_27

14. Kwon, W., Han, S.: Receding horizon control: model predictive control for state models. In: Howard, A., Iagnemma, K., Kelly, A. (Eds.) Field and Service Robotics. Springer Tracts in Advanced Robotics, vol. 62, pp. 69–78. Springer, Heidelberg (2005). https://doi.org/10.1007/978-3-642-13408-1_7

15. Köpf, F., Ramsteiner, S., Flad, M., Hohmann, S.: Adaptive dynamic programming for model-free tracking of trajectories with time-varying parameters. Int. J. Adapt. Control Signal Process. 34, 839–856 (2020)

16. Lillicrap, T.P., et al.: Continuous control with deep reinforcement learning (2019)

17. Mnih, V., et al.: Asynchronous methods for deep reinforcement learning (2016)

18. Mnih, V., et al.: Playing Atari with deep reinforcement learning (2013)

19. Petrosian, O., Inga, J., Kuchkarov, I., Flad, M., Hohmann, S.: Optimal control and inverse optimal control with continuous updating for human behavior modeling. IFAC-PapersOnLine 53(2), 6670–6677 (2020). 21st IFAC World Congress

20. Pomerleau, D.A.: Alvinn: An autonomous land vehicle in a neural network. In: Touretzky, D. (ed.) Advances in Neural Information Processing Systems, vol. 1. Morgan-Kaufmann (1988). https://proceedings.neurips.cc/paper/1988/file/812b4ba287f5ee0bc9d43bbf5bbe87fb-Paper.pdf

21. Puccetti, L., Kopf, F., Rathgeber, C., Hohmann, S.: Speed tracking control using online reinforcement learning in a real car. In: 2020, The 6th International Conference on Control, Automation and Robotics, Singapore, April 20–23, 2020, ICCAR 2020. Institute of Electrical and Electronics Engineers (IEEE) (2020). ISBN: 978-1-72816-140-2

22. Rawlings, J., Mayne, D.: Model Predictive Control: Theory and Design. Nob Hill Publishing, Madison (2009)

23. Schulman, J., Wolski, F., Dhariwal, P., Radford, A., Klimov, O.: Proximal policy optimization algorithms (2017)

24. Shah, S., Dey, D., Lovett, C., Kapoor, A.: AirSim: High-fidelity visual and physical simulation for autonomous vehicles. In: Hutter, M., Siegwart, R. (eds.) Field and Service Robotics. SPAR, vol. 5, pp. 621–635. Springer, Cham (2018). https://doi.org/10.1007/978-3-319-67361-5_40

25. Sutton, R.S., Barto, A.G.: Reinforcement Learning: An Introduction. MIT Press, Cambridge (2018)

26. Thorpe, C., Hebert, M., Kanade, T., Shafer, S.: Vision and navigation for the Carnegie-Mellon navlab. IEEE Trans. Pattern Anal. Mach. Intell. 10(3), 362–373 (1988)

27. Thrun, S., et al.: Stanley: the robot that won the DARPA grand challenge. In: Buehler, M., Iagnemma, K., Singh, S. (eds) The 2005 DARPA Grand Challenge. Springer Tracts in Advanced Robotics, vol. 36, pp. 1–43. Springer, Heidelberg (2007). https://doi.org/10.1007/978-3-540-73429-1_1

28. Van Dinh, N., Ha, Y.g., Kim, G.W.: A universal control system for self-driving car towards urban challenges. In: 2020 IEEE International Conference on Big Data and Smart Computing (BigComp), pp. 452–454 (2020)

29. Wang, L.: Model Predictive Control System Design and Implementation Using MATLAB. Springer, New York (2005). https://doi.org/10.1007/978-1-84882-331-0

30. Watkins, C.J.C.H., Dayan, P.: Q-learning. Mach. Learn. **8**, 279–292 (1992). https://doi.org/10.1007/BF00992698

31. Yu, G., Sethi, I.: Road-following with continuous learning. In: Proceedings of the Intelligent Vehicles 1995. Symposium, pp. 412–417 (1995)

32. Zholobova, A., Zholobov, Y., Polyakov, I., Petrosian, O., Vlasova, T.: An industry maintenance planning optimization problem using CMA-VNS and its variations. In: Strekalovsky, A., Kochetov, Y., Gruzdeva, T., Orlov, A. (eds.) MOTOR 2021. CCIS, vol. 1476, pp. 429–443. Springer, Cham (2021). https://doi.org/10.1007/978-3-030-86433-0_30

33. Markelova, A., Allahverdyan, A., Martemyanov, A., Sokolova, I., Petrosian, O., Svirkin, M.: Applied routing problem for a fleet of delivery drones using a modified parallel genetic algorithm. Bulletin of St. Petersburg University, Applied Mathematics. Informatics. Control processes (2022)

34. Naumann, M., Sun, L., Zhan, W., Tomizuka, M.: Analyzing the suitability of cost functions for explaining and imitating human driving behavior based on inverse reinforcement learning. IEEE Int. Conf. Robot. Autom. (ICRA) **2020**, 5481–5487 (2020). https://doi.org/10.1109/ICRA40945.2020.9196795

35. Osinski, B., et al.: Simulation-based reinforcement learning for real-world autonomous driving (2019)

36. Shi, T., Wang, P., Cheng, X., Chan, C.Y., Huang, D.: Driving decision and control for autonomous lane change based on deep reinforcement learning (2019)

37. Shalev-Shwartz, S., Shammah, S., Shashua, A.: Safe, multi-agent, reinforcement learning for autonomous driving (2016)

38. Shan, Y., Zheng, B., Chen, L., Chen, L., Chen, D.: A reinforcement learning-based adaptive path tracking approach for autonomous driving. IEEE Trans. Veh. Technol. **69**(10), 10581–10595 (2020). https://doi.org/10.1109/TVT.2020.3014628

Author Index

Printed in the United States
by Baker & Taylor Publisher Services